THE BEEKEEPER'S HANDBOOK

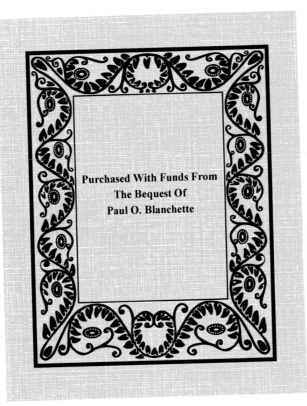

THE BEEKEEPER'S HANDBOOK

FOURTH EDITION

Diana Sammataro
Alphonse Avitabile

Foreword by Dewey M. Caron

COMSTOCK PUBLISHING ASSOCIATES
a division of
Cornell University Press
ITHACA AND LONDON

First edition published 1978 by Peach Mountain Press
Second edition published 1986 by Macmillan Publishing
    Company
Third edition published 1998 by Cornell University Press
Fourth edition published 2011 by Cornell University Press
Third edition printing, Cornell Paperbacks, 1998
Fourth edition printing, Cornell Paperbacks, 2011

Printed in the United States of America

Library of Congress Cataloging-in-Publication Data
Sammataro, Diana.
  The beekeeper's handbook / Diana Sammataro and Alphonse
Avitabile ; foreword by Dewey M. Caron. — 4th ed.
      p. cm.
  Includes bibliographical references and index.
  ISBN 978-0-8014-4981-9 (cloth : alk. paper) —
  ISBN 978-0-8014-7694-5 (pbk. : alk. paper)
  1. Bee culture—Handbooks, manuals, etc.
  I. Avitabile, Alphonse.   II. Title.

SF523.S35 2011
638'.1—dc22
                                        2010050047

Cornell University Press strives to use environmentally responsible suppliers and materials to the fullest extent possible in
the publishing of its books. Such materials include vegetable-
based, low-VOC inks and acid-free papers that are recycled,
totally chlorine-free, or partly composed of nonwood fibers.
For further information, visit our website at www.cornellpress.
cornell.edu.

Cloth printing   10 9 8 7 6 5 4 3 2 1
Paperback printing 10 9 8 7 6 5 4 3 2

Contents

Foreword to the Fourth Edition

by Dewey M. Caron

Beekeeping is different things to different people . . . for some a business, or a way to supplement income from the "daytime" job; for others, a pleasure, an intense learning experience, something to really delve into. Some bee colony owners prefer to take bees and their management more casually, although you'll discover in the pages of this impressive new edition of *The Beekeeper's Handbook* that the days of leave-it-alone beekeeping have passed. That said, you'll also find there is no one "right" way to steward bee colonies; there are many opportunities to develop and personalize your own approach.

Surprisingly, many beekeepers/consumers have no idea what honey is or how bees "make" it. Even knowledgeable consumers want to include pollen as an integral part of honey—the bees, after all, do need pollen to grow their population large enough to store surplus, and that is what allows beekeepers to gain a share of their honey. But honey and pollen are two distinct and separate products. And pollen harvesting from a bee colony is different from honey harvesting. The "key" is understanding how to estimate how much honey and/or pollen the colony can afford to give up and still survive the winter (or dry/rainy) season. This book explains the why as it shows how you can see and be part of the process.

My beekeeping mentor was Roger Morse, longtime professor of apiculture at Cornell University where I learned the basics of bees and first taught beekeeping to others. In 1998, Morse wrote in the foreword to the 3rd edition, "for two decades *The Beekeeper's Handbook* has guided thousands of beginning and advanced beekeepers in the how-to's of this entertain-ing and profitable pastime. But in recent years the science and the art of beekeeping have changed drastically, and this new, thoroughly updated edition will enable beekeepers at all levels to keep up with those changes. This third edition brings beekeeping to the threshold of the twenty-first century, with all its challenges." Challenges continue and present themselves anew, and beekeeping continues to change. With the expertise of Diana Sammataro and Alphonse Avitabile, the handbook has again been updated. It remains among only a few as the very best book to use as a tool to learn and to keep up with what is current in beekeeping.

The management of bees is clearly detailed and offered in uncluttered language, allowing beginners to readily follow colony management suggestions. Colony care options are detailed and little is assumed: the step-by-step process of colony manipulations can be followed with relative ease. The 4th edition has extensive information on Colony Collapse Disorder (CCD), Africanized honey bees, and bee mite control. New material has been incorporated throughout. Beginners will find it a joy—more seasoned beekeepers will find rereading of benefit as they continue to master the art and the science of bee colony care.

A strength of this handbook is the visual material. It is clear and used to illustrate major points of management and colony equipment. There is a good balance of text to graphics. Like the management details, the illustrations point the way clearly and patiently. The chapters are organized in a progression, and information that should be included is present and can be found relatively easily.

I do not suggest that you casually take up this handbook—it should become a favored and required reading text for your beekeeping dreams and aspirations. If you are new, *Welcome* to the world of bee- keeping! Whether you are a new beekeeper or an experienced veteran, may you *learn* and *profit* from this manual. *Enjoy!!*

Foreword to the Third Edition

by Roger A. Morse

For two decades *The Beekeeper's Handbook* has guided thousands of beginning and advanced beekeepers in the how-to's of this entertaining and profitable pastime. But in recent years the science and the art of beekeeping have changed drastically, and this new, thoroughly updated edition will enable beekeepers at all levels to keep up with those changes. This third edition brings beekeeping to the threshold of the twenty-first century, with all its challenges.

No one could do this better than authors Diana Sammataro, a noted honey bee researcher, and Alphonse Avitabile, a retired honey bee scientist and college instructor. Dr. Sammataro is also a beekeeper. She produces honey, raises queens, uses bees to pollinate crops, assembles equipment, and engages in all the other activities of beekeeping. Her intimate knowledge of honey bees is evident throughout this book. Alphonse Avitabile, also an experienced bee- keeper, is a successful gardener, nurseryman, and greenhouse manager.

The popularity of the first two editions resulted from a simple premise underlying both books: there are many ways to do things right. And this latest edition, too, unlike much of the genre, presents time-tested methods and techniques, introduces the most current ideas and concepts, and lets readers choose those which best suit their individual skills, location, and requirements. Although originally designed for beginners, *The Beekeeper's Handbook* will appeal to more advanced beekeepers as well. Rather than limit the seasoned beekeeper to traditional ways of doing things, it puts forward the newest and safest methods to deal with today's problems.

With this book, beekeeping has never been easier. Simply put, it is the best of the best of beekeeping books.

Preface and Acknowledgments

For this fourth edition, and after many folks have expressed curiosity on how the book was created, we give a short history on the development of the book and the beekeepers who inspired and helped us along the way.

Diana Sammataro acknowledges her parents here, Joseph Michael Sammataro, an architect, and Nelva Margaret Weber, a landscape architect, who guided the many interests and curiosities of their daughter's childhood with gentle kindness and encouragement. She also remembers her maternal grandfather, George Weber, who first introduced her to the world of bees at the tender age of twelve in Arrowsmith, Illinois. His two brothers, Fred and Harry, the Weber brothers, were commercial beekeepers in Blackfoot, Idaho, early in the 1900s. Diana is the only beekeeper left in the Weber family line (although perhaps a newer generation of Webers may take over).

Diana moved back to her childhood home in Connecticut after graduating with a landscape degree from the University of Michigan in 1970. It was providential that while working at the White Memorial Nature Center and Museum in Litchfield, she met Professor Alphonse Avitabile, a local teacher and beekeeper. With his guidance and charismatic inspiration, Diana was motivated and encouraged to begin her first colony using Grandpa Weber's bee hive furniture, which had been chauffeured from Illinois to Connecticut after his death. A newspaper dating back to the 1930s (*The Daily Pantograph*) was found under the metal lid of the outer cover of his hive. This first colony, along with Professor Avitabile's encouragement and patience, inspired Diana to make this fascinating insect part of her life.

After moving back to Lansing, Michigan, Diana took some classes at Michigan State University with Dr. Bert Martin, whose gentle encouragement gave her the necessary direction and creative outlet to learn more about bees. Transferring to Ann Arbor in 1973, she managed to talk the Ann Arbor Adult Education staff into letting her "teach" a beekeeping course (teaching forced her to learn). This beekeeping handbook was first envisioned when Doug Truax of Peach Mt. Press, who was taking the class, suggested making the teaching notes into a book. It was only after Jan Propst (daughter-in-law of roommate Claudia) created the original layout, with its horizontal format design, that the idea of making the rough notes into a book became a reality. *For that I will always be eternally grateful.* However, being only a novice beekeeper, Diana needed wiser, more experienced heads to help. Her first thought was to seek the guidance of Alphonse Avitabile, who had first shown her the wonders of beekeeping. Only after Alphonse corrected and added sections was the first edition realized, consisting of only 700 hard-bound copies and 1300 soft-bound copies.

The subsequent editions were produced not only to update procedures but also to address the new pests and diseases that have invaded North American shores. Throughout these changes, the expertise and dedication of Alphonse to maintain the high caliber of the book have helped make it popular and still unique in beekeeping literature.

Over the years since the first edition, many beekeepers and bee researchers have been kind enough to express their enthusiasm and honest appraisal of this book. To all of you who personally shared opinions, comments, photos, and observations, your kind words have helped more than you will ever know; thank you.

Alphonse wishes to dedicate this book to Mr. Lenard Insogna for persuading him to pursue a degree in biology; to his parents, who allowed him to spend most of his time in the woods and ponds near his home studying nature; and to his wife, Ruth, for her support throughout his studies of honey bees.

ACKNOWLEDGMENTS

Both authors wish to acknowledge Henry "Hank" Hansen and his son Jonathan for allowing the authors to share their method of installing package bees with the readers.

Diana wishes to thank the many people who were especially generous with their time, contributions, and support. Ann Harman helped tremendously by pointing out places in the third edition that needed changing and proofing the fourth edition. *Thank you, Ann.* And Dr. Dewey Caron, for all your kindness over our many years as friends, your meticulous and thorough review of the manuscript was most helpful; *thank you*, good friend. Also Dr. Nancy Ostiguy, a fellow quilter and one of the Penn State powerhouse team, had great comments that made this edition better; thank you, too! Thanks also to Bruce (currently a University of Arizona graduate student working in the Tucson Bee lab) and Linda Eckholm for the gift of the computer hardware on which the fourth edition was created.

Others who helped along the way, if not physically, then spiritually, and deserve grateful thanks include Ruth Avitabile, Carol and Ron Conkey, Eric H. Erickson, Doug and Grace Truax, Carol Henderson, Dr. Malcolm Sanford, Bob and Dorothy Kennedy, Harry and Nellie Weber, John and Gwen Nystuen, Dick and Ginny Ryan, Zander Alexander Laurie, Rob Currie, Gerry and Ginnie Loper, Gordon Waller, Judy Walker and Sabu Advani, and Maryann and Jim Frazier.

The authors wish to thank Heidi S. Lovette and Candace Akins at Cornell University Press for their extraordinary help and guidance in creating the new fourth edition.

THE BEEKEEPER'S HANDBOOK

Introduction

Beekeeping is an interesting and rewarding activity if you love nature, have a fascination with the unique social organization of insects, and are consumed with an active curiosity about how things work. And, oh, you should also enjoy honey.

This handbook is designed to help you become a good beekeeper, whether you intend to start keeping bees or already have them and need a ready guide to help you accomplish the various and often complicated tasks that you need to perform in the beeyard. It is designed to assist both new and experienced beekeepers in setting up or reorganizing an apiary and in improving the style of working with and understanding bees.

The book outlines the many colony management operations you will encounter. The text presents the key elements in keeping bees, describing all the major options available to you. It also lists the advantages and disadvantages of each important technique to help you decide which one is best for you. Also, most sections are cross-referenced to point you to more detailed information. **But remember** . . . there is **no** one correct way to keep bees. Feel free to alter any of the directions to suit your needs or situation or to try something entirely new.

Numerous diagrams and illustrations accompany the text to reinforce or illuminate the descriptions. Space is also provided at the end of each chapter so you can keep notes on your own successes and failures. Learning from your mistakes is an essential part of beekeeping.

The reference section has been updated to include as many important books, organizations, and Internet resources as possible, but in this age of instant communication and with information just a keystroke away, it is easier than ever to find what you need. Just remember, be careful what you read on the Internet; separate opinion from scientific, proven results. Experiment at your own risk with cures and treatment options. There is also an updated glossary to help beginners understand the terminology of bees.

Although considered a "gentle art," beekeeping can be physically demanding and strenuous. The typical picture of a veiled beekeeper standing beside the beehive with smoker in hand does not reveal the aching back, sweating brow, smoked-filled eyes, or painful stings. This handbook is intended to enable you to maximize the more interesting and enjoyable aspects of the art. Have fun, learn a lot, ask fellow beekeepers a lot of questions, and share your knowledge with others.

Just remember, as much as you read and learn, bees do **not** read the books and do mostly what they want and what they have successfully been doing for millions of years. That's what makes it fun. So enjoy. And welcome to the wonderful world of beekeeping.

LEGAL REQUIREMENTS

All states have laws that pertain to keeping honey bees and registering *hives* (the wooden boxes in which a colony of bees live) containing bees. Certain city and state laws limit the number of hives in urban areas. Because bees can be declared a nuisance in some cities, local laws must be studied before an *apiary* (place where beehives are located) is established.

Many states have an apiary inspection law developed to aid beekeepers by providing means for controlling and eradicating bee diseases and pests.

General requirements usually include some of the following:

- Beekeepers may have to register hives and apiaries with their state department of agriculture apiary inspection service.
- The director of agriculture and appointed deputies may be authorized to inspect, treat, quarantine, disinfect, and/or destroy any diseased hives.
- Transportation of bees and equipment may need to be certified by the bee inspector or other designated state official.
- Beekeepers may have to ascertain and comply with town or county zoning ordinances that pertain to bees and bee hives.
- All beekeepers shall have bee colonies in hives containing movable frames.
- Penalties may exist for violations of applicable apiary inspections laws.

Now, because of the introduction of parasitic bee mites, the small hive beetle, and the Africanized honey bee, some states have special laws regarding keeping bees. For specific legal requirements, check your state department of agriculture's apiary inspection law.

BEE-STING REACTIONS

An important question that you must consider as a beekeeper is your individual response to bee stings. Although most beekeepers never exhibit serious reactions to bee stings, after a few years some individuals do develop an allergy to bee venom, bee hairs, or other hive components.

When you are stung, the bee's stinging apparatus pierces flesh, and venom enters the surrounding tissues and is transported by the blood throughout the body. Fortunately for most people, a *localized* reaction results; that is, pain, reddening, itching, and swelling occur at the sting site. Sometimes the swelling can be quite alarming, but it usually subsides over a few days. Your unique body chemistry will react in its characteristic way.

On the other hand, you may experience a more serious reaction to bee stings. This is called a *systemic* reaction, a positive sign that you are allergic to bee venom.

A systemic or general reaction means that the entire body is reacting to the venom proteins. Signs of a systemic reaction may include those of a localized reaction as well as other symptoms, such as itching of the extremities (feet, hands, tongue) or all over the body (hives), breathing difficulty, swelling away from the sting site, sneezing, abdominal pain, and loss of consciousness.

Anaphylactic shock reactions are rare but can occur in a very short time in sensitized people who are highly allergic. A person who is severely allergic to bee venom will react after the second or later stings; some people (even beekeepers) can become allergic years later. Bee venom has specific protein allergens that are different from hornet or wasp venom; this means that a person allergic to wasps is not necessarily allergic to honey bee venom. Symptoms include labored breathing, confusion, vomiting, and falling blood pressure. If not treated promptly, such a reaction could lead to fainting and death.

The percentage of people who become allergic to bee venom is very small, but for those individuals, such an allergy must be considered serious. Fewer than 17 deaths per year (in the United States) result from bee stings, which is low compared with the number of deaths due to heart disease (977,700 per year), auto accidents (46,000 per year), and lightning (85 per year). If there is ever any question about whether you are developing an allergy to bee stings, bee hairs, wax, or propolis, you should consult a physician or local allergy clinic immediately!

For more information, see Chapter 5, "What to Do When Stung"; the "Venom" section in the References; and Appendix C, as well as current medical websites.

CHAPTER 1

Understanding Bees

BEE ANCESTORS

Although fossil records are incomplete, insects seem to have first appeared about 300 million years ago, during the Carboniferous period. The probable ancestors of the order Hymenoptera, to which honey bees belong, evolved some 200 million years ago as predatory wasps. Fossil insects preserved in Permian rock, dating from the close of the Paleozoic era, display hymenopteran-like structures, including the membranous wings and the antlike waists.

Approximately 50 million years later, in the middle of the Mesozoic era, the hymenopterans were firmly established in the fossil record, primarily in amber, and included primitive and subsocial ants that were mostly predatory. Bees appear to have evolved from predatory sphecid wasp ancestors, about 100 million years ago (mid-Cretaceous). The switch in diet from animal to vegetable protein and the presence of branched hairs separate bees from wasps. During the vast periods of time that followed, the flowering plants became more specialized and more dependent on mobile pollinators. Insect visitors such as bees were very important, and they, and the plants they pollinated, coevolved structures to their mutual benefit as a result of this interdependence.

It wasn't until 65 million years ago (Tertiary period) that the stinging hymenopterans became common; the land by this time was dominated by the flowering plants, or *angiosperms*, which provided plenty of *pollen* (protein source) and *nectar* (carbohydrate source). The plants that attracted bees because of their shape, color, odor, and food were pollinated and therefore set seed for the next generation. In their turn, bees developed branched hairs on their bodies to trap the pollen of flowers, inflatable sacs to carry away sugary nectars, and a highly structured social order with elaborate defense and communication systems to exploit the most rewarding of floral resources.

In the order Hymenoptera, there are over 200,000 species in 10 or 11 families and about 700 genera. The placement of the honey bee in the animal kingdom is as follows:

Kingdom: Animalia.
- Phylum: Arthropoda (many-jointed, segmented, chitinous invertebrates including lobsters and crabs).
- Class: Hexapoda or Insecta (six-footed).
- Order: Hymenoptera (*Hymen* is the Greek god of marriage; hence the union of front and hind wings [*pteron*]).
- Suborder: Apocrita (ants, bees, wasps).
- Superfamily: Apoidea (between 8 and 10 families).
- Family: Apidae (characterized by food exchange, pollen baskets, storage of honey and pollen); three subfamilies (see the figure on taxonomy).
- Tribe: Apinae (long tongues, nonparasitic, highly eusocial).
- Genus: *Apis* (bee, Linnaeus, 1758; native of the Old World, probably evolved in India and Southeast Asia).
- Species: *mellifera* (honey bearing); also called *mellifica* (honey maker), Western honey bee.

(Note: A new fossil discovered in Nevada contains a now extinct New World honey bee species, newly

3

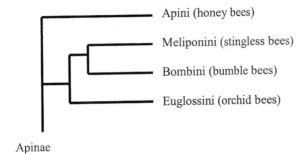

Apinae

Taxonomy of the Family Apidae. After S.A. Cameron. 1993. Multiple origins of advanced eusociality in bees inferred from mitochondrial DNA sequences. Proc. Nat. Acad. Sci. 90:8687–8691.

named *Apis nearctica*; http://www.sciencenews.org, July 2009).

EVOLUTION OF SOCIAL STRUCTURE

Social structure is defined by the degree of community living. The true, highly specialized, or *eusocial*, societies are those of ants, termites, and honey bees. The sophistication of the social structure of honey bees is indicated by a number of characteristics:

- Longevity of the female parent (queen) coexisting with her offspring.
- Presence of reproductive castes (two female castes).
- Siblings assisting in care of the brood.
- Progressive-feeding of brood instead of mass-feeding.
- Division of labor, whereby queen lays eggs and the workers perform other functions.
- Nest and shelter construction, storage of food.
- Swarming as a reproductive process.
- Perennial nature of colony.
- Communication among colony members.

A eusocial community of honey bees can be described as consisting of two female castes, a mother (*queen*) and daughters (sterile *workers*), that overlap at least two generations, and *drones*, which are the male bees. Because hornet and wasp colonies do not overwinter in temperate climates, as do honey bees, they are termed *semisocial* insects. Most insects are solitary—they neither live together in communities nor share the labor of raising their young. The 12,000

species of insects that do live in communities are the ants, termites, wasps, and bees (and include some beetles, aphids, and thrips). The origin of social insects troubled even Darwin, who could not rationalize how a special, sterile caste—the workers—could pass on their genetic information if they could not produce offspring. In other words, workers were displaying "altruism" toward their siblings at the expense of having their own children. Over the years, there have been several theories explaining the evolution of sociality. Here are a few of the more popular ones, briefly described: kin selection, mutualism, and parental manipulation.

Kin selection theory explains that workers are genetically more related to each other than they are to their parents, owing to the *haploid* (possessing only one set of chromosomes) drones. Therefore, it is more advantageous for workers (as they have no children) to rear their siblings; helping them is like workers helping themselves, enabling their genes to pass on to the next generation.

Mutualism maintains that an individual queen benefits if others, especially if they are her sisters, help in rearing her brood. If they help one another, they will be more successful and have more offspring that survive to the next generation. Such cooperation may eventually lead to members of the same species occupying a composite nest and developing some kind of communal brood care.

Parental manipulation asserts that the mother gains net survival or reproductive success by tricking or manipulating others. This theory evolved because the queen dominates her daughters by means of chemical signals called *pheromones*. Such signals reduce the reproductive potential of the daughters, forcing them to become slaves and tend their siblings instead of laying their own eggs.

Existing subsocial and primitively social insects incorporate some or all of the behaviors described by these theories. Newer scientific techniques and molecular ecology have made rapid advances in this area. Evolution of social behavior makes for fascinating reading, and you should look at other works on the complex social behavior of insects for a more complete understanding of the subject (see "Social Insects" in the References).

RACES OF BEES

There are seven to ten species of honey bees in the genus *Apis* (depending on which scientist you agree with). A *species* is a group of organisms that can interbreed, producing offspring that can do the same. Currently the species include the Western or European honey bee, *A. mellifera* Linnaeus 1758; Dwarf honey bees, *A. florea* Fabricius 1787 and *A. andreniformis* Smith 1858; and Giant honey bees, *A. dorsata* Fabricius 1793. The East Asian bees are *Apis cerana* F 1793, *A. koschevnikovi* Enderlein, and *A. nigrocincta* Smith 1861. Some bee researchers have identified new species as well as new races (subspecies) in *mellifera* (24 total), a single one in *cerana* (*A. nuluensis*), and three in *A. dorsata* (*A. binghami, laboriosa,* and *breviligula*). See "Social Insects" in the References.

By races, we are referring to populations of the same species (e.g., *mellifera*) that originally occupied particular geographic regions with different climates, topography, and floral resources. In these different regions, bees evolved characteristics that made them unique from other species. F. Ruttner, in *Biogeography and Taxonomy of Honeybees* (1988), divided bees into four groups: (1) African, (2) Near East, (3) Central Mediterranean and southeastern European, and (4) Western Mediterranean and northwestern European. From the European groups came the Italian, Carniolan, and German black bees; the Near East group includes the Caucasian bees. These four races provide the raw materials from which modern hybrid bees used mostly in the United States are derived. There are current efforts to preserve the pure races in some European countries.

The German dark bees were first brought from Europe to North America by the early American colonists in 1622 to pollinate the newly flowering orchards (e.g., apples). Equally important was their wax (for candles and waterproofing) and their honey, which provided an affordable sweetener. Then, in 1859, the first Italian queens were imported to America. This *A. mellifera* race was quickly recognized as superior to the German black bee, because it is less aggressive, has a longer tongue, and has higher resistance to bee diseases. Because the Italian honey bee was more desirable, use of the German black bee diminished, and now few beekeepers in North America have these bees. Honey bees, called "white man's fly" by the Native Americans, moved quickly across the North American continent with the early settlers—both over land and by ship—arriving in California in 1853.

Today, the Italian honey bee is the most widely distributed bee in the Western Hemisphere. The other two popular races, the Carniolan and the Caucasian, were brought to the United States circa 1883 and 1905, respectively. As with the Italian bees, they are frequently crossbred, interbred, and inbred for disease resistance, hardiness, and gentleness.

Importation of live adult bees into the United States was halted in 1922 because of the danger of introducing bee diseases and pests that did not already exist here. This action was precipitated by the discovery of tracheal mites on the Isle of Wight (in the United Kingdom) in 1919. However, in 2006 and 2007, package bees began to be imported from Australia to help pollinate California almonds.

South America did not have such restrictions, and in the 1950s Brazil imported African honey bees (*A. m. scutellata* Lepeletier 1836, one of many African races) to improve breeding stock. The accidental release of the volatile bee known as the *Africanized* honey bee (labeled the "killer bee" by the press) has led to the spread of this subspecies throughout all of South America, all of Central America, and Mexico. In 1990, swarms of the Africanized bee crossed the border and became established in the southern United States and parts of California. Go to: http://www.ars.usda.gov/Research/docs.htm?docid=11059&page=6 for the most recent updates. (For more information, see "Africanized Bees" in Appendix E and the References.) How many states this bee will ultimately occupy is presently disputed, but there is general consensus that the southern half of the United States will have them as year-round or summer residents.

Although the most common honey bee in America is the Italian, you may be interested in experimenting with other bee races or hybrids. If you raise your own queens, and do not control the drone source, crossbreeding could result in inferior queens. It is important to pay attention to the quality of both the queen mother and the drone fathers of the daughter queens you are rearing (see "Queen Rearing" in Chapter 10). Some beekeepers try to maintain only one race of bees in any one apiary, believing that pure strains are more resistant to diseases. A general overview of the most commonly available races of honey bees now used in the United States is provided below. A

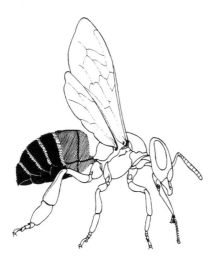

Italian Honey Bee

good reference is Graham's *The Hive and the Honey Bee* (1992).

The Italian honey bee (*Apis mellifera ligustica* Spinola 1806) originated in the Apennine Peninsula (the boot) of Italy. This race has several color types. In general, they are yellow with dark brown bands on their abdomen; the "goldens" have five bands, while the "leathers" have three. They are known for laying a solid brood pattern, producing lots of bees late into the fall, and making a good surplus of honey. On the other hand, they forage for shorter distances and therefore tend to rob nearby colonies. They also drift frequently because they orient by color rather than by object placement.

Advantages

- Good, compact brood pattern, making a strong workforce for collecting lots of nectar and pollen.
- Excellent foragers.
- Light color, making the queen easy to locate.
- Moderate tendency to swarm.
- Moderate propolizers, so hive furniture is not glued together too much.
- Resistant to European foulbrood disease.
- Relatively gentle and calm, making them easy to work.
- Moderate to high cleaning (hygienic) behavior.
- Readily build comb cells; white cappings common.

Disadvantages

- Can build lots of brace and burr comb; Italians have a slightly smaller cell size.

- Poor orientation to home hive; drift to other hives, spreading diseases/pests and causing uneven colony populations.
- Can be bothersome by persistently flying at beekeeper when worked.
- Short-distance foragers, thus have a tendency to rob weaker hives, creating a robbing frenzy in the apiary.
- Can be susceptible to many diseases and pests.
- Slow to build populations in spring, not good for early honeyflow.
- Brood rearing continues after main honeyflow has ceased, sometimes late into fall; bees may enter the winter period with too much brood and too little honey, resulting in starvation.

The **Carniolan honey bee** (*Apis mellifera carnica* Pollmann 1879) was originally brought from Yugoslavia and Austria, where the winters are cold and the honeyflows variable. They are popular in northern areas of the United States. Although they are a variety of Italian bees, Carniolans have a grayish black-brown body with light hairs; the drones and queens are dark in color. In general, they were bred for fast buildup when the spring flow starts and to shut down brood production early in the fall. They are known for their gentle disposition and low propolis and brace comb production, but they can swarm if they are not given ample expansion room. Currently the New World Carniolans are found in the United States, developed and improved by Sue Cobey (University of California, Davis, and Washington State University).

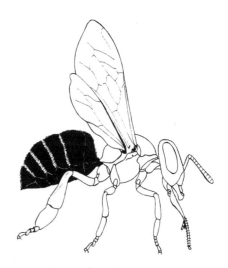

Carniolan Honey Bee

Advantages

- Rapid population buildup in early spring; good for spring pollination and early nectar flows.
- Brood rearing decreases if available forage is diminished, thus conserving honey stores.
- Exceptionally gentle; less prone to sting and easier to work.
- Few brood diseases, so less medication may be needed.
- Economic honey consumers, therefore they overwinter on smaller honey/pollen stores.
- Little robbing instinct, as they are long-distance foragers and are object oriented.
- Can have very white wax cappings, making comb honey sections attractive to customers.
- Little brace comb and propolis, making hive manipulations easier.
- Overwinter well; queen stops laying in fall and small number of bees overwinter on fewer stores.
- By comparison to other races, forage earlier in the morning, on cool, wet days, and later into the afternoon.

Disadvantages

- Tend to swarm unless given enough room.
- Strong brood population depends on ample supply of pollen; can be slow to build up in summer if forage is late or unavailable.
- Dark queen difficult to locate, making requeening operations slower.

The **Caucasian honey bee** (*Apis mellifera caucasica* Gorbacher 1916) is originally from the high valleys of the central Caucasus near the Black Sea, where the climate ranges from humid subtropical to cool temperate. Caucasian bees are black with gray or brown spots and short gray hairs. They have the longest tongue compared with the other two races described here, which could make them superior pollinators of some crops. The drones are dark with dark hairs on the thorax. These bees were introduced into the United States from Russia, circa 1905. In general, they are gentle bees, with low swarming instincts, and are good in areas of marginal forage or long honeyflows. However, sources and breeders of these bees are currently difficult to find.

Advantages

- Build strong populations, but slow to start in the spring; not good for early spring crops.
- Gentle and calm on comb, making them easier to work.
- Have a long tongue so can exploit more species of flowers.
- Little tendency to swarm, resulting in strong colonies.
- Forage at lower temperatures, earlier in the day and on cool, wet days.
- Overwinter well, shutting down brood production in the fall; conserve stores.

Disadvantages

- Maximum propolizers, making hive manipulations difficult unless collecting propolis for sale.
- Can have wet wax cappings over honey, making comb honey less attractive to consumers.
- Can sting persistently when aroused, making inspections difficult.
- Late starters in spring brood rearing; not good for early spring pollination.
- Dark queen difficult to find.
- Can drift and rob.
- More susceptible to nosema disease; may require more medication.
- Difficult to find breeders.

Hybrid Bees and Select Lines

In addition to these races, there are hybrid bees, which can be crosses between the races or between selected strains within a race. Some common hybrids were the Starline (four-way Italian cross), Midnite (Caucasians × Carniolan), and Buckfast bee lines. Many queen breeders have their own variations of these races as well, to meet the needs of their customers.

The **Buckfast Hybrid** was a product of Brother Adam (1898–1996) from the Buckfast Abbey in the United Kingdom. He crossed many races of bees (primarily Anatolians with Italians and Carniolans) in search of a superior breed that would be tolerant of tracheal mites, be gentle and productive, have high cleaning instincts and disease resistance, and possess good overwintering abilities. Currently, this hybrid is difficult to find.

The **Starline Hybrid** combined several Italian

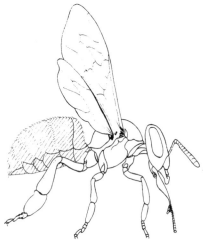

Cordovan

stocks. These bees were known for their gentle behavior and large brood numbers, which could provide a large workforce to exploit the many nectar resources. This line was ideal for commercial beekeepers who needed lots of bees for pollination and honey production. This line is not available anymore.

The **Cordovan** line is a color mutation, originally developed by Dr. Bud Cale to serve as a genetic marker in his Starline queen-breeding program. Cordovan bees were used to trace behavior and kinship relationships for research purposes (now done with molecular markers and single-drone inseminated queens). They are easily identified because the black body color comes out red in Italians and a purple-bronze in Caucasian and Carniolan bees; this latter bee is called a Purple Cordovan. They are known for their gentle behavior and pretty color, and are excellent for showcase observation hives. Some of these lines are currently being bred for disease resistance as well.

Russian bees are an example of a select line; they were carefully introduced as a Varroa-tolerant line from northeastern Russia and were imported and isolated on a Louisiana island by Dr. Tom Rinderer in 1995 (USDA lab in Baton Rouge, Louisiana). This Russian stock is now available from selected breeders. Check the bee journals for breeders of these selected mite-resistant lines.

New hybrids and crosses that are resistant to varroa mites have since been developed. These include the VSH (Varroa Sensitive Hygiene), once labeled SMR (Suppressed Mite Reproduction), and the Hygienic lines of bees, which were bred for cleanliness (pulling out diseased brood). The VSH was developed by Drs. John Harbo and Jeff Harris from the Baton Rouge USDA-ARS lab. Dr. Marla Spivak from the University of Minnesota developed a line selected for hygienic behavior in queens, and these are now widely available. Hygienic bees were originally recognized and developed by Dr. Walter Rothenbuhler (1920–2002) of Ohio State University, who was breeding them to remove diseased (foulbrood) larvae; now these bees will also remove bees killed by mites as well as disease.

Advantages

- Tolerant of tracheal and/or varroa mites (VSH, Hygienic, Russian, Buckfast).
- Low swarming instinct.
- Minimal propolizers.
- Disease resistant (Hygienic lines), including chalkbrood.

Disadvantages

- Some lines build populations slowly in spring unless good honeyflow is in progress; not good for early spring pollination.
- Offspring queens from hybrid mother may not be like the original queen; daughter queens or their progeny may not have desired characteristics.
- Poorly mated or open-mated queens may not perform as advertised.
- Super hygienic lines may not build up good populations of bees, as they are pulling out brood and mites (VSH).
- Requeening every other year may be necessary to ensure the colony is headed by hybrid queen and was not superseded.

A final note: These and new lines that are being developed should be tried with caution, as they may not perform as advertised, because of either poor breeding, hive location in your area, or the geographic region where you live. If you are planning to requeen colonies with particular lines or hybrids, try 5 to 10 queens from one breeder first, to evaluate how they will do in your area; also select different breeders to determine which give the best results. Take notes and compare with information collected by other beekeepers in your area. Open mating where the sources of drones are not controlled can lead to daughter queens lacking the desired, advertised traits.

External Anatomy of a Worker Honey Bee

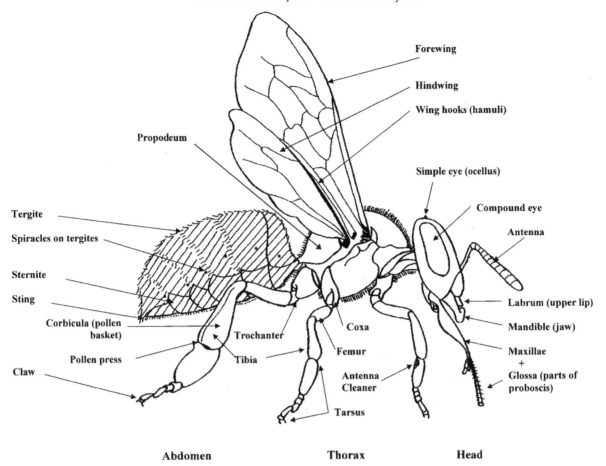

Forewing

Hindwing

Wing hooks (hamuli)

Simple eye (ocellus)

Compound eye

Antenna

Propodeum

Tergite

Spiracles on tergites

Sternite

Sting

Corbicula (pollen basket)

Pollen press

Claw

Trochanter

Tibia

Coxa

Femur

Antenna Cleaner

Tarsus

Labrum (upper lip)

Mandible (jaw)

Maxillae
+
Glossa (parts of proboscis)

Abdomen

Thorax

Head

EXTERNAL STRUCTURE OF A BEE

The anatomy of the honey bee is similar to that of other insects, except for the specialization of certain organs and structures needed by bees to carry out functions peculiar to them. Parts common to other insects include the *head, thorax,* and *abdomen*; the hard, waxy protein covering (chitin); the free respiratory system with tracheae (no lungs); the ventral or bottom spinal cord; and the open circulatory system (no veins or arteries); see the illustration on external anatomy on this page.

Located on the head are five eyes, the antennae, and the feeding structures: the tongue (*proboscis*) and the jaws (*mandibles*). The proboscis is for lapping and sucking fluids (like water, nectar, and honey), and the mandibles are used for chewing pollen and, in the case of workers, shaping the beeswax.

The thorax, or middle section of the bee, contains the muscles that control the two pairs of wings; other muscles control the three pairs of legs. The legs have specialized structures and hairs on them that assist the bee in cleaning itself and in collecting and carrying pollen. The armor-plated thorax is perforated with pairs of holes, called *spiracles*, which are part of the breathing or respiratory system. The first pair, called the *prothoracic spiracle*, is the site where tracheal mites can be found (see "Tracheal Mites or Acarine Disease" in Chapter 14). The second pair of spiracles is nonfunctional, and the last pair, although located on the thorax, is really on the first abdominal segment, called the *propodeum*.

The abdomen is the longest part of the bee and contains important organs. It is armor plated with scale-like segments, called *tergites* (top segments) and *sternites* (bottom segments), that protect the bee and keep it from drying out. It is also perforated with seven more pairs of spiracles. The bee's sting, found on only the female castes, is located in the tip of the abdomen. Wax-secreting glands, on the underside of the abdomen, and the scent gland are important abdominal glands of the worker bees.

The queen's abdomen contains ovaries for egg production, a storage sac for drone semen, many glands that produce pheromones and a sting, but no wax glands (see Appendix A for more information on internal anatomy and Appendix B on pheromones). The drone's abdomen contains the male reproductive organs but has no wax glands and no sting. Sometimes a drone can be found with both male and female parts; these rare *gynandromorphs* may actually be able to sting you!

An excellent and comprehensive book on bee anatomy is Goodman's *Form and Function in the Honey Bee* (2003).

Bee Vision

Bees have five eyes—three simple (ocelli) and two compound. The ocellus is a thick, biconvex lens or cornea that reacts only to light intensities. The compound eyes are composed of thousands of individual light-sensitive cells called *ommatidia* (singular, *ommatidium*). It is with the compound eyes that bees perceive color, light, and directional information from the sun's rays.

The color range of bee vision includes violet, blue, blue-green, yellow, and orange as well as ultraviolet light, which is invisible to humans. Because they compete with each other for available pollinators, flowers that depend on bee pollination are within these color ranges. The plants that succeeded in attracting bees with their color, nectar, and pollen gained an edge over other plants during their evolutionary development.

The structures and arrangement of the ommatidia permit polarized light to pass through certain parts of each ommatidium at any given instance. The sun's position and the bee's direction are the factors determining which section of the ommatidia will receive full, partial, or shaded regions. This pattern serves as a "compass" to the bee, giving directional information. The bee is able to monitor these shifting patterns continually as it flies and, if necessary, adjust its course.

Antennae

Most of the tactile (touch) and olfactory (smell) receptors of bees are located on the antennal segments.

These receptors are in the form of seta or hair tactoreceptors and plates and recessed chemoreceptors. These sensory organs help guide bees both inside and outside the hive and enable them to differentiate between hive, floral, and pheromone odors. If the antennae are cut off, the bee will not be able to negotiate within and outside the colony, will be unable to make comb, and will soon die. Once odors or other tactile stimulation are detected, signals are transmitted down the nerve cord to the brain. There are about 3000 plate organs on each antenna of the queen, 3600 to 6000 in the worker, and 30,000 in the drone.

Pollen-Collecting Structures

The hind legs of worker bees are specialized for collecting and carrying pollen. An inner segment on the hind leg is covered with numerous hairs, forming the *pollen combs*. Bees actively collect pollen by scraping it off of flowers with their jaws and legs; as the pollen is removed, a small amount of liquid from the honey stomach is added to make it sticky. During collection, additional pollen adheres to the bee's body by static electricity. The collected pollen is then transferred by the bee to areas on its body where it can be reached and removed by the pollen combs.

Removal of the pollen from the pollen combs is accomplished by rubbing the legs together so that the pollen is squeezed from the inner side to the outside of the legs. The pollen will be deposited eventually into a depression called the *pollen basket* (see p. 246, Appendix A). When the baskets are full, the bee returns to the hive, backs into a cell, and deposits the pollen pellets. The hive bees will add bacteria and enzymes to the raw pollen and pack it in solidly. Then the pollen, through the action of fermentation, similar to a silo fermenting cattle feed, is worked on first by bacteria, then fungi, yeasts, and molds until it eventually turns into nutritious bee bread. The formation of bee bread currently is the focus of new research to study the role of beneficial microorganisms. In the fall, cells containing bee bread are capped with a thin layer of honey and used for winter stores. Bee bread is more nutritious than raw pollen and feeds all the members of the colony. Queens and drones do not have pollen-collecting structures on their legs. See Chapter 15 for more on why bees are excellent pollinators.

Wings

The four wings of bees are designed to fold together over the abdomen while the bee is inside the colony. But for better stability during flight, the wings are held horizontally and hooked together with special wing hooks called *hamuli* (singular, *hamulus*). Wing veins help keep the wings rigid and supply blood. Honey bee wings have a distinct pattern composed of only four major veins; this is much less than what most other insects, such as dragonflies, have.

Honey Stomach, Sac, or Crop

The esophagus of the bee begins at the back of the mouth and continues through the thorax, terminating in the anterior part of the abdomen, where it expands into the crop, or honey stomach. Workers temporarily store collected nectar, honeydew, and water in this sac. The walls of the honey stomach are pleated and invaginated, allowing for it to greatly expand as the worker carries a heavy load of liquid. We now know that beneficial bacteria (*Lactobacillus* spp.) live inside the stomach and are transferred to both the nectar and the pollen when they are unloaded inside the hive. These bacteria are important in the formation of honey and bee bread, and may even provide the bees some protection against pathogens.

A muscular valve at the posterior end of the crop called the *proventriculus* controls when the contents of the honey stomach are transferred to young hive bees, which add additional enzymes to the liquid and work the nectar with their proboscis to aid in the evaporation of excess moisture. To remove further excess moisture, they place the droplet of nectar in the cells for drying and curing, so it can turn into honey.

The Sting

Stinging insects, or aculeates, belong to the order Hymenoptera, which includes both social and solitary bees and wasps. The more defensive species of stinging insects are the hornets and the yellowjackets (both of the Vespidae family); less volatile are the bumble bees (Bombini) and the honey bees (Apidae). The venoms of all these stinging insects are not chemically alike. Thus, a beekeeper who is allergic to yellowjacket venom will not necessarily develop an allergy to honey bee venom or the venom of other stinging insects.

The stinging mechanism is a modification of the egg-laying equipment (ovipositor) of female insects. Queens generally use their sting only to dispatch rival queens. The entire stinging apparatus consists of a poison sac (sometimes called the *acid gland*), an alkali (or Dufour) gland, associated alarm substances, and the mechanical equipment (muscles and hardened plates) of the sting (see Appendix C for information on sting reaction and anatomy).

The recurved barbs on the sting's lancet on the worker catch in the victim's skin and, as the bee pulls away, the entire sting structure, including the venom sac, is ripped out of the bee's body. Muscle pumps near the base of the now-detached sac force more venom into the wound for about a minute. Alarm odors are released at the sting site, inducing other workers to sting there. To minimize the amount of venom received, it is important to remove the sting **promptly** by scraping or flicking it off with your fingernail. Since the sting site is now "tagged" with alarm odors, apply smoke to the sting site to mask these alarm odors. Bees usually die shortly after stinging, but occasionally they live for hours or even days (for more information on the compounds in bee venom, see "Bee Venom" in Chapter 12). Africanized bees have the same venom as European bees but are more volatile and respond quickly to the release of the alarm odors.

THE WORKER

There are three types of bees in a colony, divided into two female castes (workers and queens) and the males or drones (see Appendix A for information on the morphology of the different bees). The most numerous members of a bee colony are the workers, the sterile female caste incapable of laying fertile eggs; in a normal hive they reach a peak population of 40,000 or more by midsummer. The workers are smaller than the drones and have a shorter abdomen than the queen.

Life Stages of a Bee

Under normal circumstances, the queen lays all the eggs in a hive. If the queen is lost and the bees are

Table 1-1 Average Development Time of a European Honey Bee

	Egg[a]	Larva	Pupa	Total	Adult Life Span	Weight[b]
Queen	Fertilized 3 days	4.6 days	7.5 days	15 to 17 days	2-5 years	178-292 mg
Worker	Fertilized 3 days	6.0 days	12.0 days	19 to 22 days	15-38 days summer	81-151 mg
					140-320 d winter[c]	
Drone	Unfertilized 3 days	6.3 days	14.5 days	24 to 25 days	8 weeks	196-225 mg

Sources: M.L. Winston. 1987. Biology of the honey bee. Cambridge, MA: Harvard University Press. E. Crane. 1990. Bees and beekeeping: Science, practice, and world resources. Ithaca, NY: Comstock.

Note: Average time between metamorphic stages, in days at 93°F (33.9°C). Conversion: 1 mg = 0.000035 oz.; 1 mm = 0.004 in.

[a] Egg dimensions: Worker and queen eggs weigh 0.12–0.22 mg, are 1.3–1.8 mm long, and take 48–144 hours to hatch, with an average of 72 hours.

[b] Weight at emergence. Weights of emerging adults vary depending on cell size, number of nurse bees, colony population, food availability and type, and season of the year.

[c] Workers in winter have well-developed hypopharyngeal glands and more fat bodies, which may enable them to live longer.

unable to rear a new one, workers often lay unfertilized eggs. Workers and queen bees hatch from fertilized eggs, and drones from unfertilized eggs; a *fertilized* egg is one formed by the union of a sperm and an egg. The egg (whether fertilized or not) consists of a nucleus, which contains the genetic material, and a large yolk reservoir, which will provide nutrients (food) for the developing embryo. The nucleus within the egg begins to divide into 2 cells, then 4, 8, 16, and so on. These rapidly multiplying cells eventually form a layer of cells called the *blastoderm*; part of this layer of cells thickens to form the *germ band*, which marks the beginning of the embryo. When the embryo becomes a completely developed larva it emerges from the egg. As the larva grows, it will gradually differentiate into the various organs and tissues that make up the adult bee.

The egg is incubated in the nursery region (called the *broodnest*) of the comb at 91.4° to 96.8°F (33–36°C). Most insects, such as the butterfly, beetle, fly, and honey bee pass through four stages: egg, larva, pupa, and adult. Insects that pass through these four stages are said to undergo *complete metamorphosis*. Eggs lose about 30 percent of their weight during incubation and, after 48 to 144 hours (average, 72 hours), depending on the hive temperature and race of bee, hatch into larvae. All honey bee eggs hatch, not by rupturing the shell (*chorion*) as in most insects, but by gradual dissolution of the membrane during hatching, a characteristic unique to honey bees. Genetics and race of bees will dictate how well

the egg hatches and the brood survives. It is essential for bees to have survival strategies that involve queens laying a large number of eggs per day as well as colony numbers sufficient to incubate those eggs, feed the hatching larvae, and build adequate space for the new population of bees.

When death occurs during any of these metamorphic stages, adult bees will clean the dead material, sometimes removing it and sometimes eating it, the latter especially during a dearth time. In either case, they are performing a crucial hygienic function, which is an important genetic trait.

Once hatched, the larva (white, wormlike grubs) becomes an eating machine, with a huge digestion system consisting of a mouth, spiracles, midgut, hindgut, salivary and silk glands, and (closed) excretory tubes. You can see a larva in its cell, a white C-shaped worm lying in the bottom. Each larva is fed between 150 and 800 times per day and will gain more than 900 times the egg weight by the fifth day (see table on development time and Appendix E). About 33 percent of the larval weight (dry weight) is made up of *fat bodies*, organs that are utilized in the pupal stage. To grow this fast, the larva has to molt six times because the skin is unable to expand sufficiently to accommodate the rapidly growing insect. Four molts take place during the first four days, one as a prepupa and the last right before the bee emerges.

Two different diets are fed larvae destined to become worker bees. First, the larvae are lavishly or

Developmental Stages of Honey Bees

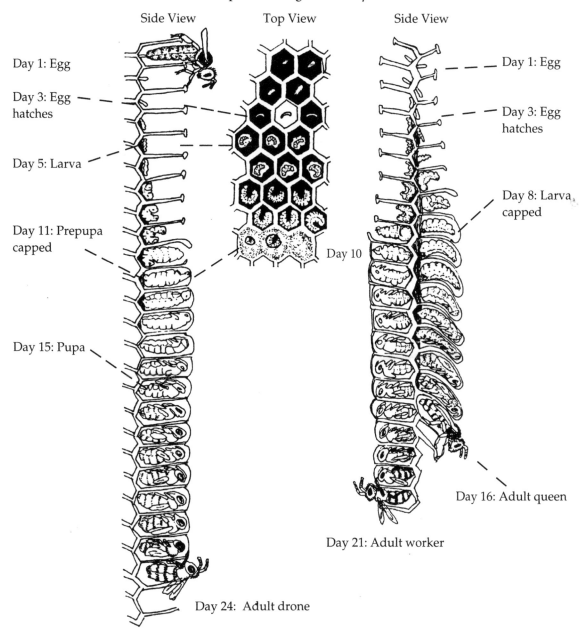

Side View   Top View   Side View

Day 1: Egg

Day 3: Egg
hatches

Day 5: Larva

Day 11: Prepupa
capped

Day 15: Pupa

Day 10

Day 1: Egg

Day 3: Egg
hatches

Day 8: Larva,
capped

Day 16: Adult queen

Day 21: Adult worker

Day 24: Adult drone

mass-fed a diet of brood food (sometimes called *worker jelly*), which is a combination of 60 to 80 percent clear fluid produced by the *hypopharyngeal food glands* of nurse bees (young worker bees whose food glands are active), plus 20 to 40 percent milky fluid produced by their *mandibular glands.* On the third day, the diet is switched to the clear component, then to bee bread and honey on the remaining days; at this time, the larvae are fed progressively, or only as needed. This diet change (which contains fewer proteins, lipids, minerals, vitamins, and sugars) and

the switch from mass- to progressive-feeding appear to be responsible for the differentiation into worker bees. Other larvae, hatching from fertilized eggs and mass-fed only royal jelly, develop into queen bees.

Between day 8 and 9 (see illustration of stages of bees on this page), the cell, which until this time is called *uncapped* or *open brood*, is capped with a wax-like cover, and is now called *capped* or *sealed brood.* This wax "cap" consists of old wax, propolis, and other components. The larva molts into a prepupa, defecates, and spins a cocoon with silk produced from the thoracic salivary glands. Normal hive temperatures of about 95°F (35°C) are necessary for normal development; if they are lower, the development time can be delayed by several days.

The next stage is called the pupal stage, where massive internal and external morphological changes take place. Recognizable parts of the bee form during this time—the legs, wings, and abdomen—and all the internal organs and muscles develop. The pupae are full of fat bodies (cell-like organs) that serve as food storage units for lipids (waxes, oils, fatty acids, and steroids) and glycogen (a stored form of glucose sugar). Fat bodies also contain essential compounds called amino acids, which assist in hardening the cuticle of young bees. In addition, they help synthesize proteins (long-chained molecules that are the foundation of all living organisms) and enzymes (proteins capable of speeding up chemical reactions). Some mitochondria are located in fat bodies; these organelles convert food into high energy molecules, specifically **adenosine-5'-triphosphate** (ATP).

The pupal skin or *cuticle* gradually darkens, and after a final pupal molt the adult is ready to emerge. Because the cuticle is so soft, this bee, called a *teneral* or *callow* bee, stays inside her cell three to four hours to harden before emerging. At the end of day 20 or 21, the emerging bee chews a hole in the cell cap sufficient to permit escape; the remaining covering is reused by other bees for other brood cappings. Teneral bees can't sting, they are relatively soft, and their thoracic hairs are light in color and matted down.

Young bees are still full of fat bodies but must receive the beneficial bacteria from the nurse bees (the so-called social stomach) and need to ingest pollen proteins. This must happen within the first few hours after emergence because these emerging bees have no microorganisms or food in their gut. Without these bacteria and the protein, their life span will be shorter and glandular development will be impaired. They will need this extra protein until they are five days old; they will also continue to beg for nutritious brood food from other nurse bees.

This young worker bee soon begins the first of many tasks she will perform during her life span. Over the next few days, glandular development, genetics, pupal temperatures, and environmental conditions rule bee activities (see Chapter 2). The worker bee's age and the needs of the colony dictate the work she is to do for the rest of her life. Generally, workers

from one to three weeks old remain within the hive where they:

- Rest.
- Feed and clean larvae and their cells.
- Tend the queen (feed, groom, and help spread queen pheromones).
- Clean the cells and the hive.
- Secrete wax, build new comb, and cap cells containing honey, bee bread, and brood.
- Guard the entrance and other areas of the hive.
- Patrol the hive; look for intruders.
- Help to heat or cool the hive as needed.
- Accept nectar from foragers, store, and cure it.
- Pack pollen.
- Take brief orientation flights to familiarize themselves with landmarks near the hive (also called *play flights*). Drones also do this, and some flight activity, such as when swarming or when the queen is on a mating flight, may also be happening.

After about three weeks of hive duties, the glands that produce the larval food and wax begin to atrophy. These workers then move away from the warm broodnest (where the eggs, larvae, and pupae are) onto broodless combs. Here they come in contact with returning foragers and are eventually recruited to food sources.

As foragers, they collect honeydew, pollen, nectar, water, and propolis. Foraging activities take a heavy toll on workers, and most of them die after performing outside duties for about three weeks. During the winter, however, many workers survive for several months. For a complete breakdown of worker activities, see Chapter 2.

THE QUEEN

Bee colonies are usually *monogynous*—that is, they have only one egg producer, the queen. The queen is the longest bee in the colony; her wasplike, slender abdomen, usually without color bands, distinguishes her from both workers and drones (see the illustration comparing the worker, queen, and drone on p. 15). Any larva that hatches from a fertilized egg is a potential queen. Thus, worker bees can raise a new queen from a larva up to three days old, either when their old queen has been accidentally lost, when she

Relative Cell Sizes

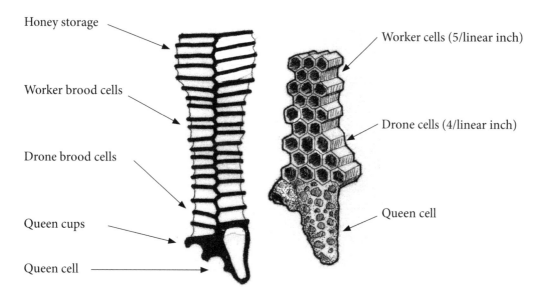

Honey storage

Worker brood cells

Drone brood cells

Queen cups

Queen cell

Worker cells (5/linear inch)

Drone cells (4/linear inch)

Queen cell

has been removed by a beekeeper, or when she is injured or too old to perform her duties.

The ability to find the queen is important, because you may need to confirm her presence in the colony or you may wish to replace her. *Requeening,* or replacing an old queen with a new one, is successfully accomplished when the existing queen is located and removed from the colony. New beekeepers need to gain the facility to find the queen among the workers and drones; once you have accomplished this, you are truly a beekeeper.

Any larva hatching from a fertilized egg is a female bee. This fact simplifies the raising of queens for commercial purposes and gives worker bees a wide latitude in selecting larvae to become new queens (see Chapter 10).

The pathway the female larvae follow is directly connected to the food they receive during their larval life. Worker bees often initiate queen rearing by constructing special cup-shaped cells. These *queen cups* are usually located on the lower edges of combs; after the queen has deposited an egg in them, they are then

Worker, Queen, and Drone Bees

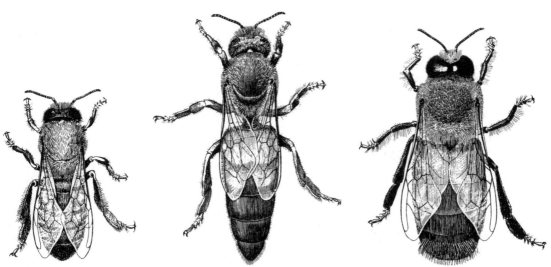

Source: E.C. Martin et al. 1980. Beekeeping in the United States. USDA Ag. Handbook 335.

called *queen cells*. The presence of cups or larvae in queen cells does not necessarily lead to the production of queens.

When queen cells are needed, worker bees construct cups or clean existing ones, or if this hasn't been done, worker cells are modified into queen cells. Larvae in these cells or cups are mass-fed royal jelly during their entire larval development. The royal diet contains several components. For the first three days it consists of white mandibular gland secretions, after which these secretions are fed in copious amounts in a 1:1 ratio with the clear component of the hypopharyngeal glands, similar to worker jelly. On the last two days of larval life there is an important addition: honey. This diet shift, with the elevated sugar content and the addition of high levels of a hormone called *juvenile hormone*, produces queen bees; this is unique to honey bees. High levels of juvenile hormone induce proteins and enzymes specific to queens, which affect the developing tissues and thus produce a queen.

Whether cells containing queen larvae begin as cups or as worker cells, as the larvae grow, the worker bees enlarge and elongate the cells, which gradually take on a peanut-like appearance. The openings of drone and worker cells lie horizontally but are inclined slightly upward on the comb; cells that cradle the queens hang vertically (see illustration showing relative cell sizes on p. 15).

Isn't it interesting that young worker bees play such an important role in a colony by selecting the next generation of queens? Remember, female larvae selected to be workers begin larval life on a diet similar to that of queen larvae, but after two days they are weaned from it and thereafter receive worker jelly, which consists primarily of proteins mixed with honey and pollen.

Queen Cell Production by Bees

Three conditions trigger queen rearing by honey bees: (1) The colony is making preparations for swarming, (2) the queen's physiological and behavioral activities are substandard, or (3) the queen is lost or dies. In each case, the purpose is to replace the existing or resident queen in the colony.

Swarming is a process whereby honey bee colonies reproduce a new colony. The existing queen departs the colony with about half of the workers and a few hundred drones, to form a new colony. Colonies preparing to swarm begin this process by constructing a great number (from 10 to 40) of queen cups. Newly constructed cups are light yellow and hang vertically from the lower edges of the honeycomb. These cups become queen cells once the queen deposits eggs in them and larvae begin to grow. If the cells are found during the swarming season—a period when colonies are casting swarms—they are called *swarm cells*.

The second condition that leads to queen cell construction occurs when bees prepare to replace a queen that is substandard; this type of replacement is called *supersedure*. Workers begin this process either by constructing queen cups or by modifying existing worker cells containing young larvae. In this case, these supersedure cells are few in number and can be found throughout the brood, on the face of the comb. Because these cups or modified worker cells are constructed of old wax, they will be brown in color.

This replacement is triggered when the queen's physiological or behavioral activities or both decline —for example, her egg production is declining or her pheromone levels are reduced (usually an aging queen) or she is injured. Worker bees are able to recognize these conditions and will rear queens to replace the resident one. After the new queen hatches, mates, and begins to lay eggs, she may coexist with her ailing mother. In time, however, only the replacement queen will be found.

The last condition triggering queen replacement, or *emergency*, occurs when the queen is absent from the colony. This can be from natural causes, beekeeper clumsiness, disease, or predation. Occasionally, the queen will fall off the comb during hive inspection and is unable to return to the hive, or is crushed between two frames as they are being removed or replaced (a process called *rolling the queen*). In such cases, unless by good fortune there are queen cells already present, bees must turn worker cells into queen cells. Worker bees will "select" and feed young larvae in worker cells and modify them into queen cells as the larvae grow; emergency cells are found on the face of the comb, but in small numbers.

Virgin Queens

While still in the queen cells, virgin queens will often *pipe* or *quack* to one another. After emerging, a

Queen mates with up to 20 drones

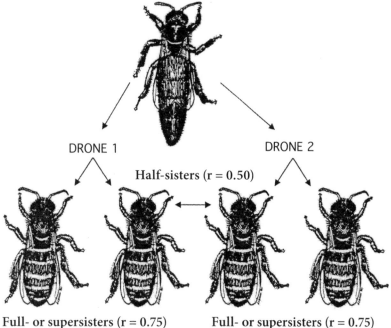

DRONE 1   DRONE 2

Half-sisters (r = 0.50)

Full- or supersisters (r = 0.75)   Full- or supersisters (r = 0.75)

If the queen's worker offspring have the same fathers (drones), they are full- or supersisters;
if they have different fathers, they are half-sisters.

virgin queen may toot or call to other virgins. She then begins to search for and partially destroy any other queen cells, leaving the workers to discard the pupae or larvae inside. Some cells may contain queens ready to emerge, in which case she will partially open these cells and sting the occupants. While performing these tasks, she may also encounter other emerged queens; fighting ensues and ultimately only one virgin queen survives.

About six days after emerging, the queen will leave the hive on a mating flight; if weather is inclement, this flight will be delayed until more favorable weather appears. During her flight, the queen's pheromones attract male bees from *drone-congregating areas* (DCAs), and she may mate with up to 10 or more drones in succession, over a few days. When her sperm sac (*spermatheca*) is filled, she will never leave the colony again, unless accompanying a swarm. Three days or so after mating, the now bigger and heavier queen will begin to lay eggs. The queen continues to lay eggs the rest of her life, pausing for a month or so in late fall. It has been reported that a queen is able to lay over 1000 eggs a day for brief periods, provided there is enough space and worker bees

to incubate and care for these eggs. Over her lifetime, a good queen will lay 200,000 eggs per year.

Genetic Traits

Because the queen mates in the open, the beekeeper has limited control over which drones will inseminate her. The few that do mate with her may be from several apiaries or from "wild," or *feral,* colonies. Now because of the presence of parasitic mites and other predators, most feral colonies have been killed off, drastically reducing the number of wild drones.

As a consequence of this random mating pattern, the queen's spermatheca may contain semen from genetically different drones. Her worker and queen progeny, therefore, will consist of individuals that are not necessarily genetically alike (that is, they will be half-sisters, full-sisters, or supersisters; see illustration on this page). The drones, hatching from unfertilized eggs (called *parthenogenesis*), are all full brothers because the queen will lay genetically similar drone eggs whether she has been inseminated or not. Only when the queen has been instrumentally inseminated with semen from recorded drone stock (or from a single drone) will a colony's workers be of known origin. Poorly mated queens do not produce strong colonies. These queens may have low sperm counts as a result of mating with weak drones or with drones of similar genetic composition; or, a queen's low sperm count may result from inclement weather that limited her window for mating flights. Other factors that influence the overall health of queens include the need to receive sufficient dietary requirements and care during their development stages. Queens parasitized by mites, afflicted with other diseases, or being reared in an environment where miticides or insecticides are present are likely to be inferior as well. Researchers have discovered that when queens mate naturally with many drones, or are artificially inseminated with genetically diverse semen, such queens will produce colonies that survive the winter in better condition, are more resistant to disease, and swarm less than queens that have mated with only a few drones.

Because the queen is the sole egg producer, she is responsible for the genetic traits of a colony. If a colony has undesirable traits, requeening should change the hive's genetic makeup and, therefore, its character.

Queens should be of superior stock to optimize desirable characteristics in her offspring, such as:

- Industry (how early or late in the day will bees continue to forage).
- Temperament (how calm the bees are on the comb).
- Handling ease (how easily aroused to sting).
- Production (how much honey is collected).
- Propolizing tendency (excessive use of propolis), unless collecting propolis for sale.
- Burr-comb building (building excessive comb between frames).
- Pollen hoarding (some strains are excessive pollen collectors).
- Plant preferences (mixed or single-source pollen loads).
- Tongue length and nectar-carrying capacity (can be measured).
- Honey hoarding (some strains store more honey than pollen).
- Whiteness of honey cappings (compared to "wet" cappings).
- Conservation of stores (bees manage their stores well).

- Total hive population (large or moderate population).
- Brood pattern (compact or scattered).
- Swarming tendency (frequently swarms).
- Winter hardiness (how many bees overwintered).
- Hygienic behavior (removing dead or diseased brood).
- Disease resistance (little or no diseases).
- Mite tolerance (overall fewer mites).

THE DRONE

Because drones are larger, beginners often mistake them for the queens. They can be distinguished from queens by the abdomen, which tapers to a point in the queen but is blunt or rounded in the drone, making him appear chunky. The number of drones per colony may be in the hundreds to thousands, usually accounting for about 15 percent of the total colony population.

Drone larvae hatch from unfertilized eggs, which, under normal conditions, are laid by a mated queen in hexagonal wax cells similar to, but larger than, worker cells (see the illustration of relative cell sizes on p. 15). On the fourth day, drone larvae are fed a diet of modified worker jelly, which contains a larger quantity of pollen and honey.

After six and a half days of feeding, the cells of drone larvae are capped with wax. The capped drone cells are dome shaped, like a bullet's tip, and are readily distinguished from the slightly convex shape of the capped worker cells. Remember, capped cells lying on a horizontal plane are either worker or drone cells; those that are peanut shaped and suspended on a vertical plane are queen cells.

Newly emerged adult drones are fed by workers for two to three days and then will beg food containing a mixture of pollen, honey, and brood food from nurse bees. Older drones feed themselves from the honey stores. Adult drones have no sting (remember, the sting is a modified female egg-laying structure) and have very short tongues, which are unsuitable for gathering nectar (see Appendix A). Drones never collect food, secrete wax, or feed the young. Their sole known function is to mate with virgin or newly mated queens; think of them as flying gametes (sperm cells).

Drones first leave the hive, about six days after emerging, on warm, windless, and sunny afternoons.

As they get older, they fly to DCAs (see p. 17). Whenever the drones in DCAs detect the pheromones of a virgin or a newly mated queen, they pursue her. A few succeed in mating with her, but the few that copulate die soon afterward.

Whenever there is a dearth of nectar (when no food is being collected), worker bees may cannibalize or remove drone brood and expel adult drones from the colony. During the summer, you can see workers dragging drones in various stages of metamorphosis out of their cells and dropping them in front of the hive. Normally in the fall, all adult drones and any remaining drone brood are gradually evicted from the hive. The evicted drones probably die of starvation (drones have never been observed foraging) or exposure. Queenless hives and those with laying workers or drone-laying or failing queens usually retain drones longer. Drones are being studied to determine if they are more susceptible to some of the pesticides commonly used both inside and outside the hive. Check the References section on current drone research.

Drone Layers and Laying Workers

An unmated queen can lay only unfertilized eggs. A failing queen is one that has mated but is no longer capable of normal egg-laying activity, laying all or nearly all unfertilized eggs. Failing queens may result from sperm deficiency, physiological impairment, disease, mite infestation, or old age. Some workers of hopelessly queenless colonies (those unable to rear another queen) undergo ovary development and start to lay eggs. All of these eggs are unfertilized. Unfertilized eggs that are laid by healthy, mated, unmated, or failing queens, or by *laying workers*, will produce mature drones, capable of mating.

Unlike a mated queen that lays unfertilized eggs in drone cells, a failing or unmated queen will often deposit such eggs in worker cells. Laying workers usually place their unfertilized eggs in worker cells as well, but though these unfertilized eggs are laid in worker cells, they will hatch into drone larvae, and as they near the pupal stage, the cappings will have the characteristic dome shape found on regular drone cells. The presence of scattered worker cells with drone cappings indicates that the colony's egg-layer needs to be replaced.

On further inspection, you may find that each uncapped cell within a scattered brood pattern contains not one but several eggs. These eggs, instead of being deposited at the bottom of the cell as is characteristic of eggs laid by queens, adhere to the cell walls. This is a result of the worker's abdomen not being long enough to reach the cell bottom. If you find these patterns in your hive, read about what to do in "Laying Workers" in Chapter 11.

The presence of clusters of occupied drone cells in the spring, summer, and early fall in a *queenright* colony (where a healthy, mated queen is present) is a normal part of the colony cycle. Because drones attract varroa mites, many beekeepers use this fact to trap the mites. They add at least one frame of drone-sized comb in each hive body to attract female varroa mites to lay their eggs. Once the drone cells are capped, the frame is frozen to kill the mites (see "Varroa Mite [Varroosis]" in Chapter 14).

 **Notes**

CHAPTER 2

Colony Activities

COLONY LIFE

The worker bees carry out the broadest range of chores necessary to maintain and promote the colony's well-being. Queens have a far more restricted range of duties and drones are confined to a single duty: to mate with and inseminate queens. No matter whether the duties are many or just one, each are key to the success of the whole.

You should be able to distinguish between the two female *castes*, the workers and the queen. The term caste in social insects is applied to individuals of the same sex that differ in morphology (form and function), physiology, and behavior. Drones, the male bees, are not members of a caste since all drones exhibit the same morphology and behavior. Because the worker caste is responsible for doing many of the tasks necessary to maintain the colony unit (see the illustration showing the cross section of a frame on p. 22), it is important to understand the role of workers in a bee colony.

Although a colony of honey bees is composed of separate individuals (250 to 50,000) including workers, drones, and a queen, you can view a bee colony as a single organism. The workers represent the somatic cells, and the drones and queen the organism's gonads. A colony of bees when viewed in this light is referred to as *superorganism*. (Ants and termites also fit into this category.)

Division of Labor

The activities of worker bees can be divided into two major categories: those that take place primarily inside the hive and those that take place primarily outside the hive. There is some overlap, but in general terms, inside bees are younger and outside bees are older. Furthermore, it is important to note that during the foraging season, the array of chores by worker bees follows a somewhat dictated age progression. This means that when a day-old bee first emerges from her brood cell, the work that she does throughout her lifetime is roughly related to her age. But age is only one factor: hormones, genetics, and pupal temperature also play a role in dictating the duties of workers. In addition, time of year and colony conditions will change the duties of the workers. For example, in northern latitudes, cues such as poor pollen forage (fewer resources), shorter days, and cooler temperatures transform workers into "winter bees." These winter bees often live longer (up to 12 months has been recorded), a direct result of fewer pollen reserves and decreased brood rearing. They also have lower levels of hormones, more fat bodies, higher levels of fats and sugars in their blood, and enlarged food glands (but with lower protein synthesis). "Summer bees" have an average life span of only four to six weeks, higher hormone levels, and lower levels of fats and sugars in their blood.

In northern latitudes, this progression in age-related duties is often interrupted or curtailed during the period when the queen's egg laying declines in the fall, which is followed by a period of complete or near complete absence of egg laying. The absence of eggs means the existing bees within the colony become progressively older, and therefore performing the duties related to age is no longer operative. By the time foraging begins in the spring, most of the bees are older than six weeks. Therefore, once spring is underway, the initial succession of age-related duties falls on bees that have survived the winter. The ability of these older bees to switch jobs is a good example of the plasticity of workers.

For example, if all the young workers were killed, some of the older bees would be able to raise brood by means of their re-activated food glands; they will even be able to build comb with wax glands that are switched back on. The opposite is also true; if all the foragers died (from pesticide poisoning), the young nurse bees are able to become foragers within a short time. In some instances, stress, contaminants, and other factors may cause nurse bees to start foraging earlier in life. This upsets the balance of the work force in a colony, which could result in fewer nurse bees raising brood; the end result is lower bee populations going into the winter, and thus, the colony could be lost during the cold season.

A functional knowledge of bee biology and an understanding of the activities performed by the members of the colony will assist you in becoming a better manager of your bees. You should be able to recognize which activities, or the absence thereof, signify the colony's overall status, such as queen egg laying, flight activity, or honey production. It is also important for you to recognize the different labors of worker bees and when (or if) they are being performed. For instance, if none of the foraging bees returning to the hive are carrying pollen loads, this could indicate that there is no brood in the colony or pollen is unavailable, or no pollen stores are adequate, and bees are instead bringing nectar as water. If there is a lack of brood, this may indicate that the queen is absent, or she has been superseded and the new queen has not commenced egg laying. It could also mean that there is no food available, and you may have to feed the colony to keep it from starving. Experience and working with other beekeepers will aid in interpreting bee behavior. Look inside and examine frames to assess the colony situation.

INSIDE ACTIVITIES

Upon emerging from its capped cell, a young worker begins to perform the first in a sequence of the many tasks she will carry out during her short but highly productive six-week life span (the average life span of summer bees). Over the next few days, the age of the

bee, its glandular development, and the environmental conditions of the colony will rule bee activities. The organizational factors determining worker bee activities are referred to as *age polyethism* (in which the same individual passes through different forms of specialization as it grows older). These tasks include cleaning cells, capping cells of brood, and caring for the brood, tending and feeding the queen, receiving nectar from foragers, making honey, removing the trash, packing pollen, building comb, ventilating the hive, performing guard duty, regulating the temperature (thermoregulation), and finally, taking orientation and foraging flights (see the illustration of the cross section of the honey bee frame on p. 22).

Researchers are able to record and document age polyethism in bee colonies by using a glass-walled observation hive containing frames of brood, adult workers, drones, and a laying queen. They paint the dorsal side of the thorax of several hundred newly hatched bees and then introduce them into the observation hive. Over the next six weeks observations of the activities of the painted bees reveal that under normal conditions bees progress from one specific activity to another. Such studies continue to this day, and each new study adds more clarity to our knowledge of age polyethism.

The progression of work "assignments," and the overall decision making of which bee does which job, suggest there is some command structure within this superorganism. However, research has shown that each worker appears to be making individual decisions and stimulates or recruits others to follow along, such as with comb building, comb use, foraging, patrolling, or even swarming. Although most young workers progress from one specific task to another, there is a great deal of plasticity in the jobs and ages of bees, which enables the bee colony to thrive during difficult or changeable environments. Age is only one factor in dictating the activity of workers; hormones and genetics also are factors. This flexibility is a key factor in the success of honey bees.

Unlike other insects, honey bees that age have higher levels of the hormone called *juvenile hormone* (JH), which is normally higher in young insects. In honey bees, JH levels increase during the worker bee's life, and could be one of the reasons why older bees learn better. Learning is needed in the outside world, to navigate, find food, and cope with the dangerous outside environment. In addition, the genetic makeup

Cross Section of Honey Bee Frame

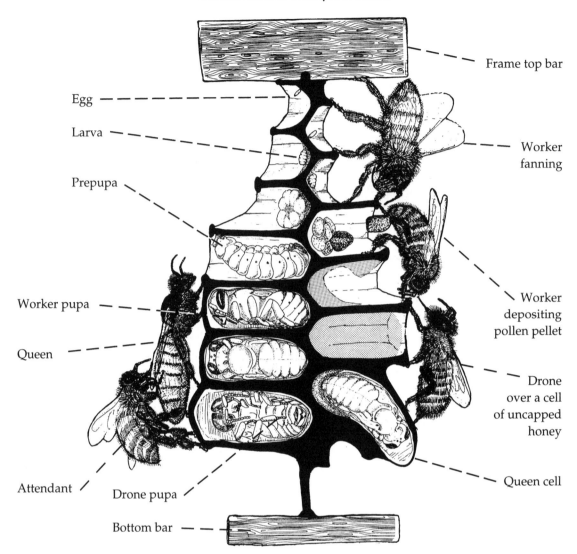

Frame top bar

Egg

Larva

Worker fanning

Prepupa

Worker depositing pollen pellet

Worker pupa

Queen

Drone over a cell of uncapped honey

Attendant

Queen cell

Drone pupa

Bottom bar

of individual worker supersisters (those that have the same father) will also influence when and how long they do particular tasks. Recent work has also suggested that the temperature at which the pupae are incubated affects the activities they will perform as adults.

As we already mentioned, the age of the worker, her glandular development, and the needs of the colony trigger which chores will require her attention. The development and maturation of her *exocrine* glands (glands that release secretions externally)—namely, the mandibular, hypopharyngeal, postcerebral, thoracic, and wax glands—as well as the conditions both within and outside the hive, provide the stimulus needed as she progresses from one task to another. At the half point of her six-week life span, the duties of a forager will take over, as her glands dry up; and

for these last three weeks she will forage for nectar, honeydew, pollen, propolis, and water.

Once a colony is in the egg-laying mode (spring, summer, early fall), the ages of bees within a colony are more equally distributed. When this occurs, workers will progress from one inside task to another, as mentioned already. Generally speaking, when you inspect a colony or observation hive, the age of various bees can be determined by the task or duties they are engaged in. For example, a bee returning to the colony with a load of pollen is a forager and at least three weeks old.

Cell Cleaning and Capping

When workers emerge from their cells, and are only a few hours old, they begin cleaning chores. Cell cleaning prepares them for egg deposition by the queen and for receiving pollen, nectar, or honeydew. Cleaning involves removal of debris and the remains of old cocoons. The cocoon, which is made up of silk thread, was spun by the larva before entering the pupal stage. Bees are unable to remove the entire cocoon, and therefore, over a period of years the interior of the cells diminish in size. As a consequence of this, emerging bees get smaller over time. Some beekeepers rotate the old combs, replacing them with new foundation in order to maintain proper cell volume. In addition, by removing old comb you also get rid of harmful residues that accumulate in the wax (pesticides, fungicides, miticides). Cell cleaning also involves the removal of fecal matter deposited in the cell before the larva starts to spin the silken case.

Once the queen perceives the cells to be clean, she will deposit eggs in them (one egg per cell). Recent studies indicate that because the inside of the hive is in darkness and the queen is not equipped with a miner's light, the cleaning bees leave a marker (pheromone) signaling that the cell is prepared to receive eggs (or nectar, pollen, etc.).

Surfaces that are difficult to clean are coated with fresh wax and/or *propolis*, a resinous substance collected from buds or the stems of trees and carried back in the pollen baskets. Often referred to as *bee glue*, it is a dark reddish to brown resin that is sticky when warm, brittle when cold. It is used to strengthen the combs and to cover any foreign matter that cannot be removed (e.g., a dead mouse). Propolis has antifungal and antibacterial properties that provide bees with some defense against pathogens. On the other hand, propolis makes it necessary to use a hive tool to pry apart the hive furniture.

These same young workers also progress to capping brood cells that are near the end of their larval stage. These cells are capped with beeswax mixed with some propolis, giving the convex caps the characteristic brownish color; honey cells with *concave* cappings are light in color since they contain no propolis. Capped brood cells are easy to distinguish from capped honey by their color.

Cleaning or *hygienic behavior* is an inherited activity and one that helps maintain the health of the colony, because workers rapidly remove dying or dead brood or mites from cells. Research has confirmed that colonies with workers prone to hygiene are more likely to be free from diseases, pests, and reduced mite levels. Dr. Walter Rothenbulher ob-

served and researched this hygienic behavior early in the twentieth century, using the following procedures: He placed frames containing sealed brood in a freezer for 24 hours, thus killing the brood, after which he returned the frames to the hives. These frames were checked after 24 hours to determine how quickly the bees had uncapped and removed the dead brood. In some colonies, removal of the brood was rapid; in others very little had been removed. The colonies that evicted the dead brood rapidly had fewer bee diseases, particularly American foulbrood (see "Hybrid Bees and Select Lines" in Chapter 1 and "Queens" in the References).

Dr. Marla Spivak (Un. Minn.) and other researchers who are looking for ways to reduce varroa mite infestation recently revisited this hygienic behavior. Drs. John Harbo and Jeff Harris developed a super hygienic line named Varroa Sensitive Hygiene (VSH, originally SMR), which removes fertile varroa mites from brood larvae, actually uncapping the cells to remove the mites (and brood) that are actively laying eggs. For more information on testing for this behavior, see Chapter 10 and "Varroa Mite (Varroosis)" in Chapter 14; also review the information on hybrid bees in Chapter 1. Colonies with strong genetic tendencies toward house cleaning are less likely to succumb to a variety of bee maladies, and it is a good policy to promote such behavior by raising queens with such genetic attributes.

In summary, hygienic workers take on the task of keeping the hive clean and can be seen:

- Removing dead or dying brood and adults from the hive. These workers are called undertaker bees and comprise about 1 percent of the worker population.
- Removing debris such as grass and leaves as well as pieces of old comb and cappings.
- Removing granulated honey or dry sugar and moldy pollen.
- Coating the insides of the hive and wax cells with bee glue or propolis.
- Propolizing cracks and movable hive parts, including frames, bottom board, and inner cover; some races use more propolis than others, making a propolis "gate" at the entrance; see "Races of Bees" in Chapter 1.

Under certain circumstances, workers remove healthy brood during nectar dearths or when the hive is lacking in food reserves. Cannibalism is not uncommon under these conditions, and usually the drone brood is the first to be pulled, followed by the worker brood. Later, adult drones are evicted in order to preserve the core of the colony—the workers and the queen.

Tending to the Brood

From cleaning and capping cells, young workers 5 to 15 days old move on to their next task, which deals with feeding the brood (worker, drone, and queen larvae) and the adult queen. Young worker bees in areas containing uncapped brood are primarily involved in feeding larvae. During this activity they are referred to as *nurse bees*. At this point in their adult life the nurse bees' hypopharyngeal and mandibular glands, located in their heads, are fully developed. Many work-related activities are correlated with the maturation of these glands. It is also imperative that these nurse bees consume large quantities of bee bread; this provides the raw material for the food glands.

The combined substance produced by these glands is referred to as *royal jelly* or brood food. All the young larvae (queens, workers, drones) are fed royal jelly lavishly; this is referred to as *mass-provisioning*. You can see adult bees sticking their heads into cells for a few seconds to determine how much food is left and then feed the larvae as needed. The brood food is placed near the cell bottom, close to the mouth of the growing larva.

While queen larvae remain on this diet, workers and drones after the third day of larvae life have their diet switched. Instead, they are provided with a diminishing diet of brood food and increasing amounts of honey/nectar and pollen (bee bread) and are supplied these substances on an as-needed basis. This is referred to as *progressive-provisioning*. Brood food therefore falls into two categories: glandular (royal jelly) or nonglandular (nectar/honey and pollen). Even though both these diets can safely be referred to as *brood food*, be sure to keep the distinction in mind.

Over the course of brood tending, one bee will rear two or three larvae. Assuming it takes one bee to rear three larvae, and on a given comb there are 1800 newly hatched larvae, these larvae will require the attention of 600 nurse bees over the course of their development. This serves to demonstrate the time and effort required to rear brood by bee colonies, from the time the nurse bees place royal jelly in cells containing eggs that are about to hatch, until the brood is capped.

Tending the Queen

In addition to caring for the brood, nurse bees tend to the adult queen and continue to maintain her diet of royal jelly, the very same diet that reared her during larval life. The ingredients in royal jelly and the amount of royal jelly available to her serve to nourish her as well as stimulate her ovaries to maximize egg production.

Again, with the use of an observation hive, researchers watching the activities of nurse bees surrounding the queen noticed that while some were feeding the queen, others were making tactile contact with her by licking and touching her with their antennae and forelegs. This group of bees (between 6 and 10) surrounding the queen are referred to as attendant bees; they form a *retinue* or circle around the queen (see the illustration of the queen's retinue on this page). By attending and caring for the queen in this manner, the attendant bees allow the queen to concentrate on one of her most important functions: lay eggs and lots of them!

At first the activities of the retinue may have been simply interpreted as a group of bees feeding and making tactile contact with the queen. However, fur-

Queen and Retinue of Workers

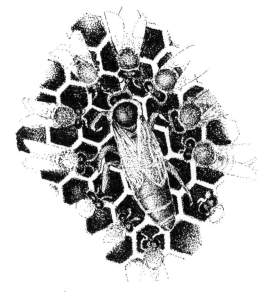

Drawing by Jan Propst.

ther observations revealed a great deal more. The retinue remained with the queen for very brief intervals—between 30 and 60 seconds. Then individual members of the court disperse, and as one or more take their leave, other nurse bees replace them. This replacement activity keeps the retinue intact. Researchers noted that as the departing bees traveled within the colony, they encountered other bees that initiated tactile contact with the former retinue members. These encounters are not like "idle handshakes," but instead play a major role in maintaining colony stability.

During these contacts, the queen substance or pheromone secreted by her mandibular glands is spread to other members of the colony. Researchers have learned that the bees departing the retinue have ingested some of the queen's pheromone(s) while they were licking her. As these bees travel within the colony they encounter other bees who solicit food from them (*trophallaxis*). By way of this encounter the queen's pheromone(s) are distributed to additional members of the colony. The former members of the queen's court can be referred to as *pheromone(s) distributors*. Remarkably, each departing attendant is physically contacted within the next 30 minutes by up to 56 other nest mates, which in turn contact other nest mates. As a consequence of this mode of pheromone distribution, thousands of bees get samples of the queen's pheromones. The effect is not only to notify other members of the colony that the queen is present, but also to provide information as to the amount or level of her pheromone output. Thus a colony of 40,000 or more bees, most of whom may never have direct contact with the queen, are fully aware of her presence!

As long as the queen's pheromone levels are interpreted as sufficient, the colony remains in a stable condition. When her pheromones are in actual decline or perceived to be so, the colony undertakes certain activities to correct this change. Reduced pheromones may trigger other activities by the workers, such as swarm preparation or queen supersedure. One of several scenarios may occur when the queen's pheromones are (1) in decline (queen is old or diseased), (2) perceived to be so, or (3) totally absent.

If the colony grows too large, some of the workers may receive only a limited amount of the queen's pheromone, and if this occurs between mid-spring or early summer in northern latitudes, bees will construct queen cups, which lead to queen cells and the

eventual emergence of a virgin queen (see "Swarming" and "Queen Supersedure" in Chapter 11). If the bees are preparing for swarming, and before the new queen emerges, the colony is likely to cast its first swarm (primary swarm), which is usually accompanied by the old queen. Thus the original colony rids itself of its mother and replaces her with her daughter, whose pheromone(s) levels return the colony to a stable unit.

If, on the other hand, the queen is perceived to be of low quality, due to disease, poor mating, or other reasons, a colony will replace the mother queen not by swarming but by supersedure. The outcome is similar in that the new queen replaces the old queen. Sometimes both the old and the new queen (mother/daughter) remain members of the same colony. It is likely that the old queen's pheromones are nonexistent, and therefore the young queen, her replacement, does not seek to immediately eliminate her mother.

If a queen is accidentally killed or lost (i.e., her pheromone is totally absent), within 24 hours the entire colony will know that it is queenless. Bees will construct *emergency queen cells* and rear a replacement. Occasionally, a queen dies for any number of reasons. For example, while a beekeeper is inspecting a comb, the queen falls to the ground and is crushed when the beekeeper inadvertently steps on her, or the queen, after falling to the ground, is unable to return to the hive.

In another instance a queen dies and the bees are unable to replace her because there are no eggs or young larvae available to rear a new queen. In this case, a given number of worker bees undergo ovary development resulting from the absence of worker brood pheromone, which, when present, suppresses ovary development. Workers that are able to lay eggs are referred to as laying workers. However, they are unable to mate, and any eggs they lay are haploid and will produce only male bees (drones). Such a colony, unless discovered early and successfully requeened, is doomed.

Comb Building

The wax comb (and surrounding cavity) is the nest and abode of the honey bee. Although comb is referred to as honeycomb, which implies that it contains honey, it has many more functions than solely the storage of honey. The hexagonal cells are used to store pollen, which is converted to bee bread; they are also a place to store nectar as it is being processed into honey, and the place for depositing water droplets needed to cool the hive when rising temperatures would cause problems for the colony. In addition, the cells serve as a depository for the queen's eggs as well as incubators, cradles, and transforming sites while the developing bees move through the four metamorphic stages. The comb is also a communication system, especially the area in the vicinity of the hive entrance. This area is referred to as the *dance floor*, where returning scout bees perform the important round and wagtail dances. The purpose of these dances is to recruit other bees to nectar, pollen, propolis, and water sources. Furthermore, vibrations produced by the dancers travel to more distant areas of the honeycomb, where they are detected by other bees that then move to the dance floor to receive information from the dancers. Honeycomb also provides the colony with its particular odor, and defends the colony against pathogens (in the form of propolis).

A prerequisite to the construction of honeycomb by bees is that they first must locate themselves, or be located by humans, within a suitable cavity. An ideal cavity includes the following elements:

1. Low interior light.
2. A dry interior, protected from adverse weather.
3. Sufficient capacity to hold a reasonably large bee population.
4. Ample room for combs (for brood rearing, honey and pollen storage).
5. An entrance/exit that allows bees easy passage, while restricting, by various means, entrance by predators or parasites.

In the wild, the comb is usually confined within a dark enclosure such as a hollow tree, a small cave, or an opening in a house, although some nests can be found in the open. In the desert Southwest, for example, where trees are not abundant, feral bees nest in small caves formed by rock outcrops, as well as in electrical boxes and under manhole covers.

Once bees move into a cavity, either on their own (as in the case of swarms that have not been captured) or placed by a beekeeper, then the process of comb building begins. Occasionally, honey bees will construct comb outside of a cavity; such undertakings are suicidal in higher latitudes and altitudes.

As we have previously noted, many work-related activities of bees parallel the maturation of specific bee glands. In the case of wax glands, they become fully developed when workers are between 12 and 18 days old, and it is bees of this age that not only extrude the wax scales, the building blocks for honeycomb, but also become directly involved in the construction of the comb. The wax for honeycomb is produced by four pairs of wax glands, located beneath the ventral segments (sternites) of a worker bee's abdomen. The wax is secreted as a liquid from each of the paired glands, and as it passes to the outside, it solidifies on contact with air into thin oval scales. The ventral surface of the sternites is referred to as the wax mirrors. The hardened wax flakes are referred to as *wax scales.*

Each wax scale is removed from its wax mirror with special hairs located on the bee's hind legs. From there, each scale is passed forward by the next two pairs of legs and on to the bee's mouth. Here the bee's mandibles (jaws) masticate or knead the scale and mix it with a secretion from the mandibular glands. The end result is a pliable product that bees can employ to construct comb from scratch or augment existing comb. Bees in the process of producing wax can be observed suspended from one another in formations that resemble strings of beads or chains. These formations are referred to as *festooning* (see illustration of worker bees festooning on p. 27). The function of festooning still is not entirely understood. Workers in a festoon will stay there for a time and then move off to feed brood or do other tasks, thus allowing the wax glands to recharge.

During the construction process, bees are also warming the wax to more than 109°F (43°C) in order to construct the perfect hexagonal cells. When you see a series of parallel combs constructed by bees, you may find it difficult to believe that so many combs can be produced from tiny wax scales. Each comb is separated from an adjacent facing comb by a space (the diameter of two bees) that permits bees on the surface of opposite facing combs to go about their chores independent of each other. The size of one bee (about 3/8 inch or 8–10 mm) is called a *bee space*. Any space that is smaller or larger than a bee space will be filled in with comb (burr comb). The discovery of the bee space in 1851 enabled L.L. Langstroth to develop a beehive with removable frames.

The cells of the honeycomb do not lie on a com-

Worker Bees Festooning

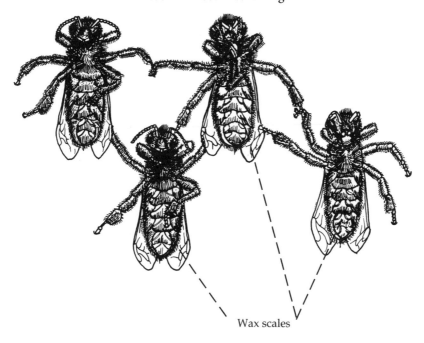

Wax scales

Appearance of festooning bees on a frame

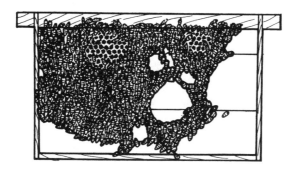

pletely horizontal plane: the openings are slanted slightly upward by 9 to 13°. This inclination prevents nectar, which has a low viscosity, from spilling out and serves to concentrate the brood food at the base of the cells holding the young larvae. The cell walls are 0.003 inch (0.07 mm) thick and the combs hang vertically. Bees have special sensory hairs in the joints of their legs that enable them to detect gravity, and they measure the cell thickness by using sensory hairs on the tips of their antennae.

The production of beeswax is economically very expensive; it takes 50 pounds of honey (22.7 kg) to produce 10 pounds (4.5 kg) of wax. Therefore, bees need to engorge on large quantities of honey, nectar or sugar syrup, and pollen in the form of bee bread, to stimulate wax production. A hive placed on a scale provides important information in this regard, because a hive beginning to gain weight indicates that

a honeyflow is in progress. This is the ideal time to provide bees with foundation (see p. 42), which they will convert into wax comb. This is precisely why the best time to have foundation "drawn" into honeycomb is during a honeyflow and/or by feeding sugar syrup to bees or by capturing a swarm. As a colony prepares to cast a swarm, worker bees will engorge on excess amounts of honey or nectar and store the same in their honey sacs. This activity serves three vital purposes: (1) it provides a reserve of food in their honey sacs while the swarm is clustered and scouts are searching for a home site; (2) it stimulates their wax glands, which will be needed to construct honeycomb at the new home site; and (3) any remaining honey or nectar can be regurgitated into the new comb, providing food for the new colony.

Beekeepers often hive swarms on foundation. Why? Ten days before a swarm issues from their hive, worker bees begin engorging on honey and continue to do so until the swarm departs from the hive. During this time they are digesting the honey as well as filling their honey sacs with it. One of the results of this activity will be the stimulation of the bees' wax glands. Thus, bees in a swarm are primed to construct comb. For the same reason swarms are often used in the production of comb honey. The bees will draw the foundation, fill the cells with honey, and then seal them with clean white cappings. Clean sections of comb honey are pleasing to the eye and will command a premium price.

Honeycombs once constructed consist primarily of hexagonal cells, each connected to one another and forming a large comb containing thousands of cells on each side. The hexagonal cells are of two sizes: the smaller worker cells, which are by far the most prevalent, and the larger drone cells.

Exceptions to the normal hexagonal cells found in colonies are queen cups and queen cells, which are fabricated of wax and are suspended vertically from comb. *Queen cups* (similar in shape to an acorn cap) can be found in colonies at any time but are most prevalent during the swarm season. Most cups are suspended from the bottom edge of the comb; by tilting up a deep hive body, you can usually observe these cups. Once a queen deposits an egg in a cup they are then referred to as *queen cells*. As the queen larva grows, bees add more wax to these cells, enlarging and elongating them. Eventually, they take on the shape and form of a peanut. When these cells are found in the spring and early summer and are located along the bottom edges of comb, the colony may be preparing to swarm.

On the other hand, when bees are preparing to supersede their queen (or need to replace her in an emergency situation), bees convert worker cells containing eggs or larvae into queen cells by reshaping them. Thus, in supersedure or in the event of an emergence, queen cups are not involved. As the queen larvae develop, the reshaped worker cell will begin to take on the shape and form of "regular queen cells." Note: supersedure and emergence queen cells almost never look as robust as do queen cells developed during the swarm season. In addition, these queen cells are found in the brood area and are usually brown in color rather than the medium to light yellow color seen on swarm cells. (For more information, see "Queen Supersedure" in Chapter 11.)

The ability of bees to organize and execute the many tasks needed for colony survival, including nest design and construction, is indeed a marvel of the natural world.

Nest Homeostasis

The maintenance of temperatures and other environmental factors at a constant level inside the colony despite external conditions is termed *homeostasis*. Colony homeostasis is sustained by cooperative living or social behavior and is found for all social insects (ants, termites, wasps, and bumble bees) and some mammals (naked mole rats). The ability of a colony to survive temperature extremes and dearth or abundance is an obvious advantage over a solitary life, common to most insects.

Recent work with infrared thermographic cameras (which are able to determine temperatures) has revealed that there are "incubator" workers who generate heat in their thoraxes. They can be seen in a "heating posture" over capped brood with their thorax pressed down onto the cappings. They have been timed to hold this posture for 30 minutes with a body temperature of 109°F (43°C). In addition, some of the empty cells within the broodnest are also used by "heater" bees to keep the brood warm. These bees formerly were called "resting" because all that had been seen was the pumping tip of the abdomen. We now know better.

After about 30 minutes, the heater bees have used up their energy stores and need to eat. Other "filling station" bees feed these exhausted heater bees; in fact, they will search for the heater bees in order to feed them, especially in the cooler winter months. The amount of heat energy needed to maintain broodnest temperatures has been estimated to be the equivalent of a continuously powered 20-watt bulb. This amazing work is clearly illustrated in the entertaining and very readable book *The Buzz about Bees* by J. Tautz (2008).

Food Exchange, Handling, and Hive Odor

Bees within a hive exchange honey or nectar. Foragers returning from the field pass nectar (honeydew) to the hive bees, who then pass it to other bees. This food exchange not only communicates what the colony is receiving from the foragers but also indicates the availability and quality of incoming food.

Beneficial Microbes in Honey Bees

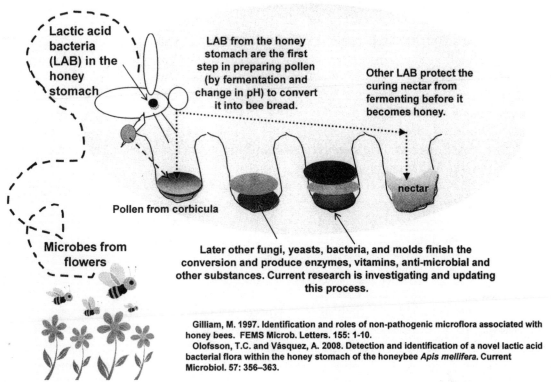

Lactic acid bacteria (LAB) in the honey stomach

LAB from the honey stomach are the first step in preparing pollen (by fermentation and change in pH) to convert it into bee bread.

Other LAB protect the curing nectar from fermenting before it becomes honey.

nectar

Pollen from corbicula

Microbes from flowers

Later other fungi, yeasts, bacteria, and molds finish the conversion and produce enzymes, vitamins, anti-microbial and other substances. Current research is investigating and updating this process.

Gilliam, M. 1997. Identification and roles of non-pathogenic microflora associated with honey bees. FEMS Microb. Letters. 155: 1-10.
Olofsson, T.C. and Vásquez, A. 2008. Detection and identification of a novel lactic acid bacterial flora within the honey stomach of the honeybee *Apis mellifera*. Current Microbiol. 57: 356–363.

Schematic by D. Sammataro

A returning forager will offer a drop of nectar to two or three house bees, who will then move off to a quiet corner to work the nectar by mixing it with enzymes and bacteria that help keep the nectar from fermenting. Later these bees will extend their tongues to expose the droplet to the warm air. This helps further cure the nectar into honey by evaporating some of the 80 percent water that is in nectar (similar to the sap from sugar maple trees; the water must be removed to create maple syrup). As the water content is reduced, the nectar is placed in cells, where further curing takes place. This ripening is finished in the honey cells and will take from one to five days to complete, depending on the amount of humidity in the air, the water content of the nectar, ventilation, and the amount of nectar being cured. Through this ripening process, the final product, honey, will have a finished water content of less than 18 percent.

Honey bees also process the pollen pellets that foragers bring back to the colony. Pollen is the colony's only supply of protein, providing essential amino acids (the building blocks of proteins and enzymes), lipids, vitamins, minerals, and sterols to the diet of honey bees. Once the pollen foragers return to the hive and deposit their pollen loads (raw pollen) into the cells, house bees begin their work by ramming their heads against the pellets to pack them tightly into the cells. As the foragers add more pollen and the packing progresses, the house bees moisten the pollen with nectar and saliva that contain enzymes and bacteria, a process referred to as *seeding*. Some of the lactic acid bacteria belong to the genera *Lactobacillus* spp. and *Bifiobacterium* spp. The addition of these organisms acidifies the pollen, making it habitable to other yeasts, fungi, and molds (see illustration on this page), which finish the conversion of raw pollen into bee bread. Bee bread provides bees with amino acids, vitamins, and preservatives not found in raw pollen. When the cell is nearly filled, a light film of honey is placed over the pollen bread. The honey acts as a preservative (similar to adding a wax seal over jelly in a jar to keep out harmful organisms). It is now believed that bees continually work these cells to keep the beneficial microorganisms active.

Given that pollen is the only source of protein and that it is ultimately converted to bee bread, we are faced with the fact that some agricultural practices require that certain crops be sprayed while in bloom (the time when pollen is collected by bees). Scientists are concerned with the effects these chemical sprays may have on raw pollen and the organisms that convert the pollen into bee bread. Ultimately, the contaminated bee bread is eaten by bees to manufacture brood food. Stay current with research on this subject by reading journals and searching for information on scientific websites.

An additional function of food transmission is the spread of the hive's unique odor. Each colony has its own characteristic odor, which may aid the bees of one hive in distinguishing bees from other hives (such as robbing bees) and foreign queens (see "Queen Introduction" in Chapter 10). To keep foreign bees out, guard bees patrol the colony entrance and challenge any incoming bees that may be intruders. Guard bees are older workers that have very high concentrations of the alarm pheromones.

The needs of a colony will dictate how quickly incoming foragers will be relieved of their liquid loads by the house bees. For example, as temperatures rise inside the hive and fanning alone is not sufficient to lower them to acceptable levels, forager bees transporting water back to the hive will be given unloading priority by house bees over those bringing in nectar. As long as the need for water remains critical, incoming water foragers are rapidly unloaded. When foragers are unloaded quickly, it has the effect of attracting recruits (idle foragers or novices) to their dances, which provide information to the location of water sources.

As the number of bees returning to the hive with water increases and the hive's temperature returns to appropriate levels, the number of water foragers and the attention given to them by the house bees declines. (See more below in "Foraging and Communication.")

Ventilation

Bees can often be seen fanning their wings on the extended deck of the bottom board with their heads facing toward the hive entrance. In this position, warm air is pulled out of the hive. This fanning also takes place inside the hive, including on the portion of the bottom board within the hive that is hidden from view. The best time to observe fanning behavior is on warm days when abundant amounts of nectar are being collected. Workers of all ages perform this task, but many young bees, less than 18 days old, are often the ventilators, especially on hot days.

Fanning circulates air through the hive and helps to:

- Regulate the hive's humidity at a constant 50 percent.
- Reduce or eliminate accumulations of gases, such as carbon dioxide (CO_2).
- Regulate brood temperature.
- Evaporate water carried into the hive to reduce internal temperatures (see the figure on temperatures on p. 99).
- Evaporate excess moisture from unripened honey (nectar with a high percentage of water); as this moisture evaporates, it too will cool or humidify the hive.
- Keep the wax from melting as the temperatures climb.

Another type of fanning helps spread workers' pheromones. In this case, the fanning bee's abdomen is raised, and a gland called the scent gland (Nasonov gland), located on the last two dorsal segments of the abdomen, opens and releases a mixture of pheromones. These chemicals, which have a sweet-smelling, lemony scent, guide other bees toward the fanners. It is often called the "come hither" odor and is most often detected in swarms, where it helps to keep the bees together.

This type of scent fanning is commonly seen:

- When a swarm or package of bees is emptied at the entrance of a natural homesite, hive, or inside a hive body, to guide stragglers.
- When bees are shaken off a frame or otherwise disoriented.
- When scout bees from a swarm mark the entrance to a new homesite.
- When a swarm is moving to a new homesite.
- When a hive that is queenless or has a virgin or newly mated queen or laying workers is opened.
- When a swarm begins cluster formation.

Guarding the Colony

Worker bees between the ages of 12 and 23 days defend their colony by flying at and often stinging an intruder. Each bee does guard duty for only a few hours or days in their entire life. They can be recognized by their posture: they stand on hind legs with the antennae held forward and the first two legs (forelegs) upraised. These guard bees often inspect incoming bees, which will not be admitted if they do not smell or behave "correctly." Stray or foreign drones, young workers, and foragers carrying a full load of nectar are generally allowed to enter. During strong nectar flows, foreign bees pass easily into a hive; during a dearth, however, guard bees closely inspect strange bees.

If a large animal approaches the colony, some of the guards will often fly out to challenge it. This defensive action should not be interpreted as "meanness" or "aggression" but rather as a defensive action. When an intruder approaches and enters or begins to open a hive, some bees raise their abdomens, begin fanning, and thereby disperse the alarm odor being released by a gland at the base of the sting. This pheromone has an odor similar to that of banana oil. It incites other bees to defend the colony. Once some of the attacking bees sting, some alarm odor remains at the site, tagging the victim. Thus tagged, the victim may become the target of further defensive acts as long as the odor remains on the clothing or skin.

Many factors influence the temper of a colony (see "Bee Temperament" in Chapter 5). Africanized bees, a recent invader from South America, are known for their volatile defensive stance, which distinguishes them from European races (see Appendix E).

OUTSIDE ACTIVITIES

Flight

As we have already discussed, except for occasional orientation flights, worker bees generally remain within the colony for the first three weeks of their adult lives, cleaning, feeding, building comb, ripening honey, and packing pollen. These routines are more or less discontinued at the end of the third week as bees turn to tasks that require flight. An ability to recognize the different kinds of flying patterns near the hive entrance will enable you to discern the purpose or function of each pattern.

Orientation Flight. Young bees take orientation flights to familiarize themselves with landmarks surrounding their hive as well as void feces. Small to large numbers of these bees hover near the hive entrance. Their first flight will last only five minutes, with successive flights over the next few days lasting longer. This is a common sight on warm afternoons, when young drones and workers are seen hovering in front of the entrance.

Recent research has now reexamined this behavior, and by marking bees, it was discovered that mass orientation flights take place at all times during the day, that many of the bees are in fact older foragers, and if there is no queen in the colony, there are no orientation flights. It now appears that these mass orientation flights coincide with the queen nuptial flights; what role these orientating workers play is still not understood.

Drone Flight. On warm sunny afternoons, mature drones fly out of the colony to gather at the Drone Congregating Areas (DCAs; see p. 17). Virgin and newly mated queens fly to these DCAs where they come in contact with the drones who rush to mate with them. Drones in these mating areas have a limited amount of food to sustain them; when it is nearly exhausted, the drones return to the hive to refuel. Queen breeders who want to collect drones for instrumentally inseminating queens use this time of day (late afternoon) to collect drones.

Foraging Flight. This is the final task of a worker bee. First time foraging bees fly out and away from the hive in a random direction in search of nectar or honeydew, pollen, propolis, and water. Because their brood food and wax glands have degenerated, these bees look smaller and later the edges of their wings will become torn and ragged. The characteristic pattern of returning with pollen, or flying straight into the hive or onto the extended deck of the bottom board, will distinguish them from orientating bees. After all of her trips, each forager will have traveled about 500 miles (800 km) before she dies (see the illustration on forage areas on p. 32 and the "Foraging and Communication" section).

Robbing Flight. Unlike orientation flights, which are short in duration, robbing activity is a form of foraging. On first approaching a hive, the robbers sway to-and-fro in front of the hive to be robbed in a manner somewhat similar to a figure eight. Robbing bees can also be seen checking out cracks and other

Forage Areas for Honey Bees

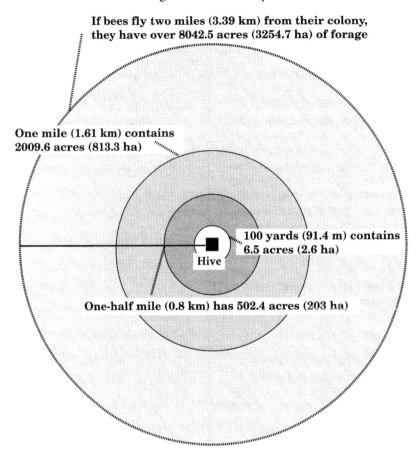

If bees fly two miles (3.39 km) from their colony, they have over 8042.5 acres (3254.7 ha) of forage

One mile (1.61 km) contains 2009.6 acres (813.3 ha)

100 yards (91.4 m) contains 6.5 acres (2.6 ha)

Hive

One-half mile (0.8 km) has 502.4 acres (203 ha)

openings in a hive (such as an ill-fitting super over a weak colony) in order to gain access without detection. Once the hive has been invaded, other robbing bees are "recruited" to it when the initial robbers return to their home and dance the location of the new "food" source.

Robbing often takes place during a dearth, when little or no nectar is available; abandoned and weakened colonies are the first targets for robbing bees. Beekeepers who expose combs, take off honey supers, or attempt to feed weak colonies at this time may inadvertently initiate robbing. You must minimize your activities in the beeyard during a dearth or you can start a robbing frenzy (see "Robbing" in Chapter 11). It is important for you to recognize robbing so you can intervene to stop it.

Cleansing Flight. Individual bees usually release their body wastes or fecal material outside the colony during flight. Evidence of this activity is the yellow or brownish spots or streaks found on the snow, plants, cars, or laundry hung out to dry. (The so-called yel-

low rain discovered in Southeast Asia during the Vietnam War turned out to be bee excrement instead of chemical warfare.) After bees have been confined for long periods—during the winter months, for example, or during wet and cold weather, and even in a package—the yellowish brown droppings are easily seen because of the large number of bees defecating as soon as flight is possible. In such circumstances, the flights are referred to as *cleansing flights*. If bees are confined for too long or they have dysentery or are infected with *Nosema apis*, defecation may take place inside the hive or just outside on the hive furniture. If you see dark stains on the outside of the hive in the late fall or during the winter, your bees may not survive the winter.

Foraging and Communication

The gathering of water, propolis, and food for feeding larvae and for storage requires a high degree of cooperation and communication among the members

Dance Language of Honey Bees

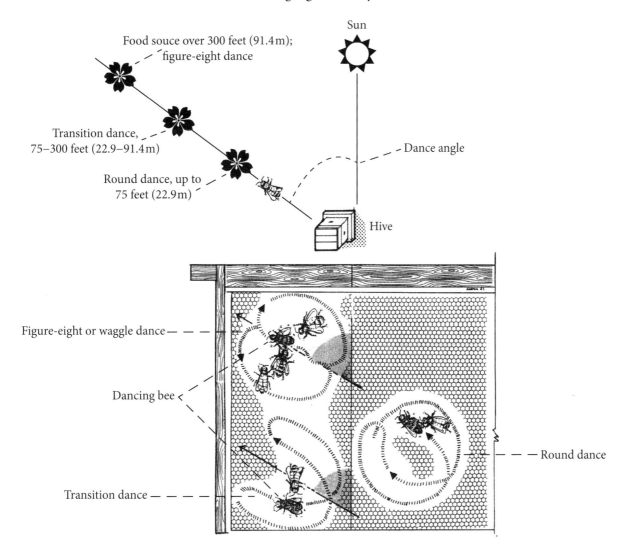

of a colony. Haphazard searches for food by the older worker bees would require too much energy and time and could not be sustained over long periods without adversely affecting the well-being of the colony. Communication among the bees dramatically increases the efficiency of food-gathering activities by directing bees to known water, food, and propolis sites.

A worker bee orients herself as she goes out to locations according to the following external stimuli:

- The sun's position and polarized light.
- Landmarks, both horizontal and vertical.
- Ultraviolet light, which enables her to see the sun on cloudy days.
- A "visual odometer" of images passing over the

compound eyes resulting in an "optical flow" of images; this helps the bee determine flight speed and distance.

A forager is able to inform and recruit other bees as to the location of a food source through a series of body movements called *dances*. Dr. Karl von Frisch, an Austrian scientist, received the Nobel Prize for his discovery of the function of the bee dance in 1973. He and others observed that the movement of the dancer on the surface of honeycomb could be outlined as circles and figure eights. He was able to demonstrate what the dance movements meant by marking bees at a food source and observing the marked bees back at the observation hive. Bees following the dancer(s) can learn the distance and the direction to a food, water, and propolis source. During the dancing, the dancer rubs her wings together, causing vibrations on the comb, which may provide additional information to the recruited bees following her. Other clues may include samples of the provision being sought and its associated odors as well as other chemical signals the dancers release. This information, as well as scent markers at the source, minimizes the time and energy needed by the recruits to locate the provisions.

Dancing bees utilize a particular area of the comb, or "dance floor," on which to recruit other foragers. This area is near the entrance of the hive and is chemically marked by pheromones. When the dance floor is cut and moved to another part of the colony, the dancing bees move to the new location.

Several dance configurations have been recognized, but the two most easily identified are the *round dance* and the *wagtail dance*. The round dance communicates that the provisions are 160 to 330 feet (50–100 m) from the hive in any direction and the wagtail or figure-eight dance is used for provisions over 330 feet; the pattern formed by this dance is similar to the figure 8. This dance conveys to recruits the distance and direction to a specific location. Different races of bees (see chapter 1) convey distances with different numbers of waggles per run; these differences are referred to as *dialects*. Bees being recruited to locations beyond 330 feet obtain distance information from the waggle portion of the wagtail dance and directional information from the position in which the bee's head is pointing in relation to the sun when waggling. The number of waggles within a given "run" conveys the distance. The wagtail dance includes three components: comb vibration, frequency of waggles, and the number of waggles (see the illustration on dance language on p. 33). During this time, the bee transmits the waggle vibrations to the surface of the comb, where they are felt by other bees attending the dancer, especially if the dance is performed on uncapped cells. This vibration is thought to be picked up by the recruited workers via special organs on their legs or possibly their antennae. Since the wagtail dance resembles the number eight, the middle section where the dancer waggles is referred to as the *straight run*. Distance is communicated by the number of straight runs or waggles in the dance. At the same time, the dancing bee is covered in the odor of the flowers she visited, and these scents are also being picked up by

the bees following the dancer. Dance language is still being studied, and development in new recording technology is being used to decipher this enigmatic action.

A transition or sickle dance is done by some races when distances to the food are between those communicated by the round and the wagtail dance. Workers have been observed doing other dances, such as vibration and migration signals; these are still being studied. The most current information is nicely summarized in Tautz's *Buzz about Bees*.

Other Behaviors

Washboard Movement

Beekeepers can often observe bees, usually in the late afternoon, on the front wall of the hive with their heads pointed toward the entrance. These bees are standing on the second and third pair of legs and seem to be scraping the surface of the hive with their mandibles and front legs, as if to clean it. As they scrape, their bodies rock back and forth in a motion similar to someone scrubbing clothes on a washboard. This is called the *washboard movement*. The exact purpose of this activity is not currently understood; it usually occurs only in very populous colonies.

Bee Bearding

The phenomenon known as *bee bearding* refers to the clustering or hanging of older bees outside the front face of the hive during very hot weather, giving the impression of a beard. This behavior is common on hot, humid days when temperatures reach the high 80s and 90s (27–33°C). This activity is viewed by some as an indication that bees are preparing to swarm (in northern latitudes, swarming usually takes place from late April until the middle of June).

Yet, bee bearding takes place even when it is not during the swarming season, or when temperatures are far greater than those from mid-spring to early summer. In most cases, a portion of the bees in a hive preparing to swarm hang from the bottom board in a conical cluster rather than spreading themselves over the front facing portion of the hive in a beard-like formation.

Bee beard forms any time bees need to cool the inside of the hive. When summer is in full swing, the colony populations are nearing their peak and combs are filled with open brood—the act of bringing in water by forager bees to cool the hive down may not be sufficient. As a consequence, a given number of adult bees vacate the inside portion of the hive and move outside to assist in moderating the internal temperature in order keep the brood from being killed by excessive heat.

When bearding occurs, and weather forecasts indicate a sustained period of high temperatures, the beekeeper needs to take some action to ameliorate the situation. Increase the ventilation by adding supers to reduce crowding. Screened bottom boards will also help resolve the problem. Propping open the outer cover will provide additional ventilation. In extremely warm climates, hives should be located in areas that receive shading during the afternoon. In addition, the following tips were outlined by Khalil Hamdan (*Bee-World*, 2010) and can be useful:

- Nothing beats the screened bottom board for ventilation.
- Increase the entrance space. Larger entrances are good for hot weather.
- Place shade boards to shade the hives, especially if the sun is beating on them.
- If you have 10 frames consider going to 9 to allow more space for ventilation.
- Provide an upper entrance to improve ventilation in warm humid conditions during the summer.

- Place a ventilation box with screened openings on top of the inner cover, then place the outer cover on top of the ventilation box. This allows for good airflow through the hive. Ventilation boxes can be made out of unused honey supers: drill holes in the sides on a slant so rain cannot enter and screen them on the inside to prevent robbing.
- Painting hive bodies with white paint or some other light color helps to reduce overheating. White reflects the sun's heat.
- Use an inner cover with either a wider or two openings.
- Provide a source of water for the bees in a partially shaded position near the apiary (within half mile or less) or use Boardman feeders. On a hot day, a strong colony will use more than one-quarter of a gallon (1 liter) of water to cool the hive and to prevent overheating.
- Sliding a super back about one inch (2.5 cm) to improve airflow is not a recommended approach. This may encourage robbing and allow rain to get inside the hive.
- Remove a few frames of honey or harvest supers when sealed and add new boxes of frames to provide space and ventilation.

Check the References for more information under "Bees and Beekeeping."

 Notes

CHAPTER 3

Beekeeping Equipment

A *hive* is a structure that houses a full colony of bees (workers, drones, and a queen). A man-made hive consists of a bottom board; hive bodies (with 8 or 10 frames; see below) for the broodnest and, if honey has been stored, one or more honey supers (the number depends on the abundance of nectar); and an inner and outer cover or a migratory cover. The depth of the hive bodies varies.

Some beekeepers use only standard or deep supers; others use the shallower supers for housing both the brood and the honey. If only the standard 10-frame deep hive bodies are used, lifting off the honey will be very strenuous because a deep super full of bees, brood, wax, wooden frames, and honey can weigh over 100 pounds (45 kg).

If, on the other hand, you choose only the medium supers to house your colony, the weight will be less, as one full 10-frame medium super weighs about 65 pounds (29 kg). Finding the queen in the broodnest, however, becomes much more time-consuming and disruptive to the colony. If two medium supers are used as the broodnest, the queen can move more rapidly from frame to frame when you are in the hive than if she were on deep frames; in this case, it may take you longer to locate her and thus you will be keeping the colony exposed longer.

The number of hive bodies needed to house bees varies throughout the season; in general the brood-nest occupies the equivalent of two deep supers. In regions with cold weather, the extra honey supers are removed, and the bees winter in two deep and one shallow super; the top deep and the shallow are used for winter stores of honey and pollen. But this can vary; other combinations are one deep and one shallow, two deeps, or sometimes even three deeps or three shallows (see "Wintering" in Chapter 8). Whatever configuration is used, an ample supply of provisions must remain or be supplied before the onset of cold weather.

It has been traditional to paint the exterior of hive bodies white to reflect the sun's heat in the summer months, thus keeping the colony cooler, especially in southern parts of the United States. Remember to paint the metal top of the outer cover white; this will keep it cool to the touch during the hot summer months. Although white is most favored in southern climates, beekeepers in northern areas might consider painting hive bodies darker shades to retain the heat longer. For hives located in highly visible sites where vandals or thieves may be a problem, other colors serve to camouflage hive equipment.

Whatever color is used, the outer sides or rims of the wooden hive parts should be painted to extend the life of the equipment. Because bees produce moisture as part of their metabolic activity, a latex paint would be least likely to blister as the moisture leaks out; lead-based or other toxic paints should never be used. Paint the bottom and sides of the outer cover and bottom boards (especially underneath) to extend their life and make sure to paint underneath the ridge of the handholds.

To save money, you can buy off-color latex paints from hardware or paint stores; by mixing the paints together, you can create your own unique hive colors. It is unnecessary to paint or coat any interior hive parts with any substance. Some beekeepers paint the rabbets (see the diagram of a beehive on p. 38), or apply a thin layer of petroleum jelly (such as Vaseline), to reduce propolizing in this area.

Some pieces of equipment, notably the inner and

outer covers and the bottom boards, are available in plastic, but their suitability compared to wooden equipment can be disappointing. Visit with other beekeepers to obtain information about using plastic equipment in your area.

Beekeepers who live in damp and humid regions should consider other additional measures to extend the life of the hive furniture. Immersing wooden parts in boiling paraffin is an excellent way to preserve them (see *Hot Wax Dipping of Beehive Components for Preservation and Sterilization* [2001] in the References). This method works well but takes significant physical effort to heat the wax and can be dangerous, thus requiring great care. It is economical if you have a lot of equipment to dip; or you could combine your efforts with other beekeepers. Many wood preservatives used for lumber can be very toxic to bees and should not be used.

If your hives are in an area where loss through theft is a concern (or if you keep many colonies in various locations), you could brand the wooden hive parts with individual identification. However, branding your name or symbol on hive parts merely alerts the thief that eradication of such marks is necessary. A less expensive alternative is to leave your name and address under the tin roof of an outer cover, or with a permanent marker, write your name **inside** your supers and even if the bees cover your name with propolis, it can be easily scraped away. Or you can add or leave out an extra nail, or otherwise individualize equipment to assist in proof of ownership in the event of theft. If you have just a few hives, the extra time may not be worth the effort. New microchip technology now exists, and this option should be explored; check recent bee journals for the latest in identification technology.

BASIC HIVE PARTS

Outer Cover

Two types of outer covers, a *telescoping* cover and a *migratory* cover, are placed on top of a beehive to act as a roof. The telescoping cover is usually made of wood covered with some sheet metal, like aluminum. It "telescopes" over the rim of the inner cover and uppermost hive body. If you live in an area with long, hot summers, some form of insulating material (newspapers work well) should be placed between the metal sheet and the wooden roof during construc-

tion. The illustration of a beehive shows a telescoping outer cover (p. 38).

The flat California or migratory cover does not telescope over the sides and is used in drier areas because it will not last long in extended wet weather. The lack of sides telescoping over the edge of the hive allows beekeepers to strap hives together to move them to different locations for pollination services and to particular or abundant honey sources. It also permits colonies to be packed together on pallets and trucks for migratory operations. Whichever outer cover you use, place a heavy weight on top of it to prevent it from being blown off by strong winds. If a colony loses its roof, exposure to robbing, heat, snow, rain, and cold winds could weaken or kill the bees.

Inner Cover

The *inner cover* is a wooden, Masonite, or plastic board with 1/2-inch (13 mm) rims and an oblong hole in its center. Some inner covers come with an additional half-circle hole notched in one end of the rim to provide an extra entrance for bees. This hole, as well as the center opening, helps to vent moist air, especially over winter when the inner cover is turned rim-side down. When migratory outer covers are used, inner covers are not necessary.

If a device called a *bee escape* is placed in the oblong center hole, the inner cover becomes an *escape board*. By placing this board below a honey super or supers, the worker bees passing through the one-way bee escape are unable to reenter the honey supers above, and thus the boxes are cleared of bees; see "Removing Bees from Honey Supers" in Chapter 9 for other removal methods.

Whenever it becomes necessary to feed a colony, you can invert a jar or pail of syrup over the oval hole of the inner cover. Bees can collect the food without venturing outside. Place an empty super on top of the inner cover enclosing the feeder and the outer cover on top of that super (see Chapter 7). This will protect access to the feeder by robber bees.

Shallow or Honey Supers

The shallow supers, which are usually used for honey storage, come in various depths and with frames and foundation of corresponding size. In the United States, three different honey supers are used:

A Beehive

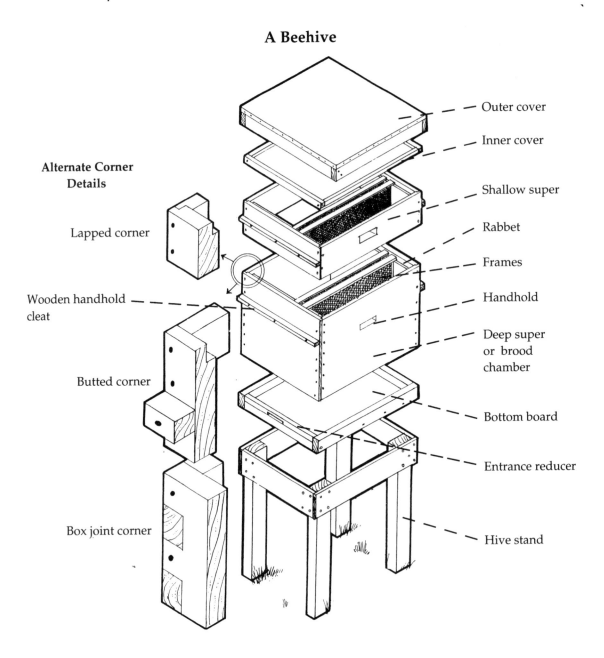

Alternate Corner Details

Lapped corner

Wooden handhold cleat

Butted corner

Box joint corner

Outer cover

Inner cover

Shallow super

Rabbet

Frames

Handhold

Deep super or brood chamber

Bottom board

Entrance reducer

Hive stand

- 4 ¹³⁄₁₆ inches (12.2 cm), referred to as the comb honey or section super.
- 5 ¹¹⁄₁₆ inches (14.5 cm) or shallow super.
- 6 ⅝ inches (16.8 cm) or medium super (also called Dadant, Illinois, or western super); this can also be used as a brood super.

The different super sizes are illustrated in "Super Sizes" in Chapter 9.

The number of honey supers per hive will vary during a honeyflow, depending on the amount of honey collected and the number of supers you have. Although supers are designed to hold 10 frames, this configuration is generally used only in the brood supers. Experienced beekeepers put 8 or 9 frames in honey supers, so the bees will draw out wider comb. When filled with honey, such thick frames are much easier to uncap when harvesting the honey.

Queen Excluder

A queen excluder is a perforated plastic or metal sheet or a metal- or wood-framed metal grill. *Queen excluders*, as the name implies, exclude the queen from going beyond where the device is placed. Because of its small openings, only the worker bees can squeeze through; the larger drones and queens cannot. The device is placed on top of the broodnest to prevent the queen from entering the honey supers above. It is also used on two-queen colonies or for any other reasons that require the exclusion of the queen. Drones are often trapped by excluders, and if they die in large numbers, they can clog it with their bodies. This can be prevented by leaving (or boring) an escape hole in the honey supers and by eliminating adult drone or drone larvae from the supers. Be careful, robber bees could also enter, so be alert.

Deep Super or Brood Chamber

The deep or standard brood box is 9 5/8 inches (24.4 cm) in depth. Ten full-depth frames (9 1/8 inches) are used in this super. The deep super is typically used in the United States to contain brood and winter stores. The general understanding is that since these deep, heavy boxes are not lifted or moved much, they are used as brood chambers; but to collect honey, the shallower supers keep the weight more manageable. At one time, bigger supers called the Jumbo and the Modified Dadant deep supers were common. Commercial beekeepers tend to use only one size of super for both brood and honey production.

Some bee supply companies are reintroducing the smaller 8-frame hive equipment, which has the same vertical dimensions as the other supers but holds 8 instead of 10 frames. This super is lighter in overall weight and has smaller dimensions, and some beekeepers say that the bees do better in it; again, chat with beekeepers who have this equipment to see if they like it. The only problem is that this equipment may be difficult to find for purchase and cannot be mixed with the standard 10-frame bodies.

Bottom Board

The floor of a hive is a wooden structure called a *bottom board*. Hive bodies are stacked on top of the bottom board, which should be placed on a firm and level foundation to keep honey-heavy hives from tipping over. Never place the bottom board directly on the ground as it will rot quickly; use a hive stand, bricks, cinder blocks, or flat rocks (see the illustration of the beehive on p. 38). Treating the bottom board with wood preservatives may extend its life, but many chemicals not only are toxic to bees but also can contaminate honey and wax. Bottom boards are a good candidate for paraffin dipping (see text above for the reference).

Most bottom boards have two rim heights—a short winter rim and a deeper summer rim—and these are called *reversible bottom boards*. Many beekeepers, instead of reversing the bottom board (which is difficult to do), use an entrance reducer to restrict the hive opening in the fall. By reducing the entrance, weak colonies are protected from robbing bees and other insects, and cold winter weather is likewise blocked. Mice guards are also available that double as entrance reducers. Plastic bottom boards are available, but they sometimes buckle if the hive is very heavy, rendering them useless.

Commercial operations often use handmade pallets, on which bottom boards are permanently attached. This system serves as a convenient way to move bees (via forklifts) and provides a solid and safe hive stand.

Screened Bottom Boards

With the invasion of parasitic varroa mites and the small hive beetle, new equipment is available as a passive control method. The screened bottom board is an effective and noninvasive way to control for these mites, but it needs to be used with other mite control options because it does not keep mite populations from reaching peak levels. However, the screened board is a means to determine the mite population in your colonies when it is used in conjunction with a sticky board to collect and count mites. Check the current research on these boards and speak with other beekeepers about using them. Also, special bottom boards or traps for the small hive beetle are also available and should be researched if you have this pest in your area (see Chapter 14 for more information). Check bee supply catalogs.

Commercial Hive Stand or Alighting Board

Most bee supply companies sell an on-the-ground hive stand and alighting board combination. Although it is sufficient for most situations, this type of hive stand is not recommended because unless it is made of cypress (or rot-proof material such as plastic) it will rot in a few years. It will also not protect bees from ants, mice, or other predators. Therefore, most beekeepers eventually make their own hive stands from materials at hand in order to keep the bee hives elevated. For more information, see "Hive Stands" in Chapter 4.

Bee Space

Bees do not space natural honeycombs at random in a wild colony or in the wooden beehive. They adhere to a strict code and do not construct comb in spaces less than ⅜ inch (9.5 mm). This fact was published by the Philadelphia minister L.L. Langstroth over 100 years ago. It was the basis on which he designed the prototype beehive used today. The ⅜-inch space enables beekeepers to remove frames without having to cut the combs from the walls and covers. A ⅜-inch gap separates the frames from the hive walls, the top bars from the inner cover, and the bottom bars from the bottom board; the space between any two adjacent frames is the equivalent of 2 bee spaces, which allows bees on the vertical surface of the comb to move freely (see illustration of the bee space on this page).

By utilizing this natural spacing, beekeepers are assured that the bees will not attach comb to the walls or to other sections of comb, and that the frames can be easily removed. If the frames are spaced farther apart, or if you neglect to return a frame to the hive after examining it, the bees will fill the gap with a perfectly formed piece of comb; you will have to cut this comb out in order to put in a frame, making it rather messy, especially if the comb is full of honey.

Any space less than the bee space will be filled

The Bee Space

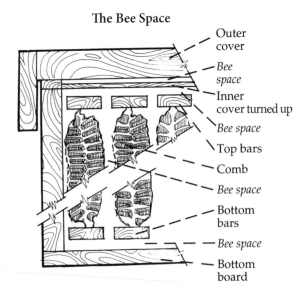

Outer cover
Bee space
Inner cover turned up
Bee space
Top bars
Comb
Bee space
Bottom bars
Bee space
Bottom board

with propolis or bee glue. This helps to keep cold air drafts out of the hive and kill any microbes that may live in these tiny spaces. See more about propolis in Chapter 12.

It is illegal to use beehives with fixed honeycombs such as straw *skeps* or log gums. Up until the mid-1800s, colonies of bees were kept in skeps or straw domes. To collect honey, the bees were killed and the honey and wax were cut out and separated. Today, all bee equipment must have removable frames so bees can be inspected for diseases.

Frames

Bees in the wild attach honeycomb to the ceiling and often to the walls of a cavity. Such combs can be removed only by cutting or breaking them from their attachments. In modern beekeeping equipment, hive bodies contain *frames*, which are rectangular structures made of wood or plastic, and hold *foundation*, which is a sheet of beeswax or wax with a plastic base, embossed with either hexagonal worker or drone cells. When the foundation has been drawn out and completed with finished wax cells it is called *drawn comb*. There are now one-piece plastic frames with embossed plastic "foundation" available in several colors; darker colors make it easier to see eggs.

The vertical parts of frames, called *side bars*, are designed to space the frames in the hive; see the illustration showing frame sizes on p. 41. Thus, when bees have fully drawn wax into honeycomb, the natural "bee space" is created between combs. This space allows the bees to move freely from comb to comb and the queen to lay her eggs.

As illustrated on p. 41, frames come in many sizes and styles (European countries have more sizes, so be careful if you are ordering hive furniture from international sources). The top bars of frames will have either a *wedge* that is removed and then nailed back once the foundation is in place, or grooves that allow you to slide the foundation in. The bottom bar may be either solid, grooved, or two pieces. Before the foundation is set in, you should string the frame with wire. This will add support for the foundation, which is especially important for frames that will be placed in a honey extractor. Because the extractor spins frames at high speeds, stress is placed on the comb and may cause it to buckle or break. The wire support will prevent breakage in most cases. It is also

**Frame Sizes
(inches)**

Nailed-in top bar wedge

1.   2.   3.   4.   5.

11¼   9⅛   6¼   5⅜   4½

Slotted bottom bar

1. Jumbo or Modified Dadant
2. Deep frame
3. Medium frame
4. Shallow frame
5. Section super frame

important to make sure the foundation is perfectly flat in the frame; otherwise uneven or warped honeycomb will result.

A simple jig can be made or purchased to facilitate wiring frames. It is a good idea to hammer eyelets in the holes on the wooden side bars before wiring; otherwise the taut wire strung through these holes may cut into the wood, which can cause the wire to slacken. The wire used for stringing frames is no. 28 tinned, although any thin wire that won't rust will do. Some beekeepers use 40-pound fishing line or monofilament instead of wire, in which case there is no need to embed the line in the foundation. However, there are reports that bees can cut the monofilament, especially if there is no strong honeyflow when these frames are installed.

If you decide to use plastic-based foundation, or plastic sheets sprayed with beeswax or plastic molded frames that already have the hexagonal honeycomb, wiring is not necessary. Check with your bee supply dealer or with other beekeepers to compare notes on which is the best to use in your area (see "Foundation" in this chapter).

With wired frames, the wire is embedded into the foundation by lightly heating the wire. Two tools can be used for this purpose. One is called the *spur embedder*, which is heated in boiling water and then rolled over the wire. The heat melts the wax around the wire; when hardened, the wax holds the wire-foundation combination in place. Another tool, called an *electric embedder*, heats the entire wire with

an electric current, which melts the surrounding wax enough for it to become embedded.

When the frames are fitted with sheets of wax foundation, start with 10 frames in the hive so the bees will draw out the foundation into even combs. Some beekeepers later remove one frame to allow easier manipulation of the remaining 9 frames or to allow the honey frames to be "fat"—making it easier to extract. To evenly space these 9 frames, use special spacers or follower boards, or merely space the 9 frames evenly by hand. If you do not space the frames evenly, bees will fill in any larger gaps with additional comb or expand the comb to fit the empty space. Make sure to add all 10 frames of foundation to a new super so the bees will draw out the comb correctly. This is especially true for the plastic foundation.

Empty frames—those that contain no foundation or comb—should not be placed in a colony, because the comb may be attached in an unsatisfactory way, such as at right angles, which would fasten all the frames together and make their removal and inspection impossible. Some beekeepers will use washed hardware cloth or plastic screen instead of foundation or, for reasons of economy, will put only a small strip of foundation (wax or plastic) attached to the top bar of a frame. Although straight, even comb may be achieved by this method, the practice of putting in small foundation strips is not recommended: with deep frames, the bees may construct drone-sized comb rather than worker cells, or fill the empty

Frame Parts and Foundation

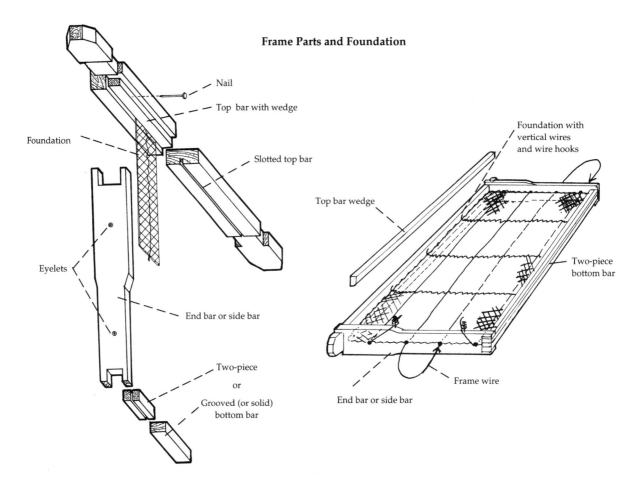

space with crooked comb, and it will not be wired, further weakening the comb.

With the advent of the varroa mite, some beekeepers practice mite control by encouraging drone production, because this mite prefers to infest drone larvae. You can now purchase either drone foundation or plastic drone frames. Another technique is to remove a deep frame and in its place substitute a shallow frame. Bees will draw drone comb beneath the bottom bar. Once the varroa mites are attracted to the drone brood, the infested pupae can be removed and frozen to reduce the population of mites (see "Varroa Mite [Varroosis]" in Chapter 14). **Make sure** to label the top bar of this frame so you can find it easily. The drone comb can always be removed in the fall; this method should be used only for a few colonies; otherwise, buy drone foundation and, again, label those frames.

A word about plastic frames and plastic foundation: Some suppliers sell one-piece frames. While beekeepers report good success with such frames, and they may help with mite control, others do not use them at all. If you wish to experiment, try plas-

tic frame units on a small scale; they may or may not work well in your area. Remember, if you find foulbrood in your colonies and need to burn frames, the plastic frames will not burn cleanly, and it may be illegal to burn them because of the toxic gases released. You will then have to find another way to destroy or otherwise sterilize plastic equipment. Irradiation chambers exist in some areas for sterilizing diseased equipment.

Foundation

Foundation is a thin sheet of beeswax, or wax with a plastic base, embossed with the hexagonal shape of honeycomb cells. Sheets of plastic embossed with the hexagonal comb are also available; they come with or without a wax coating. They are sold with worker-sized or drone-sized cells. This sheet is set in the rectangular frame and wired or pinned in place. When put into a hive of bees that are gorged with honey, nectar, or syrup (a state that induces worker bees to produce wax), the bees will use the foundation as a base and draw out the walls of the cells. They do this by adding secreted wax to the base, then by drawing up the sides of the cell walls. When completely filled out, the frame is said to contain drawn comb.

The kinds and combinations of foundation can be confusing. Basically, there are three different categories:

- 100 percent beeswax (with or without vertical wires).
- Thin beeswax sheets with a plastic core (with metal edges).
- Plastic embossed sheets with beeswax sprayed on top.

Different thicknesses of wax foundation also exist, but basically the thicker sheets of 100 percent beeswax with inlaid vertical wires are used for brood frames and in the extracting supers. The thinner 100 percent unwired beeswax foundation is used in honey supers for producing cut-comb, round, or section comb honey. As we already mentioned, unwired frames should **not** be placed in a honey extractor. Although most foundation sold today contains worker-sized cells, drone-sized-cell foundation, now in plastic, is also available for varroa mite control (see below). Drone foundation is also used to increase drone populations in queen mating yards to ensure there are enough drones to mate with virgin or newly mated queens (see Chapter 10). It is also used in honey supers.

To set the foundation firmly in the frame, especially in the brood frames, horizontal wires (or monofilament fishing line, 40-lb. weight) should be first strung across the wooden frame (see the illustration of frame parts and foundation on p. 41). Use wired foundation, deep or shallow, in these wired frames that will go through a high-speed extractor. When a wired frame contains unwired foundation, the resulting comb will break apart when it is run through an extractor, and pieces of wax will "blow out," leaving large pieces of comb in the extractor. The embedded wires also help keep the comb from sagging, keep the cells from stretching in warm temperatures, and strengthen the honey-filled frames against breakage during extracting. Distorted cells in sagging combs are unsuitable for raising brood and decrease the number of cells in which the queen can lay eggs, which results in spotty brood patterns. Such frames should be replaced with new foundation or newly drawn comb.

The plastic-core foundation sheets are easy to install and require no wiring. Drawbacks include the bees not drawing new comb on the plastic sheet, should the wax separate from it. Bees prefer the 100 percent beeswax foundation.

Smaller worker-sized-cell foundation has become the focus in controlling varroa mites. Some beekeepers have tried the smaller, 900-count foundation, but converting bees to the smaller cell size will take several generations, and most bees don't like to work the small foundation. Africanized bees, however, work this smaller comb very readily. The most recent research has found that small cell size does **not** reduce the varroa mite populations.

Foundation can be stored for a long time but should be kept sealed and away from heat or freezing temperatures. If kept in large plastic bags in a cool, dry place, the wax sheets will remain fresh and soft. Never store foundation in hot or sunny areas, as the wax will melt. Also, the plastic bags will protect it from wax moth predation. Also, foundation may now contain pesticide contaminants. Check Chapter 15 "Products of the Hive Beeswax" for more information.

FOR THE BEGINNER

To the beginning beekeeper, the plethora of equipment available from the bee catalogues may seem somewhat confusing. The basic equipment listed in the "Beginner's List" below provides a good starting point.

It is not a good idea to keep just one hive of bees, because if the colony fails to develop properly, or the queen becomes injured or dies, you will lose an entire year's experience at working and learning the art of beekeeping. Two to five hives is a better number for the beginner, or go in with a partner, each getting one or two hives.

While used hive bodies and frames are less expensive than new equipment, they could be contaminated with brood diseases that are not readily apparent. If equipment is questionable, it should be sterilized. The most economical way to sterilize hive bodies, bottom boards, and inner and outer covers is to scorch them with a propane torch, but this is not always reliable or effective. Frames are too difficult to sterilize and should be discarded.

A more effective sterilization technique is to use

a boiling paraffin dip. This should be done with only hive bodies, not frames (see above); see "Fumigation Chambers" in Chapter 13.

Other, more expensive equipment, such as honey extractors, can be shared by several beekeepers on a cooperative basis. Some hobbyists take their honey supers to a sideline or commercial beekeeper for extraction. Work out a fee, usually a percentage of the honey extracted, beforehand. The joy of harvesting your own honey is part of the lure of becoming a beekeeper, but the actual process of extracting is a messy and sticky affair. Letting someone else who has the right equipment and setup to extract is a better alternative for most beginners.

Beginning hobbyists are cautioned not to buy every gadget on the market. When in doubt about the usefulness of a particular piece of equipment, seek the advice of more experienced beekeepers.

Beginner's List

Most hive parts come disassembled, and some require painting. Some bee supply companies sell plastic or wooden equipment made from pine or cypress. More durable hive bodies are made from cypress, which has a longer life than pine. Check prices and available equipment from various bee supply houses, or from your local bee club. Generally, when you are first starting, it is best to buy all new equipment from one source. Hive dimensions vary just enough to cause problems when you mix and match hive furniture from different suppliers.

The current trend is to purchase hive furniture completely assembled and sometimes painted. This is certainly a time saver, but it will cost more, and you will lose out on the fun and education of putting together your own equipment.

While different bee supply companies have a variety of beginner's kits, a complete kit for one hive should include:

- 1–2 Standard deep hive bodies, inner and outer covers, bottom boards, and frames (unassembled or assembled)
- 10–20 sheets of wired foundation, deep
- 1 smoker
- 1 veil
- 1 hive tool
- Gloves

- Entrance reducer
- Bee brush
- Feeder
- Beginner's book and assembly instructions
- 1–2 medium (6 5/8 in.) or shallow supers and frames
- 10–20 sheets of wired foundation, medium or shallow
- Queen excluder
- Smoker fuel
- Subscription to bee journal

TOTAL COST $300–$600

Optional Equipment

- Paint (latex)
- Mouse guard
- Bee suit, zip-on veil, and gloves
- Or helmet and veil separate
- Division board or top feeder
- Bee escape or escape board
- Roll of duct tape
- Uncapping knife, electric
- Extractor (two frames, hand powered)
- Jars for honey and labels
- Another good beekeeping book
- Year's subscription to another bee journal
- Bee club membership

TOTAL COST $300–$500

The prices listed are approximate for 2011, and they will vary somewhat depending on make, supplier, and so on. Shop around to get the best deals.

BEEWEAR

There is no stigma to wearing protective equipment; it will save you and your visitors a lot of worry if you are properly attired to look through your colonies without the fear of being stung. You will learn to gauge bee temper, handle equipment properly, and understand basic bee management by working bees. Listed below are the essentials.

Veil

A bee veil is a must. Although you will see photos that show beekeepers working without veils, such practice is discouraged. Stings on eyes, lips, scalp, or inside the nose or ear canal are extremely painful; it is downright foolish to risk them. All sensible beekeepers wear veils. Veils can be purchased separately or attached to helmets.

There are several kinds of veils:

● Wire-mesh veils (square, folding, or round) worn with a helmet
● Sheer veils (tulle or nylon) worn with a helmet
● Alexander-type veils, worn without a helmet, held in place by an elastic head strap
● Veils attached to coveralls, with or without helmets
● One-piece bug bag that fits over the entire torso and arms

Check with other members of your bee club and try on different veils and suits to determine which type you like best.

Helmets

Helmets are worn to support veils, to keep the veils away from your face, and for shade. They are available in plastic, cloth, or woven mesh styles, with or without attached veil material. Some helmets have ventilation holes or louvers, but remember, if you live in an area that has the smaller Africanized bees, they can get through these holes. You will have to tape over or stuff cotton in these holes to keep them out.

Most helmets have bands inside that allow you to adjust it to your head size, but none fit very well. Helmets have a tendency to slip to one side or onto your face at inappropriate moments (such as when you bend over or are looking up at a swarm), and can create problems for the inexperience beekeeper.

Felt or straw hats are sometimes used as helmets but are not usually strong enough to hold up to rough use. Veils and helmets are typically worn with a coverall; if veils are attached to the coverall with a zipper, the outfit becomes a *bee suit*.

Coveralls and Bee Suits

Bees are less likely to sting people wearing light-colored attire. They are more prone to sting dark, furry objects (such as skunks and bears). Wearing lightweight, breathable material is preferable, because the best part of the day for working with bees is usually also the hottest part of the day. However, fabric that does not "breathe" well, such as nylon, may cause you to sweat more than usual, and heavy perspiration may aggravate the bees. Rule of thumb: fabric light in color and weight is best.

The easiest, most foolproof clothing to wear is a one-piece bee suit (with a zip-on veil) either jacket-length or full-length. Bee suits come in handy if you are working a lot of colonies. Even though bees are able to sting through them, they rarely do. These suits keep your clothes cleaner and protect them from wax, honey, and propolis (the latter is difficult to remove thoroughly from clothing). If you are purchasing a bee suit, buy the best one you can afford, for cheaper ones tend to wear out quickly. The suits should incorporate all of the features discussed in this section. Bee suits or coveralls, whether purchased or homemade, should be made of white cotton or a white polyester-cotton blend and have pockets and pouches to carry hive tools, matches, and the like. All wrist and ankle cuffs should have elastic in them to fit tightly with no gaps. Some suits have zippers on the legs, which make getting into the suit easier, especially if you are wearing boots. Make sure the pockets of purchased bee suits or coveralls do not come with openings that access your inside pockets. Bees could get in.

If you are not wearing a bee suit, turn up your shirt collar before putting on a bee veil and close the trouser and sleeve cuffs with tape or elastic to keep bees from getting beneath clothing. You can also wear gauntlets. If using a jacket-style suit, some beekeepers tuck their trousers into their shoes or socks. If clothing is not closed tightly, bees can crawl inside unnoticed, and when a bee is pressed between clothing and skin, it will sting. Once a bee gets inside the clothing, you may attempt to release it, but it is easier to crush the bee before she stings you.

If bee suits do not come with a helmet, wear a visor cap or some other hat underneath to keep the sun off your face. Remember, veil mesh does not shade your face from sunburn. Sunscreen is always recommended.

Some bee supply companies sell child-sized suits. Check online or in bee journals for other styles of suits. If you still can't decide, have your local bee club put on a "beauty" contest and show off the various bee wear. Wash your bee suit periodically to prevent the accumulation of dirt, disease spores, and venom from stings.

Gloves

Many old-time beekeepers disdain using bee gloves, but for the beginner it is a good idea to start with them. Again, there should be no stigma to wearing gloves while working with bees. As you gain experience, you will probably not wear your gloves, at least most of the time.

Gloves sold today at bee supply outlets are made from cloth or canvas, plastic-coated canvas, leather, or some kind of plastic. Wearing gloves that do not fit well will make handling frames more awkward and may even invite more stings than not wearing gloves at all (yes, bees can sting through leather). All bee supply houses carry men's sizes, but the smaller women's sizes may be harder to find; check the Internet or other bee supply houses. Buy gloves that fit snugly. Even dishwashing gloves work well, and they can be replaced easily. A double layer (two pairs) of medical latex gloves, now widely available, does not interfere with dexterity and can reduce or even eliminate the risk of stings; however, your hands will sweat a lot. These gloves are disposable, thus reducing the risk of moving spores and other diseases among colonies. Work with other beekeepers to observe how they use (or don't use) gloves; you can learn much by watching others work their bees.

Gloves are a great help in keeping wax, honey, and propolis off your hands, but wash them periodically, especially after working with diseased bees. One disadvantage of gloves is that they may retain the alarm odor long after bees sting them, which is another reason to wash them regularly. Most times, however, gloves are not necessary and you should not make a habit of using them—you can get stung more wearing gloves than bare handed. And never work someone else's bees with your used gloves, for they may contain foulbrood spores. In a pinch, latex or rubber gloves are a great alternative.

Some people wear leather gloves when applying miticide strips for varroa control. These are **not** recommended. Chemical-resistant rubber gloves must be used when applying miticide strips (see "Varroa Mite [Varroosis]" in Chapter 14).

If you wear a black plastic or leather watch band, remove it during bare-handed apiary work, because such bands seem to incite bees to sting your wrists.

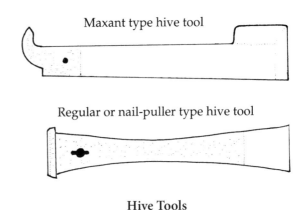

Maxant type hive tool

Regular or nail-puller type hive tool

Hive Tools

HARDWARE

Hive Tools

Several types and styles of hive tools are available from bee suppliers, including Teflon-coated tools. The scraper-nail-puller type can usually be found in most hardware stores (see the illustration of hive tools on this page) and comes in 7-inch or 10-inch lengths—buy both. Hive tools are an invaluable aid to the beekeeper when prying apart hive bodies and frames that have been propolized.

It is a good idea to have several hive tools on hand because they are easy to misplace. Hive tools should be periodically heat-sterilized and scoured clean of excess propolis and wax; use steel wool and sandpaper. Paint your hive tools bright colors (or stripes) such as a neon red, orange, or yellow, but do not use blue, green, or brown (which are camouflage colors). Painting your tools bright colors will help keep them from being lost in the grass and is a way to identify your tools from someone else's. The ends should be sharpened at least once a year, and the sharp corner edges blunted.

Try several kinds of hive tools, and see what kind beekeepers like to use in your area.

Smoker

The second most important piece of equipment you will need, besides the veil, is the smoker. The smoker is a metal cylinder, in which a fire is lit, with attached bellows. Start the smoker by following these steps: (1) ignite some paper, (2) drop it to the bottom of the smoker, (3) pump air through the smoker by way of the bellows, and add more fuel once the lower layer is well lit. Smoke blown from the nozzle (see the il-

Bee Smoker

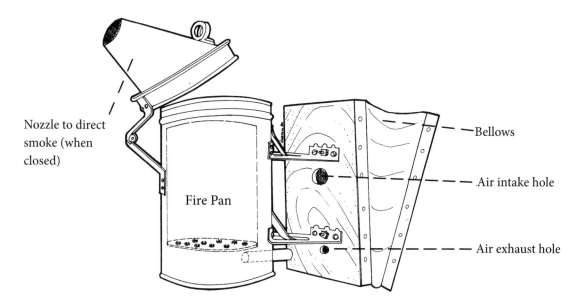

Nozzle to direct smoke (when closed)

Fire Pan

Bellows

Air intake hole

Air exhaust hole

lustration of the smoker on p. 47) is directed into the hive and between the frames to encourage bees to gorge honey. Once engorged, bees are more docile and less prone to sting. Also, smoke masks the alarm pheromone bees release when disturbed.

Smokers come in 4 × 7 inch and 4 × 10 inch sizes. The small size generally does not hold enough fuel for more than a two-colony inspection trip; when purchasing a smoker, get the largest available size. For long life, buy a stainless steel smoker with an attached heat shield and front hook. See Chapter 5 for more information on lighting and using your smoker.

OTHER EQUIPMENT

Frame Grip

When held in one hand, this spring-loaded metal tool enables you to grip the center of the frame's top bar and pull out the frame in one motion. This is especially helpful to those who find it difficult to pull out frames. The frame must first be loosened with a hive tool, to break the propolis seal. However, if the grip is not used correctly, you can scrape or injure clinging bees or the queen if she is on the frame. You must lift the frame slowly, carefully, and straight up out of the hive.

Bee Brush

This soft bristle brush is used to clear bees off a frame, especially frames with honey that you want to extract. It is not a practical way to remove bees from many honey frames. It does come in handy, though, when you need to exchange frames (such as honey) from strong colonies to weaker ones (without the clinging bees and mites). In a pinch, you can use a handful of grass, soft evergreens, ferns, feathers, or leaves for brushing bees from a frame. Remember, moving frames around the apiary will also move diseases and mites around, so be careful not to spread problems.

Top Bar Hives

There is a recent trend to "go organic" again, which now includes using the African-style top bar hives instead of conventional equipment. The top bar hive (TBH), also called the Kenyan Top Bar Hive (KTBH), was designed to replace fixed comb log hives commonly hung in trees and is used in many parts of Africa. The KTBH has sloping sides and wide top bars or slats that fit together, forming a roof or top of the hive. Some beekeepers add an extra cover to make it rainproof. The sloping sides of the hive box allow the bees to naturally draw comb, and they normally will not attach the comb to the sloping sides. Top bar hives became popular first in developing countries because the materials needed for their construction are inexpensive (local lumber), and assembling the parts can be done with low-tech tools. They can be made as long (up to 25 top bars) or as short as needed, and they are designed to hang, not rest on the ground.

From its original intent, the use of this hive has become increasingly popular. Why? Many beekeepers feel it is more natural for bees to construct their own honeycomb rather than being given wax foundation enclosed in a frame. When bees construct comb without foundation, they are making it to fit the natural size of the bees. An even more important advantage is that by making their own wax, the comb will not contain the contaminants now found in beeswax, even in purchased foundation. This type of hive uses little if any commercial foundation and forces the bees to draw fresh wax, thus creating a very clean and uncontaminated wax. There are many websites on the top bar hive, so be sure to look them up (see References, "Equipment").

Advantages

- Materials for construction are inexpensive, often can be obtained at a local saw mill.
- The hive is a single horizontal unit, eliminating the need to lift off hive bodies which are stacked one on top of the other (such activity is necessary for removing honey and carrying out normal hive inspections).
- Reduces storage requirements; no empty honey supers and unused deep supers to put away.

- Can be made to hang off the ground, which protects colony from ants, skunks, and other ground pests.
- Can be adjusted to fit honey supers.
- No queen excluder needed.
- Honeycombs can be used as cut-comb honey or chunk-comb honey.
- The top bars are moveable; hence, the hive conforms with laws that require moveable frames.

Disadvantages

- Honey from comb cannot be extracted in the same manner as with conventional frames.
- To extract honey from these combs they need to be crushed.
- The top bar needs to be inserted with a small strip of wax foundation, from this base the bees will construct the rest of the comb.
- Occasionally, comb may be attached to the side walls of the hive.
- During strong honeyflows you cannot add supers to a top bar hive (unless you have made a modified KTBH); this means you have to remove the filled combs and additional top bars need to be added to continue to take advantage of any major honeyflow.
- Unlikely to be used in commercial operations.
- There are good websites on the subject, check the References under "Equipment." Before trying this, check with other beekeepers who are using this method to see if it is a good option for you.

Notes

CHAPTER 4

Obtaining and Preparing for Bees

Whether you are about to become a beekeeper, or are a novice or an established apiarist, there are a number of options available for obtaining your first bees and/or augmenting the number of colonies you wish to maintain.

- Buy package bees.
- Buy or produce your own nucleus colony or established colonies.
- Retrieve colonies nesting in natural or man-made cavities or bait boxes.
- Collect bees that have started building exposed combs.
- Collect swarms.

BUY PACKAGES

Package bees come primarily from the southern states, Hawaii, and California and are shipped in the spring by mail or are picked up by dealers and trucked to their final destination. To order packages, look for advertisements in the bee journals (such as the *American Bee Journal* and *Bee Culture*). Most bee organizations also have a website or a state newsletter with local ads. Bee supply dealers and searches on the Internet will provide the names of other package producers as well.

Call several dealers to compare prices and delivery dates and order packages early in the winter months (December and January) to obtain the desired number of packages and choice of shipping dates. Request early delivery of your packages (e.g., three to four weeks before the dandelions and fruit trees bloom in your area) to ensure that bees will have enough time to develop. All necessary bee equipment should be ordered and fully assembled well in advance of the arrival dates of your packages. For more complete information on packages, see Chapter 6.

You can also contact local dealers or beekeepers within your state that purchase packages and queens for resale. Local dealers are usually experienced beekeepers who sell not only packages but also nucleus colonies, established hives, and an assortment of bee equipment. By attending meetings of your state or regional beekeeping organizations, you will meet dealers from your area as well as other beekeepers.

Generally, a 3-pound package of bees (about 1.4 kg) plus a mated queen will provide an ample number of bees needed to begin a good colony. The approximate cost of such a package in 2010 with one laying queen was $80 to $100 or more. Prices are subject to change and will reflect current conditions.

Advantages
- Easier for beginners to work (fewer bees than in an established hive).
- More adult bees than in a nucleus colony.
- Certified as apparently healthy (and mite-free) and from healthy stock.
- No brood diseases.
- Easy to obtain replacement queens.
- Available in 2-, 3-, 4-, and 5-pound units (there are approximately 3500 bees per pound) with or without queens.

Disadvantages

- Queen could become injured or lost due to stress in shipment.
- Drifting is common (bees fly into other hives or become lost), especially during installation.
- Success is dependent on weather; if it is too cold, bees may not feed and even may starve.
- There are no eggs or brood until queen starts to lay; time until new adult workers emerge is about 21 days.
- Diseases, such as virus, may be present.
- Bees must be fed heavily to draw foundation, since feeding stimulates wax glands to produce wax; feed until the first major honeyflow to keep bees from starving.
- Bees must be fed a protein supplement until there are enough flowers blooming.
- Bees may not forage if weather is too cold or wet, which will delay colony development.
- Bees should be medicated.
- Bees could be infested with pests such as mites and small hive beetles (see Chapter 14).

On occasion a package will arrive with bees that are soaking wet, along with several layers of "wet" dead bees in a similar condition. The surviving wet bees, even after being released, will not live for long. **Do not accept such a package** from a local dealer; if the package was received by mail, notify the sender immediately (see Chapter 6, "Inspecting the Packages").

BUY OR PRODUCE NUCLEUS COLONIES

Nucleus hives (or *nucs*) consist of three to five frames in a small, nucleus hive box. Nucs contain adult bees, comb(s) with eggs and brood, usually a laying queen, and various amounts of stored honey and pollen. Some nucs contain a virgin queen, or a queen cell, or are left to raise their own queen (not recommended). Nucs and nuc equipment can be purchased from dealers or beekeepers.

Advantages

- Usually cared for by an experienced owner.
- May be available locally, with the owner nearby to answer questions; assistance and information obtainable directly from the seller.

- Easy to transfer frames and clinging bees from a nuc into a standard hive.
- Includes all ages of bees and brood, when nucs are purchased during the spring/summer months.
- Quicker turnaround than packages in that the queen has drawn combs to deposit her eggs, young bees are emerging from brood combs, and a fairly good age distribution of adult bees is present.
- Are easy to transport.

Disadvantages

- Nuc boxes, frames, if not brand new, may contain disease spores and pesticide residues.
- Queen could be old or of poor quality and stock.
- Bees may require additional feeding and medication.
- Nuc may be infested with mites and/or small hive beetles.
- Additional equipment is needed if nuc grows fast and when the contents of the nuc are transferred into a standard hive body.

Installing a Nuc

If you are buying a nuc, make sure there is no sign of disease, beetles, or mites and that the frames have new wax; if in doubt, contact an experienced local beekeeper or bee inspector for advice. Merely install the purchased frames of bees and queen into one deep (if the frames are deep) hive body and fill the rest of the space with foundation or clean drawn comb. Feed or medicate as needed and provide additional room as the colony grows. This is an easy way to get a hive going, provided the queen is young, healthy, and of good breeding stock. Be careful not to buy problems, such as varroa mites or foulbrood, nosema, or viral disease.

BUY OR SPLIT AN ESTABLISHED COLONY

Established colonies when purchased in the spring, summer, and early fall contain adult bees (workers and drones), brood of all ages, eggs, a laying queen, and various amounts of stored honey and pollen. These colonies are usually purchased in a minimum of one but usually at least two deep supers (each containing 9 or 10 frames) and may also include any number of shallow supers. Retiring beekeepers often sell complete colonies and extra equipment. If you need to have expert help, call an experienced bee-keeper in your area or your local bee inspector to assess the health and condition of such colonies. You may be buying more trouble than you can afford. Before moving any bees and used equipment, check and comply with all legal requirements (see "Moving an Established Colony" in Chapter 11).

If you already have colonies of bees and wish to add more, an alternative to purchasing new ones is to divide some or all of your existing colonies into *splits*. The advantage of dividing your own colonies (making splits) is that you are familiar with the condition of the bees and the equipment (since they are your own) and you can save money. For more information on splitting colonies, see Chapter 11. The problem with making splits is that you will need additional equipment. Also, the split will require a queen, which you will have to rear or purchase.

As with nucs, purchased colonies should be **inspected and certified free of diseases and pests.**

Advantages
- Hive is fully assembled.
- Seller is available for assistance and information.
- Established colony contains full complement of bees.
- An established populous hive usually produces surplus honey the first year.

Disadvantages
- Combs could be old and may have to be replaced.
- The hive including supers, frames, bottom board, and inner and outer covers may be old and need to be replaced.
- Old combs have reduced cell size and may harbor pesticide residue, thus requiring replacement.
- Some of this pesticide residue may result from the application of miticides in bee colonies.
- Comb (or honey) may harbor American foulbrood or *Nosema* spores and virus.
- Equipment may not be uniform (different-size brood chambers, shallow supers, and frame).
- Queen's age is unknown unless marked and record of age is provided, and she may be of inferior stock.
- Success is dependent on time of year; colony may be preparing to swarm or supersede the queen.
- Honey may be crystallized (if so, it cannot be extracted from comb).

- If hive has been left unattended, frames may be heavily propolized (stuck together) or wooden equipment may be in poor or otherwise unusable shape.
- Large established healthy colonies would be very populous and thus difficult to work, especially for a beginner. Moving established hives can be a challenge; see "Moving an Established Colony" in Chapter 11.
- Established hives may be heavy and require a minimum of two adults to lift them. All hive bodies, the bottom board, and the cover need to be tightly secured before being moved. If anything comes apart during any phase of the operation, the situation will become problematic.
- Bees may be infested with mites and other parasites or viruses.

A final note concerning the purchasing of established or nucleus colonies: obtain from the seller assurance that the bees are free of disease and mites. Most states have bee inspectors, so request their assistance in evaluating the bees, or ask a fellow beekeeper to accompany you when you inspect the hive and the equipment before making a purchase. In addition, comply with all state regulations dealing with moving bees from one site to another, especially if you are crossing state lines.

RETRIEVING WILD COLONIES

A *colony* of bees—consisting of thousands of workers, usually one queen, and sometimes drones—can live in a man-made cavity, such as a bait box and a building, or in a natural cavity in a tree. Wild bee colonies can often be obtained free of charge from the property owner with appropriate permission. They should not be confused with a swarm of bees: a *swarm* of bees is a portion of a colony between homesites that has **no** honeycomb.

Bee colonies living in buildings are difficult to remove and can cost much in time and stings. Their removal is only worth it for the experience of getting bees out of buildings or for collecting the honey. Removing the entire colony and its combs successfully involves tearing off the outer portion or the inner portion of the building containing the colony. Leave this to the professionals.

Removing bees and combs from *bee trees* usually

involves felling the tree and splitting it open. Much of the comb and many of the bees, even the queen, are often crushed when the tree hits the ground. Similarly, removing bees from caves or cavities in the ground can be challenging and require other skills and equipment. In addition, such colonies may be Africanized and therefore more difficult to work with.

In general, the method of retrieving a wild colony involves cutting the combs and fastening them to the frames (with elastic bands or wire). This arrangement is only temporary, and eventually the frames will have to be removed and replaced; the final step is to lure the wild colony to go into a new, modern framed beehive. You can place a second deep box above the makeshift frames with a frame of larvae to attract the bees to move up. Consult an experienced beekeeper before deciding on any method of removing bees from trees or buildings; see "Bees, Beekeeping, and Bee Management" in the References. Alternately, check with local professional bee removers (maybe you could help a local bee removal expert and learn the tricks of the trade) to learn how they do it.

Advantages

- Retrieval would be interesting and educational (and could be profitable).
- Bees could be used to augment weak hives, make nucs, or start new hives.
- Extra wax and honey are available from removed combs.

Disadvantages

- Bees could be diseased or infested with mites.
- Queen might be injured or killed.
- A great deal of labor may be required, with little reward.
- Bees could be of inferior stock.
- Queen is often difficult to find and capture.
- Owner may expect you to repair dwelling.
- You may be liable for others getting stung.
- After the bees have been removed and repairs made, if the cavity is not eliminated, bees from another swarm may reoccupy it and the owner may attempt to hold you responsible.
- Transferring comb from a cavity onto frames without comb is time-consuming, requiring cutting the combs to fit into the frames.
- Bees could be Africanized.

COLLECTING COLONIES THAT HAVE ESTABLISHED RESIDENCY ON EXPOSED COMBS

Occasionally a swarm of bees fails to move from its cluster site to a natural homesite (cavity). Such a colony remains anchored to its homesite and will initiate comb construction. With the presence of comb, bees begin collecting nectar, pollen, propolis, and water, and the queen commences egg laying.

A swarm that reverts to a "normal" colony in this situation has its nest exposed to the elements. Two explanations have been proposed for this behavior: (1) scout bees fail to find or to agree on an adequate homesite or (2) inclement weather, accompanied by a reduction in daylight levels, stimulates bees to initiate comb construction. Once set into motion, the swarm remains wedded to its cluster site. Over time, as the colony's population increases, more comb is constructed. The problem is that bees on exposed comb are unable to survive the winter months in northern latitudes. See below ("Retrieving Wild Colonies") on how to collect these established bees.

Advantages

- As long as the exposed comb is within easy reach, retrieval should not be difficult.
- This is an educational and interesting way to start a colony.

Disadvantages

- Exposed comb may be entangled in bushes and shrubs, making it difficult to remove.
- Queen may be difficult to find, and lost in the process.
- Special bee equipment may be required to properly hive the colony because the existing comb will have to be removed and fit into a standard hive body.
- Bees require medication.
- Colony is not as gentle as "normal" swarm.
- The colony may be Africanized.

COLLECTING SWARMS

Swarms represent a portion of a bee colony that has departed its hive and is temporarily clustered on a tree branch or any location that provides a place to land. Once the cluster has settled (a minimum of bees flying around), scout bees fly off the cluster and begin searching for a suitable home site (a cavity) for the swarm.

Collecting swarms and wild colonies was a common way for people to obtain bees free of charge or to augment one's holdings. Swarms come from standard hives (or feral colonies) seeking a new home in any unoccupied cavity. Today, because of possible liability exposure, we caution you from obtaining bees in this way. We encourage you to retrieve some swarms, however. Not only is catching a swarm a learning experience but also you are helping a friend or neighbor by collecting "nuisance" bees.

Swarms can be a joy to collect under the right circumstances! Unfortunately they are not as common as they were prior to the introduction of the varroa mite in North America. This mite has decimated both feral and "domesticated" colonies, and the number of beekeepers and the number of the colonies they maintain have declined appreciatively as well.

Usually when a swarm alights, most property owners seek a beekeeper to remove them. Most swarms are calm and not prone to sting and therefore can be retrieved with a minimum of effort, provided they are within easy reach. Occasionally a swarm will exhibit aggressive behavior; therefore, as a precautionary step you should wear a bee veil and have a working smoker at your side.

The major problem with swarms is that not all of them are within easy reach. Ironically it is usually the larger ones that are beyond reach (or so it seems), creating a temptation to retrieve them at all costs. Any risk to life and limb far outweighs any swarm's value.

Before you embark on a swarm-collecting expedition, it would be prudent to bring along some bait hives, which contain pheromone lures. In the event that the swarm is out of reach, set up the bait hives to tempt scouts from the swarm to select one for their homestead (see "Bait Boxes" in Chapter 11).

Generally, swarms are easy to collect, but as a precaution you should treat the bees as if they are diseased (or infested with mites) (see "Catching Swarms" in Chapter 11).

Advantages

- If the swarm can be retrieved with ease, this is a great way of obtaining bees at no cost.
- Swarms captured in late spring may have ample time to develop into full fledged colonies. An old ditty in support of this statement says, "A swarm in May is worth a load of hay, a swarm in June is worth a silver spoon, and a swarm in July isn't worth a damn."
- Educational and interesting, swarms usually collect a lot of onlookers and on occasion the news media, providing the beekeeper with the opportunity to educate the public about honey bees. Media stories publicizing the activity of the beekeeper usually generate additional interest from people wishing to rid their property of swarms.

Disadvantages

- Swarms need to be medicated as a precaution, owing to the possibility that they are diseased or carrying the varroa mite.
- Occasionally a swarm is unable to locate a homesite and its food reserves run out. Such a swarm is referred to as a "dry swarm" and can become defensive.
- Another circumstance in which swarms exhibit defensive behavior is when someone, out of fear or malice, throws rocks or other objects at the swarm; bees under these circumstances release an alarm odor (pheromone), which permeates the area so that when a person approaches the swarm, the likelihood of being stung is real.
- Swarm could be Africanized.

THE APIARY

It is becoming increasingly difficult to obtain apiary (beeyard) sites, in this day of urbanization, changed farming practices, and public awareness of "killer bees." Many who keep bees as a hobby put them in their own backyards, much to the concern of their neighbors; but this need not be a deterrent if you practice a "Good Neighbor Policy" (see p. 55 or Appendix G "Urban Beekeeping"). Maybe you can find a farmer or a friend with some rural property willing to take bees in exchange for some bottles of honey. In addition to offering honey, you can relate to a farmer the advantages of having bees located on the farm—namely, crops that require pollination will benefit immensely from the presence of bees. Keep in mind that apiary locations may require registra-

Ideal Apiary Site

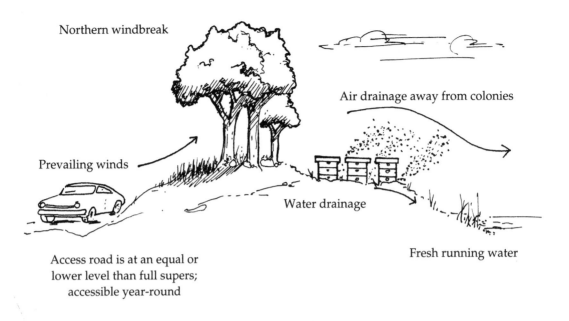

Northern windbreak

Air drainage away from colonies

Prevailing winds

Water drainage

Fresh running water

Access road is at an equal or
lower level than full supers;
accessible year-round

Poor Apiary Site

Wet, stagnant air could keep
honey from curing properly;
could develop frost pockets

Access road is higher than full,
heavy honey supers

Floodplain; colonies could be washed away

Drawing by J. Propst

tion with appropriate town, city, and state agencies.
Check your state's agriculture department.

Choose a site to optimize these conditions (see
the illustration of ideal and poor apiary sites on this
page):

● Close to freshwater—a stream, pond, or lake or
a faucet or other device. In the absence of such
sources, water must be provided; otherwise, the
likelihood of bees visiting neighboring swimming pools, birdbaths, etc., becomes a reality.
(Bees appear to be more attracted to *saltwater*
swimming pools.) Such visits will annoy your

neighbors and often lead them to seek some
relief.

● Easy, year-round vehicle access, which is at or
lower than supers, to avoid carrying heavy supers
uphill.

● Near dependable nectar and pollen sources
(within a 2-mile [3.2 km] radius).

● On upper sides of slopes to improve air drainage
away from hives.

● Along the edges of open fields.

● With a northern windbreak for winter protection
and noontime summer shade to keep hives cool.

● Near owner or neighbors (with clear commitments on both sides) to discourage vandals and thieves and to encourage visits from the beekeeper.

- At least 3 miles (4.8 km) from other beeyards, to diminish the spread of diseases and pests.

As a consequence of modern technology, you may be able to search for apiary sites by using mapping search engines on the Internet, such as Google Earth. You may also zoom in to determine where bees might obtain good forage, such as swampland, fields, wild meadows, or other areas covered with flowering plants. In addition, when searching for apiary sites using the computer, you may be able to locate other apiaries (depending on how recent the image is) that could be close to your apiary.

Apiary sites to *avoid* include:

- Bottom lands where air tends to stagnate.
- Fire-prone regions or flood areas (check soils or land-use maps).
- Areas where vandalism and thievery may be prevalent.
- Areas where the flight path of bees will be an annoyance (public parks).
- Areas where bears are prevalent (in recent years bear problems have increased, and bears have moved into states after an absence of decades).
- Areas that are infested with Africanized bees.

Identification

You should post your name and address at each *out-yard* (an apiary that is not near your home). Information about the location of each of your yards should be kept in an accessible place in your house, in case someone needs to find you quickly. Maps, GPS data, all phone numbers, and other pertinent details to help locate you should be included.

The property owner may need to reach you for several different reasons: a swarm is available, a bear has overturned your hive or been seen lurking near your hives, suspicious individuals have been seen observing the hives, or the owner is about to sell the property and the new owner wants the bees removed or wants to lay claim to them. Written agreements between you and the property owner clearly stating that the bees are *your* property will protect you should the current owner, kin, or new owner claim

the bees. In a litigious society, everyone needs agreements in writing because the gentleman's handshake is a thing of the past. These agreements should also be kept in an accessible place. You may have to have an agreement drawn up by a lawyer.

The property owner may also wish to be protected from liability through your insurance policy. This process is usually routine and adds no additional cost to your homeowner's policy. Certainly the fact that you can present this additional assurance to the property owner may influence the decision to grant you permission to place hives on the property. Should the property owner inform you that he (or she) needs more time to make a decision, leave a jar of honey that contains your name, address, and telephone number. By this act, you are providing the owner with a way to get in touch with you. Once the apiary is set up, post information (name, address, phone, etc.) enclosed in a waterproof plastic bag stapled to a hive or a nearby post to facilitate the bee inspector or anyone else who needs to get in touch with you.

Good Neighbor Policy

To keep on good terms with your neighbors, landowners, or other people likely to come in contact with your bees, here are some simple rules:

- First, make sure it is legal to keep bees at a particular location, especially if inside city limits. Most municipalities have zoning and planning commissions that can provide you with information concerning honey bees.
- Keep your bees out of sight by planting tall shrubs in front to hide the beehives and to force bees to fly higher than your neighbors' heads.
- You can place hives behind evergreens, a stockade fence, or a stone wall and paint the hives in colors that blend in with their surroundings; locate them in less traveled areas.
- Use a gentle race of bees and keep them gentle by requeening.
- By working bees only on sunny, warm days and during a honeyflow, bees will remain calm.
- Erect a sturdy fence to keep out curious children and pets.
- Do not work bees during inclement weather or during a nectar dearth.

- Have no more than two or three hives at a home property site and not more than 20 to 30 in an *out apiary.*
- Don't work hives if the neighbors are outside working in their yard or holding outdoor events.
- Initiate practices to minimize swarming so as not to alarm your neighbors.
- Provide and maintain a water source so your bees are not in a nearby pool or birdbath or hummingbird feeder (use a bird or livestock fountain or Boardman feeder). A four-gallon plastic fount normally used to provide water for fowl (such as those found at www.mcmurrayhatchery.com) can be used as an outside source of water for bees. Watering should begin as early in the year as possible; otherwise, by the time you provide the bees with water, they may have discovered other sources and as a consequence it may be too late to redirect them. Do not underestimate the water requirement for your bees!
- Give honey freely to next-door neighbors and a magazine or book explaining how beneficial bees are to agriculture.
- For keeping bees in urban areas, see Appendix G for more tips.

HIVE SCALE

A hive scale is a device that is placed under a strong colony and from which accurate records of weight gains and losses can be made. These scales can be a valuable piece of equipment, and for the serious beekeeper, it is important to have at least one scale per beeyard. Some times, marked changes in weight are recorded from day to day (e.g., when a colony is strong and a heavy honeyflow is in progress), as well as over the course of one day. A substantial weight gain or loss would be considered a change of one-half pound to several pounds (0.2 to 0.9 kg) daily. At other times, a hive scale may record subtle changes or no changes. The scale provides essential information to the beekeeper such as when the hive continuously gains weight over days or longer, likely indicating that a honeyflow has started. If the scale shows that the hive has grown heavier daily, it means a strong honeyflow is in progress and the hives can be *supered,* that is, have extra supers (shallows or mediums) placed on top of the broodnest to collect

honey. If you miss doing this, the colony can become "honeybound," which will reduce the space where the queen can lay eggs.

The type of scale often used in the past was a farmer's grain scale, but now several different scales, especially designed for bee colonies, have become available. These scales may be obtained from companies or dealers that sell bee equipment, as well as some Internet sites. Check with other beekeepers for suggestions and "Equipment" in the References for more information.

When a honeyflow is in progress, the hive gains weight because of the incoming nectar. You should be alerted to do certain tasks, depending on the season:

- Add frames and/or supers full of foundation, since the worker bees' wax glands are stimulated during honeyflows. Timing supering to honeyflows almost guarantees that the foundation will be drawn into full honeycomb, so take advantage of such opportunities.
- Interchange the locations of weak and strong hives (see "Prevention and Control of Swarming" in Chapter 11) as long as the weaker colonies are not diseased or mite infested.
- Check hives for swarming preparations, especially during or shortly after a spring honeyflow (see "Swarming" in Chapter 11).
- Requeen or make splits when there is a good honeyflow in progress (see Chapter 10).
- Divide strong colonies to increase the number of colonies in your apiary.
- Start queen-rearing operations. Colonies used to raise queens do so with greater success during honeyflows. See Chapter 10.

Whenever the scale shows a hive gaining weight, check and note which flowers are in bloom in order to anticipate nectar flows in future years. Occasionally, hives gain weight when no major nectar plants are in bloom. If you have been keeping good records and this weight gain fails to correspond to previously recorded ones and the weather is extremely hot, suspect that bees are harvesting water. Good beekeepers become good detectives. Another occasion when hives gain weight and it isn't hot (and there is no honeyflow in progress) is when *honeydew* (a sugary liquid excreted by insects that are feeding on plant sap) is available. Don't forget: the possibility exists that a honeyflow, the availability of honeydew, and the need for water may occur simultaneously. Weight gains may also be attributed to the times of rapid population expansion in a colony (3500 bees = 1 lb or 0.45 kg), or that bees are robbing from other colonies. Check for robbery or unusual places from which bees may be obtaining syrup (e.g., a nearby dump or landfill).

On the other hand, when the scale records no change or shows the colony is losing significant weight, it's time to open up the hive to see what is taking place. During late fall, winter, and early spring a slow but steady decline in weight should be anticipated, because with the exception of water, no other bee products are available. In these cases, colonies should be checked because they may be almost out of food and near the brink of starvation. Colonies in this condition need to be provided with supplemental food (sugar syrup, fondant, candy, dry granulated sugar, pollen, or pollen supplement; see chapter 7 on feeding bees). Naturally, colonies will show a weight loss when a swarm is cast, is diseased, is queenless, contains laying workers, is weakened by parasites or other maladies, or is being robbed out.

NASA is interested in tracking climate change as it relates to the flowering times of bee plants and weight gains or losses in an apiary; check out its website (honeybeeNet.gsfc.nasa.gov) for more information.

HIVE ORIENTATION

In most apiaries, the hives are placed in rows or are paired on hive stands in rows. The hives in a pair (on a shared hive stand) should be 6 to 8 inches (15–20 cm) apart, and the pairs spaced 5 to 8 feet (1.5–2.4 m) apart. This minimizes vibrations and jostling while you are working a colony, but keeps the hives close enough together to make your work more efficient.

When the hives are in long rows (more than six hives), there is a tendency for some bees to drift to the hives at the end of the row. Honey production will be less in the middle hives under these drifting conditions. This drifting may be due to prevailing winds, which continually push returning bees toward the end of the rows. Drifting is undesirable, and may lead to misinterpretation of hive scale readings. But the more critical problems that result from drifting are

Hive Orientations That Reduce Drifting

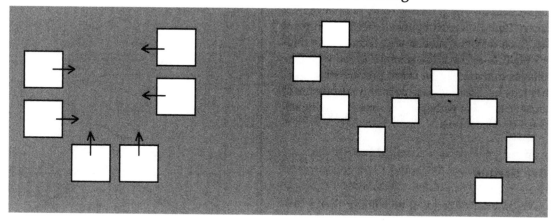

Horseshoe configuration   "S" configuration

Hive Orientation That Promotes Drifting

Hives lined up in a straight line disorient bees. Those at the end (dark boxes) will be stronger than those in the center (light boxes).

the spread of diseases, dissemination of mites, and robbing (and killing) of weaker colonies.

For all of these reasons, placing colonies in rows should be avoided whenever possible. In addition, during periods of nectar dearth, bees from strong colonies drifting into weaker ones are likely to seize the opportunity to rob honey and then return to their own colony and recruit other bees to continue the robbing. We cannot emphasize enough that the practice of placing colonies in rows, no matter how aesthetic or convenient, needs to be curtailed.

Reduce drifting by placing the hives in a horseshoe configuration (entrances facing in or out), by putting up a windbreak, or by shortening or staggering the rows (see the illustration of hive orientations on p. 57). If hives must be placed in a row, you should alternate entrances front and back along the row to reduce drifting, or if this is inconvenient (because you can't drive "behind" the colonies), paint the hives or bottom boards different colors or patterns or place landmarks near the hives, such as rocks or bushes. In areas where there is a flat horizon, bees may not be able to find their way back to a colony because of the lack of vertical landmarks. Consider erecting a snow fence or planting trees or shrubs to help the bees orient themselves.

HIVE STANDS

The amount of bending and lifting that a beekeeper must do while working a colony can be minimized when the colony is placed on a stand about 18 inches (46 cm) above the ground (see the illustration of hive stands on p. 59). Such a stand, in addition to saving your back, will keep the hive dry, extend the life of the bottom board, help keep the entrance clear of weeds, and discourage animal pests, such as skunks. Wood that is continuously wet or damp, such as a bottom board, can quickly rot, and pests, such as carpenter ants and termites, are likely to nest in damp bottom boards (see "Minor Insect Enemies" in Chapter 14). Other pests, such as skunks and mice, have less easy access to hives that are placed on some sort of hive stand. Hive stands are also good in drier areas where fire ants, rodents, and snakes are a year-round problem. If ants are a continuous problem, paint the legs of a hive stand with oil or grease. When working a colony on a stand, you can set the hive bodies that are temporarily removed from the hive onto an empty super or extra hive stand to reduce the amount of lifting and bending.

Some stands are constructed to sit on the ground in areas where dampness is not a problem. Low stands help to create a dead air space underneath the hive, thus providing extra insulation and enhancing the bees' wintering success (see illustration of low hive stand on p. 59). Make sure the colonies are level (or tilting slightly toward the hive entrance) so water will not collect inside the bottom board. It is also important that the hives not tip over in high winds, when loaded with honey, or when the ground is saturated with a lot of rain.

Types of Hive Stands

Hives can be kept off the ground by placing them on any one or a combination of these materials:

- Cinder blocks covered with tar paper or shingles or painted with waterproof paint.
- Bricks or drain tiles.
- Wooden railroad ties, pallets, or 2 × 4 lumber, painted or covered with tar paper.
- Wooden hive stands made of durable lumber (but be careful, some wood preservatives are harmful to bees).
- Permanent cement platforms, metal stands, or even flat, level rocks.

Pallets, unless with solid tops, are not recommended; you can easily twist an ankle or be easily injured especially when the platform for the pallet is rotten or weak.

R. D. Wynn, a California beekeeper, offers these helpful suggestions. He uses 2 × 4 × 8 pressure-treated lumber for the high hive stand. To put it together, he uses deck screws so they don't work out or become loose from temperature changes or from high winds buffeting the stand. He also sets the stand on a stepping stone purchased at a building supply store, to keep the wood from contacting the ground.

RECORD KEEPING

Careful record keeping is key to successful beekeeping, whether you raise the bees to make your own honey or to use for pollination services. Maintaining accurate records will allow you to refine and tweak

High Hive Stand

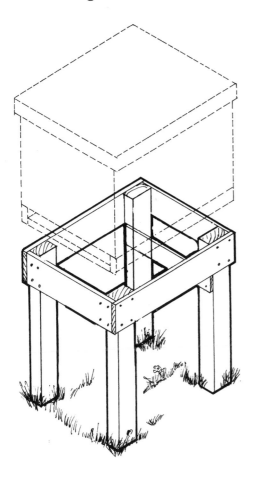

Parts List

High stand (pressure-treated lumber):
- (4) 1 × 4 × 16.5 inches (42.9 cm)
- (4) 2 × 4 × 18 inches (46 cm)
- (2) 1 × 4 × 24 inches (61 cm)

Low stand (pressure-treated lumber):
- (2) 6 × 20 inches (51 cm)
- (2) 6 × 48 inches (122 cm)

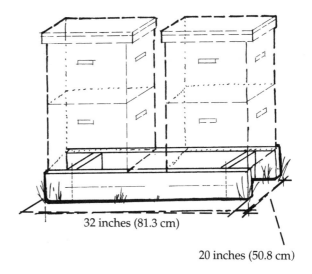

32 inches (81.3 cm)

20 inches (50.8 cm)

Low Hive Stand

your activities and considerably reduce repetition of previous mistakes or misinterpretation of honey bee activities. Although record keeping is tedious and burdensome, in the long run the information gleaned from records should and will make you a master of your hobby or business. Record keeping does have its limits though, because beekeeping is weather dependent; for example, dry weather affects plant-growing conditions, with flowering plants reducing or curtailing nectar output, and a prolonged rainy period could suspend foraging activities.

Meticulous records are absolutely necessary for those who desire to continually upgrade their stock. So, keep track of your queens—where and when did

you purchase them and from whom (see more in Chapter 10). The goal that most beekeepers strive for is gentle bees that overwinter well, remain disease free, and produce a surplus of honey whenever weather and floral conditions permit (see "The Hive Diary" in Chapter 5).

Financial records should also be kept for income tax and loan purposes and to determine the amount of income lost or gained in a season. During any given year you will have a certain amount of expenses, such as the purchasing of equipment, bees, vehicles, and the time employed in caring for the bees and assembling the equipment. All represent monetary expenditures. On the other hand, income represents the money you take in from pollination services, sales of bee equipment, sales of bee products (honey, pollen, propolis), and other services related to beekeeping (swarm or colony removal, if you were paid). If the expenses exceed the income, you have sustained a loss; if the income is greater than the expenses, you have made a profit.

On the actual beekeeping side, records of the condition of each hive will help guide you as to when queens need to be replaced, which colonies have mite infestation and the treatments provided, the times of major and minor honeyflows, and the weight changes in hives during mild and not-so-mild winters, to name a few. A diary of the blooming times of major nectar- and pollen-producing plants (and crops) will help you anticipate the time important sources of forage will become available. This information makes it easier to plan wisely the activities necessary for successful beekeeping. Check "Bees, Beekeeping, and Bee Management" in the References.

In summary, keep records for each apiary, if not each colony, and note such things as:

- Dates of all beekeeping purchases.
- Equipment bought, destroyed, stolen, or sold.
- Mileage to apiary.
- Dead or stolen colonies, queens, or packages.
- Vendors where queens, packages, or equipment were purchased.
- Colonies killed by pesticide sprays.
- Medications for beekeeper and for bees.
- Organizational memberships; journal subscriptions.
- Lectures, talks, shows, and fairs attended or entered, with associated fees.
- Books and conference fees.
- Equipment for selling honey (labels, bottles, etc.).
- Amount of honey extracted, bottled, and sold.
- Amount of comb honey packaged and sold.
- Other related income/expenditures.

 **Notes**

CHAPTER 5

Working Bees

It is important to know the purpose of the inspection before opening a hive. Be prepared so you can keep the amount of time spent in each colony to a minimum—no more than 15 minutes, unless you are doing some specific task that requires more time. Each time a colony is examined, the foraging activities of the worker bees and activities of house bees are disrupted and it may be a few hours before normal activity resumes. During a major honeyflow, this disruption could result in a measurable drop in the quantity of honey collected.

It has been estimated that an average of 150 bees are killed every time a hive is worked. Bees that are killed or injured release their alarm pheromone, which may excite other bees to become more defensive. Careful handling of the bees and the hive furniture can minimize the release of the alarm by bees, thus reducing the number of stings you might receive.

Avoid quick movements when working with the bees and do not jar the frames or other equipment. By proceeding slowly and gently, you allow time for the bees to move out of the way. Although killing some bees is unavoidable, the beekeeper who works slowly but precisely can keep the number of squashed bees to a minimum. Remember, work in slow motion.

WHEN TO EXAMINE A HIVE

A precise timetable for checking hives cannot be given because conditions vary from colony to colony throughout the year, and some hives will require more attention than others. Here are some general guidelines explaining when to open your hives:

- In the spring, when temperatures first reach over 55°F (12.8°C), to briefly check the general conditions and determine if the colony has an adequate food supply and a laying queen. You can do this in cooler temperatures, but don't keep the colony open too long (less than 5 minutes).
- After the first fruit bloom, to check hives periodically for growth, strength, pests (such as small hive beetles), diseases, and signs of swarming (presence of queen cells).
- After a major honeyflow, to remove or add supers.
- Periodically after a honeyflow, to check condition of queen and brood.
- Before the winter season sets in.

Certain changes to a hive will require that you follow up to see how that colony responded to your previous visit. For example, check a hive:

- 14 days after installing a package or swarm.
- 7 days after introducing a queen.
- 7 days after dividing a colony.
- Whenever pesticide damage, disease, mite infestation, queenlessness, or similar conditions are suspected.

A hive should **not** be examined:

- During a major honeyflow, unless absolutely necessary—for example, if disease is suspected, to requeen, if pesticides sprays are imminent, or to add or take off supers.
- On a very windy or cold winter day.
- When it is raining.
- At night.

BEFORE GOING TO THE APIARY

The following equipment and supplies should be available before you depart for the apiary. Although some of these items will not be needed during every trip to the apiary, it may be prudent to keep them near at hand, such as in the vehicle toolbox or in the apiary shed. It is safest to house the smoker in a separate nonflammable container at all times, such as a metal can with a lid. All other bee supplies and tools can be stored in the back of a truck in a special box.

If you are going out for a quick look at several colonies, your checklist should include:

- Two hive tools.
- Matches, lighter, or propane torch.
- Dry fuel for smoker kept in a waterproof container, such as a small metal can or plastic bag.
- A container of water to wash sticky hands or to extinguish smokers.
- Newspaper to unite hives, start smokers, or wrap comb samples for disease identification.
- Hive diary and extra markers, pens, and pencils.
- Duct tape and screen to close holes and cracks.
- Bottle of water for you.
- A nonflammable metal container for the hot smoker.
- Cell phone

If you have many colonies (over 50) and are going out to visit several apiaries for the entire day, you may want to include these items:

- Can or jar of fresh syrup or dry sugar for emergency feeding, and extra feeders.
- Plastic spray bottle containing sugar water or medication if needed for swarm capture (to dampen the bees' wings and keep them from flying off).
- Sample jars or plastic lunch bags to collect bee samples for mite or disease testing.
- Extra frames, in case you find broken ones.
- Extra hive bodies, outer covers, and inner covers.
- Division screen or empty hive, for emergency colony division (e.g., should many queen cells or queens be found).
- Container to collect scrapings of wax or propolis; this keeps you from throwing away valuable hive products and spreading disease.
- Queen excluders.

- Burlap or cotton sacking, to protect uncovered supers from robbing bees (if you are taking off honey). You can find these in fabric stores, or you can use plastic or nylon covers.
- String, rope, or hive straps to move colonies, or for other uses.
- Bee medication, in case it may be required.
- Empty queen cages, for superseded queens, loose queens, or for emergency queen introduction.
- Marking paint (model paint), to paint queens.
- First-aid kit including a sting kit or other medications for the beekeeper.
- Pruning clippers, saw, or sickle (or weed trimmer) to keep weeds and brush under control.
- Skunk/mouse guards, entrance reducers.
- Hammer and nails for repairs.
- Spray can of herbicide to kill poison ivy or other noxious weeds growing on or around your hives.
- Your lunch with extra water.
- Extra gloves, veil, and maybe even a bee suit.
- Extra smoker.
- Robbing cloth (burlap or cotton sack) to cover exposed super if bees start to rob (see p. 153).

THE HIVE DIARY

Methods of keeping track of the condition of each colony vary. Some beekeepers use a system of bricks or stones placed on top of the hives in some code to tell the queen's age, swarming tendencies, or the like. But because the stones can be removed by bee inspectors or others, or the code forgotten, other methods giving more precise information should be used.

A sheet of paper stapled or tacked to the underside of the outer cover is a good place to keep records. However, this is only a temporary solution, because the bees will chew up the paper and all your notes will be lost; you can stall this process by placing your notes in a plastic sandwich bag and staple or tack it on the cover or outside the hive. Another technique is to write on the underside of the outer cover, or outside the hive with a grease pencil or permanent marker. Number your hives and be sure to note important events in your hive diary (such as if the colony swarmed or is queenless).

A better way is to keep some sort of a hive diary that can be filled out each time you work your hives. Or you can purchase some inexpensive calendars and keep track of when you medicated, requeened, supered, and so on, for each apiary. Later, you can enter the information into a spreadsheet in your computer. Some folks take their laptops out and enter information while they are in the apiary; just don't get honey on it. Other beekeepers make special bar codes for each colony and keep track of them in this manner. No matter which method you use, be sure to keep good notes in some form. A good diary is also useful for tax purposes, as discussed in "Record Keeping" in Chapter 4. Check Internet sites as well.

Diaries are important if you plan to rear queens and select breeder queens from your best colonies. Queen characteristics should be recorded so you can tell whether or not a queen is up to your standards (see "Breeder Queens" in Chapter 10). In addition, by referring to the diary before going to the apiary, you will be less likely to forget any needed supplies or equipment. Every time a particular hive or group of hives is worked, note down some pertinent facts in the hive diary; add others as you go along. Note down:

- Layout of apiary, including windbreaks, water source, and GPS and compass directions (so other people can find your apiaries if you are incapacitated).
- Date of visits.
- Weather conditions (wind, temperature, humidity, etc.) the day you worked bees; if you want to be fancy, get a weather station for your apiary.
- Colony strength, that is, the number of frames with sealed brood and frames covered by adult bees (do this each spring and fall).
- Characteristics of the colony—defensive, gentle, productive.
- Swarming record—how often, dates swarms found, what colony the swarm came from (colony with a low worker population and lots of swarm cells or marked queen missing).
- Manipulations made that day—reversing, supering, medicating, etc.
- Effects of last manipulation and time elapsed— after requeening, package installation, or other activities.
- Hive weight gained or lost since last visit.
- Amount of honey harvested for each colony, in case you want to raise queens from a good honey producer.
- Requeening schedule—age and origin of queen.
- Disease, pests, or mite record.
- Wintering ability—amount of stores consumed.

- Medication—what type, when, and for what reason.
- Mite controls used (e.g., miticides, powdered sugar, etc.).
- Number of stings received and reaction.

DRESSING FOR THE JOB

If you absolutely do not want to get stung (although bees can sting through leather gloves), you can lessen your chances considerably by dressing appropriately for the job, much of which depends on reading the temper of the bees and dressing accordingly (see "Bee Temperament" in this chapter). First, bees are attracted to some odors, so do not wear strong perfumes or hair sprays. Then, depending on the degree of armor you want to put on (and the outside temperature), use the checklist below to make sure you or your friends, curiosity seekers, or other visitors are adequately dressed to work or observe you working bees (see also "Beewear" in Chapter 3).

- Bee veil, a necessity; getting stung on the face is dangerous and painful. **Always wear a veil**. Protect your eyes.
- Some sort of hat or helmet to keep off the sun, usually worn under the veil.
- A light shirt, bee suit, or coveralls to cover up gaps through which bees can and will crawl. A bee suit is best as it keeps your other clothes clean; propolis and wax stains will not come out in the wash. Avoid floppy shirt sleeves or loose material that could get trapped between hive furniture.
- Long pants, with socks pulled over the cuffs or with some other form of closure (duct tape, bicycle straps), or a long bee suit with elastic cuffs. The same goes for the shirt—elastic closures around the wrists are important to keep bees from crawling up your arms.
- Boots or sturdy walking shoes; many apiaries have poison ivy and other "goodies," not to mention wet grass and brambles, snakes and spiders.
- Gloves are optional, but handy to have if the bees decide to get fractious. They are essential to have if you are doing work that will irritate bees, such as taking off honey or setting upright a colony turned over by bears or preyed on by skunks.

If you find yourself in a situation where smoke will not quiet the bees and they continue to sting, here are some simple guidelines:

- Don't swat at bees as swatting motions will only irritate them and attract attention to yourself.
- If you are without a veil, push your glasses (if you wear them) close against your face to protect your eyes, and cover your mouth; keep your head down and walk calmly (or run) into some shrubbery or shady forest canopy.
- If a worker lands on you or in your hair, kill her quickly, by smashing or pinching her.
- If you are attacked by Africanized bees, get out as fast as possible, and get inside a car or a building; do not jump into a lake or pond.

SMOKING

The use of smoke while working bees is essential. No hive should be opened or examined without first smoking the bees. A few periodic puffs of smoke will help keep the bees under control, but bees that are oversmoked might become irritated. In a pinch, pipe or cigarette smoke can be blown in, but smokers work better.

Smoke works in several ways to keep bees from stinging. When bees are smoked, they seek out and engorge on honey or nectar in the hive, and bees with full stomachs are less prone to fly or sting. Also, when the hive is first opened, guard bees—which are sensitive to hive manipulations—release an alarm pheromone to alert other bees. When many bees are releasing this pheromone, you too can detect the odor (especially if you open a hive in winter). It smells like banana oil. The alarm pheromone causes the bees to react defensively to protect their home from "intruders." If you direct a puff of smoke from your smoker into the entrance of the hive, it will mask the initial release of the alarm odor by the guard bees, allowing the other bees to continue their routine hive duties rather than assume a defensive stance. Smoke may also reduce the sensitivity of the bees' receptors to the alarm pheromone; if they can't detect it, they won't release or react to it.

Smoke can also be used to drive bees away from or toward an area within the hive. Additionally, it can be used to mask the alarm pheromone left after you have been stung. Because the gland that releases the alarm pheromone is at the base of the sting, some of this pheromone marks the area where you are stung. Other bees that detect this signal may also sting the tagged area. Hands, clothing, and bee gloves that have been stung should also be smoked (and washed regularly) to clean off the alarm odor, dirt, venom, honey, sweat, and disease spores.

Purchase the larger smokers—they are easier to light than the smaller ones, burn longer, and are less likely to fail when needed most. Smart beekeepers carry a lidded five-gallon pail in which dried fuel, extra matches, and hive tools are kept. Metal boxes are available in which to store hot smokers. Clean the smoker of soot when it builds up in the nozzle and thoroughly scrub it with steel wool and hot soapy water, especially after working a diseased (especially foulbrood) colony. Keep your smoker in a fireproof container and **never** place a lit smoker in any vehicle.

Lighting the Smoker

You should become thoroughly familiar with the smoker before using it at the apiary. It is a good idea to practice lighting it a few times to get the hang of it; nothing is worse than having the smoker go out when you are in the middle of examining a colony. All beekeepers have their favorite fuel and may use it exclusively. The best fuel to use is the fuel that works best for you and is readily available in your locale.

Some commonly used fuels are:

- Straw or dried grass mixed with something else (wood chips or bark).
- Dry, rotted, or punk wood, which produces good smoke but can burn too hot; use with other fuels.
- Sumac bobs.
- Dry evergreen (pine) needles and cones that are old and open.
- Cedar bark or other bark chips or bark mulch.
- Peanut or other nut shells.
- Rice hulls.
- Burlap, untreated and dry (purchased from fabric stores); do not use old feed sacks as they may have pesticides in them.
- Wood shavings, sawdust mixture.
- Rags, 100 percent cotton only.
- Dried horse or cow dung, which burns well without much odor.
- Cotton stuffing (can be purchased from bee supply stores).

To cool the smoke, put a thin layer of green grass on top of the burning fuel. This layer catches the hot embers and prevents the bees from getting burned.

Some smoke, including that from burning dried corn cobs and leaf tobacco, is toxic to bees, so be careful that you are burning clean, nontoxic fuel. In addition, synthetic materials, petroleum starters, or rags treated with pesticides should **never** be used, since they may also give off a toxic smoke. Newspaper and corrugated cardboard boxes should not be used as the only fuel; the embers are too big and could burn the bees, and the cardboard may have been treated with pesticides or other toxic compounds (e.g., glue). Some beekeepers use binder twine (from hay bales), but since this is usually treated with rot-retarding chemicals, which may be toxic to bees, we do not recommend its use. Generally, any organic material can be used, but be careful not to mix in dried poison ivy vines with the smoker fuel; this can be found in wood chips from country roadside tree trimmings (sometimes the workers leave the chips or give them away).

Here is how to light a smoker:

Step 1. First clean the smoker and scrape out any clogging soot from inside the nozzle.
Step 2. Use a match or lighter and ignite a small piece of newspaper or punk dry wood and drop into smoker.
Step 3. Drop on top of this a small amount of dry fuel.
Step 4. Puff the smoker bellows and slowly add more fuel; puffing in air will help ignite the fuel as you pack the smoker. Do not overpack, and make sure the fuel is well lit and smoking well. **Be careful not** to get burned! Smoker bodies get hot.
Step 5. If your fuel is damp, add bits of beeswax, or start your fuel with a small hand-held blow torch.
Step 6. Puff the bellows hard until the smoker stays lit.
Step 7. Once it is going, put a handful of grass or green leaves on top of the fuel to cool the smoke and catch hot embers; make sure you don't smother your fire.
Step 8. Do not pack the fuel cylinder too tightly and keep filling it periodically with fresh fuel. A well-stocked large smoker should last 30 to 45 minutes.

After finishing work in the apiary:

- Place the hive tool(s) in the opened smoker and puff a blaze to sterilize the tools. This will make them very hot, so be careful when removing them; douse them with water to cool before storing them.
- Empty the remaining fuel and ashes onto dirt or pavement and drench them with water (which you should have carried in your truck). Always have on hand some water to drench old, smoldering smoker fuel. Some beekeepers stuff a cork or green grass into the nozzle of the smoker to suffocate the fire if water is not available. Another way to smother a smoker is to place a sheet of paper over the top and close the lid tightly over the paper. This also will help keep your nozzle from filling with soot.
- Make sure the fire is out and the smoker is cool before putting it away, and never leave a lighted smoker in a vehicle, even the back of a truck. It can be fanned into a blazing fire during just a short drive down the road. Buy a metal box designed to hold hot smokers from a bee supply store.

EXAMINING A COLONY

The general method used by most beekeepers to open and examine a hive is outlined below. The procedure may vary somewhat, depending on the number of supers on the hive and the purpose of the examination. Work with many different beekeepers and pick up their good habits on working bees.

Step 1. Approach the hive from the side or back.
Step 2. Do not stand in front of the hive at any time, since you will be blocking the flight path of outgoing and incoming bees.
Step 3. Puff some smoke into the entrance (be sure it gets inside) and wait 30 seconds so the bees can begin to gorge on honey.
Step 4. Gently pry or remove the outer cover and direct a few puffs of smoke through the oblong hole of the inner cover; again wait 30 seconds for the bees to gorge honey. Then gently pry off the inner cover, puffing some smoke between the frames. If an inner cover is not used on the hive, puff some smoke under the outer cover as you take it off and wait 30 seconds.
Step 5. Lay the inner cover at the entrance so clinging bees can reenter the hive; do not block the entrance.

Examining a Colony

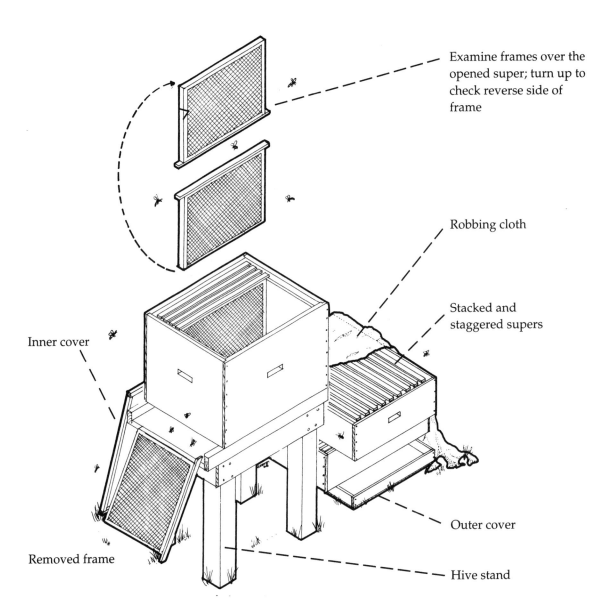

Examine frames over the opened super; turn up to check reverse side of frame

Robbing cloth

Stacked and staggered supers

Inner cover

Outer cover

Removed frame

Hive stand

Step 6. After the covers have been removed, smoke the bees down from the top bars of the frames; use the smoke judiciously. Too much will cause the bees to run in every direction, making your work more difficult and decreasing the likelihood of finding the queen. In addition, too much smoke can damage brood cappings. Smoke bees just enough to make them move; experience will teach you what amount is right.

Step 7. Use the outer cover (underside up) or a spare hive body or stand as a base for stacking supers (hive bodies) as they are removed from the hive

(see illustration on examining a colony on this page).

Step 8. As you begin to remove frames and set the supers aside, avoid quick body movements or jarring or bumping the equipment; such actions tend to increase the defensive posture of bees. Slow, deliberate actions and the gentle handling of equipment will keep bees calm.

Step 9. Throughout the examination, smoke the bees as needed to keep them out of your way and to keep them from getting squashed.

Step 10. The purpose of the inspection will dictate

whether to first remove all supers above the bottom ones (to inspect the broodnest) or whether to work from the top down (to see if nectar or honey is being collected). Most of the time you will want to inspect the broodnest, which is where the queen, eggs, brood, drone larvae, and queen cells are; therefore, start at the bottom-most super first, and move upward in the stack.

Step 11. Each time a super is pried off, puff a bit of smoke onto the super below and to the bottom of the one you are moving.

Step 12. If the hive is very populous, it is best to start by examining the bottom-most hive body, after stacking all other supers on the upturned cover nearby (give them an occasional puff of smoke as you work). If you were to begin by working at the top, many of the bees smoked from successive operations on the upper supers will crowd to the lowest super, making it very full by the time you reach it and making it almost impossible to find the queen.

Examining Frames

Now you have made it to the brood chamber hive body, where you want to start examining the bees. Here's how you should proceed:

Step 1. Whenever you decide to begin your examination, smoke the bees off the top bars and down between the frames. Before removing frames, choose the one closest to the hive wall and push all other frames away from it with the hive tool; this will create sufficient space for easy removal of the frame. Avoid removing frames from the center of the hive first, as the queen may be crushed in the process of pulling out the frame.

Step 2. Once the first frame is removed, you have created more space to remove subsequent frames. Lean the removed frame against the bottom hive body or some other object, out of the sun and where it won't be kicked or jarred, or place it in an empty hive body.

Step 3. As each frame is examined, hold it vertically over the hive; in this way, if the queen falls from the frame, she may drop back into the hive.

Step 4. Continue to examine each adjacent frame until your objective is completed.

Reading the Frames

For each frame you inspect, quickly check for these items:

1. Sealed brood: It should be compact, in a concentric semicircle at the bottom half of the frame. If there are many open cells, it may mean that the queen's eggs were not viable and/or the larvae were pulled out.

2. Ratio of eggs to open larvae to capped pupae: A ratio of approximately 1:2:4 is ideal. This means there are twice as many larvae and four times as many capped pupae as there are eggs. It indicates that the queen is laying continuously and the bees are of sufficient numbers to incubate the eggs.

3. No eggs found: If no eggs are found in the open cells, you can estimate how long ago the queen stopped laying by opening up some capped worker brood. Young pupae with white eyes will emerge in about seven days; if the eyes are purple, they will hatch in two to three days.

4. Queen cells: If you find supersedure cells, the queen is failing for some reason. If you find queen cells with larvae, lots of drones, sealed brood, but no eggs, the colony may swarm in about a week. If you find sealed swarm cells, some with holes in them, sealed brood, and few bees in the honey supers, the colony swarmed and a virgin queen is emerged. You should see eggs in about a week.

5. Other observations: Note any changes in the behavior of a colony since your last visit, especially if the bees are more volatile; this could indicate lack of forage, pesticide use, pests, mites, queenlessness, or disease. Observe the amount of incoming honey and pollen in case bees are starving (no honey, dead brood on bottom board) or are becoming honeybound (honey filling all available space, even into the brood combs). Also note the physical condition of the combs and frames, including any wax moth damage and uneven comb or foundation; fix any broken frames.

Step 5. Frames should be returned to their original positions and spacing unless you are adding frames of foundation, honey, drawn comb, brood, or eggs. If frames with brood and eggs are separated from the broodnest and placed elsewhere, those frames might become chilled, because the bees will have a hard time maintaining the proper temperatures in a scattered broodnest. This can result in brood chilling, making them more susceptible to chalkbrood disease; if chilled too long, the whole frame of brood could die. When adding frames with eggs or brood, be sure the colony has enough bees to care for them.

Step 6. If while working bees you see bees landing on the top bars, or bees fighting on uncovered frames, supers, or at the entrance, robbing may be in progress. This happens when there is no honeyflow. You must quickly cover the exposed equipment with a robbing cloth (wet cloth or blanket) or better yet, curtail hive examinations for the day (see "Robbing" in Chapter 11).

Step 7. When replacing hive bodies, the bees in the super below will be milling on the top bars and rims; smoke the bees down so they will not get crushed as you replace the hive furniture.

Step 8. Whenever possible, scrape excess propolis and *burr comb* (comb not in the proper place) from the frames with a hive tool. These materials should be placed in a closed container; the wax can later be melted down in a solar extractor (see "Beeswax" in Chapter 12). Never discard propolis and wax around the apiary or anywhere else. This material not only will attract pests such as skunks and bears, but also could promote robbing and transmit diseases. Remember, wax and propolis are marketable products.

WORKING EXTRA-STRONG COLONIES

Working a colony more than two deep supers (hive bodies) full of bees can be a challenge, even to experienced beekeepers. Some colonies occupy three or even four deep supers, making their examination a daunting task.

To handle such a colony, it may be best to break it down into a more manageable size. This can be done by taking off the top deep bodies without looking at the frames. First, smoke the entrance as before, then take off the outer cover. Smoke the inner cover hole but do not remove the cover. Now, using your hive tool, insert it into the place between the two deep bodies, and lift the top deep super. This breaks the seal between the two boxes. If you lever the top box up a few inches, blow smoke into the area between the upper and lower deep supers; continue breaking the propolis seal and remove the top super. Place it on the upturned outer cover, and then proceed to remove the second super (see also "Requeening Defensive Colonies," Chapter 10).

This method will keep the house bees on the frames in the upper super; they will not have been smoked down into the lower box. The end result is the bee population is now almost equal in all the supers. Next, begin your examination, starting at the bottom-most hive body. If the colony is especially populous, you may find swarm cells, in which case you may want to divide the colony into one or two new colonies, each with queen cells, keeping the original queen with the parent hive. You don't have to use queen cells if you are concerned about the quality of the queen that may result; instead introduce a purchased queen of known characteristics to these splits (see "Relieving Congestion" in Chapter 11).

You may have to use more smoke on a populous colony, but the rules are the same, work slowly, try not to kill too many bees, and avoid knocking or bumping the equipment.

WHAT TO LOOK FOR

In the spring the colony must build up strength in order to achieve the peak population of 40,000 or more; such numbers are needed to secure a good honey crop.

You should be able to verify that:

- A queen and/or eggs or young larvae are present (see "Reading the Frames" on p. 67).
- There are adequate food stores (pollen, honey, or stored sugar syrup). If not present, you must provide these to ensure colony survival.
- The brood pattern is compact for both uncapped (larvae) and capped (pupae) brood; if the brood is scattered or there are a lot of empty cells, determine the cause.

Also check for and take measures to correct, the following adverse conditions:

- Queenlessness: add a queen or queen cell or unite queenless colony.
- Queen cells (supersedure or swarm cells): manage appropriately.
- Presence of a failing or a drone-laying queen: replace queen.
- Presence of laying workers: unite with a queen-right colony.
- Leaking feeders: replace and check to see if bees are not harmed by leaking syrup. If too much syrup leaks out, robbing could be initiated.
- Crowded conditions: add extra honey supers or brood chambers.
- Overheated conditions: provide shade or additional ventilation.
- Diseases, mites, and other pests: treat accordingly.
- Robbing activities: reduce entrance, seal cracks and other openings.
- Bottom board filled with dead bees, debris, wax moth, or propolis: check condition of colony, then clean off or replace bottom board.
- Wet, damp, or rotting equipment or carpenter ant infestation: replace.
- Dwindling populations: determine cause, and if disease and mite free, unite.
- Broken combs or frames: replace.
- Cracked or broken equipment: replace.
- Obstructions in front of the entrance (grass, weeds): clear and cut down.
- If there is a mouse nest inside, remove it, and replace damaged frames.

Queen and Court

Finding the Queen

The queen's presence and her reproductive state can be determined indirectly, without actually finding her. Brood frames with a concentrated pattern of capped worker cells, frames mostly filled with eggs or larvae (uncapped brood), or a combination of both indicate a healthy colony headed by a fertile and productive queen.

If it is necessary to find the queen, use as little smoke as possible, open the hive gently (as outlined in "Examining a Colony"), and remove the outermost frame. She will seldom be found on frames with just honey and pollen or on frames with capped brood; she will **most likely** be found on or near frames containing eggs and uncapped larvae.

The queen can often be spotted in the midst of her encircling attendants or retinue. When a queen moves slowly along the frame from cell to cell, the other bees will begin to disperse, but the circle will re-form when she pauses (see the illustration of the queen and court on this page).

If the queen must be found—whether before requeening, to kill her before uniting colonies, to mark her, or just to satisfy the need to see her—but cannot be located within 15 minutes or without disrupting the entire hive, it may be helpful to use the following method:

- Place a queen excluder between the two brood chambers (usually the two lower hive bodies).
- Five days later, the queen will be in the hive body whose frames contain eggs. Because all eggs hatch in three days, the brood chamber from which she was excluded will have no eggs.

If you do not see the queen or eggs, or the colony is not behaving normally, refer to Chapter 11 for potential problems with your colony, or have another beekeeper or bee inspector check it.

BEE TEMPERAMENT

Good Disposition

To minimize the likelihood of being stung and to encounter fewer defensive foragers, work the hive on days when most field bees are foraging.

Generally, it is best to work bees:

- In the spring, when populations are low or honeyflow is in progress.
- During a good honeyflow.
- On warm, sunny, calm days.
- When populations are low, as with package bees.
- When bees are well gorged with food, as with a swarm or package bees that have been fed.
- Between late morning and early afternoon (roughly between 10:00 a.m. and 1:00 p.m. depending on season and time zone).

Irritable Disposition

Bees are more prone to sting when most of the foragers are in the hive. Conditions outside the hive (usually weather) are the reason for the foragers not being out. Other factors that can cause bees to become more defensive include:

- Queen temperament, genetically passed on to her offspring.
- The effects of pesticides.
- Disturbance by skunks, bears, or other pests.
- Poor honeyflow, when there is little food coming in.
- Autumn, after the honeyflow has ceased.
- Impending thunderstorm.
- Cool, wet, cloudy days.
- Hot, sultry, humid days.
- Windy days.
- Early morning or late afternoon or evening.
- Improper handling resulting in the killing of many bees.
- Jarring of the hive or a hive part.
- Disease/mite infestation.
- Examination of the colony without using smoke.
- Removal of honey or leaving supers or frames exposed during a dearth, stimulating robbing activities.
- Pungent hair oils, lotions, deodorants, or perfumes.
- Queenlessness.
- Presence of laying workers.

To minimize your getting stung, remember these rules:

- Work in slow motion; avoiding rapid, jerky movements around the bees.

- Don't swat at flying bees; ignore them and they may do the same.
- Don't drop, bang, or bump hive parts as vibrations upset bees.
- Don't stand in front of the flight entrance of the colony; stand to one side or at the back.

UNEXPECTED OCCURRENCES

When working with bees, situations may sometimes arise for which you are not prepared. These are some of the more common unexpected happenings:

A bee gets in your veil. Squash it quickly, before it stings your head; or get inside or in a car. You can also walk behind a tree or bush outside the immediate area, trying not to let other bees follow you, and remove the veil quickly to release the trapped bee. Replace veil.

Your smoker goes out. Cover exposed supers with extra outer cover(s) or cloth to prevent robbing and relight the smoker.

The queen flies away. During package installation, or other times when the queen is directly released or handled, she may fly off if she is not first clipped (something to do if you have a very expensive queen). Virgin queens are more prone to fly than are mated, laying queens. In either case, do not panic. Shake a frame or two of bees at the front entrance of the hive she flew out of. Many of the bees will start to fan and leave scent at the hive entrance, which will hopefully attract the queen to land. Watch for a cluster of bees to form at the hive entrance, or on a nearby branch. If this latter occurrence happens, the loose queen will likely be in that cluster and you should collect it like a swarm. Lay it down in front of the hive and let the bees walk in (see "Swarming" in Chapter 11).

The queen is balled. If the queen is surrounded by a "ball" of workers, they may be trying to kill her. This can happen when she is released directly into a colony formed from a package or when she is introduced into a hive that is being requeened. Balling may also happen when you are requeening a colony that already has a queen or when the hive is roughly handled. In these cases the bees consider the queen to be "foreign" and commence to surround or ball her. Balling will either suffocate or kill a queen by raising her body temperature too high, thus "cooking" her, or by tearing her apart. Do one of the following:

- Cover the hive quickly and hope for the best (not usually the best choice).
- Break up the ball with smoke or water and cage the queen; reintroduce her using the indirect-release method discussed in Chapter 6.
- Break up the ball and spray the queen with syrup, then place her on a frame of uncapped brood.
- Requeen the colony using a queen cage; the one they are balling may be a virgin or invading queen.

You are chased by many bees. Blow smoke on yourself and walk casually behind bushes or trees. Be sure your smoker does not throw off flames; otherwise, your clothing might ignite. Bees see movement (like fleeing bodies) very easily but are confused if many objects like branches or leaves are between them and their target.

The colony is exceptionally defensive. Close the hive as quickly as possible and wait for another day. Try to determine the reason for the bees' unusual behavior. Check other colonies in the same apiary to see if they exhibit the same behavior. A skunk or bear may be bothering your hives (see "Animal Pests" in Chapter 14). If only a particular colony is defensive, it may be a genetic trait or it may have become Africanized. If such is the case, requeen that colony (see "Requeening Defensive Colonies" in Chapter 10).

BEE STINGS

What to Do When Stung

If a worker bee pierces your skin with the barbed lancets of her sting, she cannot withdraw them once they are embedded (see "Analysis of Bee Venom" on p. 184 and Appendix C on sting reaction physiology). As the bee struggles to free herself, the poison sac attached to the lancets is ripped from her abdomen. This means that the bee will ultimately die, and having left most of her sting in your tissue, she will obviously not be able to sting again before she does.

Other stinging insects have either smooth lancets or lancets with ineffectual barbs; they can, therefore, withdraw that portion of the sting and repeatedly reinsert it. The queen honey bee, hornets, and yellowjackets (superfamily Vespoidea) have such a sting. Even when handled, queen bees rarely sting beekeepers; they use the sting against rival queens.

Scrape off the sting with a fingernail or hive tool as soon as possible to minimize the amount of venom pumped into the wound. Start to scrape the skin about an inch away from the sting and continue scraping through the sting; which will pull out easily. **Speed of removal, not method, is important to reduce the amount of venom injected and thus minimize the swelling and itching afterward.**

Because an alarm pheromone is associated with the sting, other bees are likely to sting in the same vicinity as the first sting. Apply smoke to the area of the sting to mask the alarm odor.

Treatment

Local Reaction

Bee venom contains enzymes (hyaluronidase) and peptides (melittin) that cause the pain (see "Analysis of Bee Venom" on p. 184). For local reactions, there is very little an individual can do except to relieve the itching. Since the sting barbs are so tiny and the puncture so small, no treatment will be effective in reducing the amount of venom other than the prompt, proper removal of the sting structure.

Every beekeeper has a favorite treatment for bee stings. Although treatment does not **cure** the wound, it does give a different sensation to the area and thus takes your mind off the momentary pain. The following items are often used to relieve the pain and itching of bee stings:

- Bee-sting treatment kits.
- Vinegar.
- Raw onions rubbed on the area.
- Toothpaste.
- Honey.
- Juice from the wild balsam, jewelweed, or touch-me-not (*Impatiens pallida*).
- Baking soda.
- Ammonia.
- Meat tenderizer, as a paste.
- Mud.
- Hemorrhoid treatment cream.
- Ice packs or cold water.

The best course is to put an ice pack on the sting. Whatever treatment you choose, it works best if you apply it immediately after you are stung, but this is not usually possible if you get stung away from home or through a bee suit. To give relief to the itching red welt that appears following a bee sting, apply calamine lotion or other insect-bite or poison ivy preparation, or hot water.

Systemic Reaction

A good summary of bee stings and allergies is in Graham's *The Hive and the Honey Bee* (1992) as well as at some online sites. In general, if you break out in a rash (hives) or have difficulty breathing after being stung by a honey bee, you are probably having an allergic reaction to bee venom.

In such cases, call for emergency medical help or take the person who has been stung to the hospital immediately! Time is critical.

Medication for an allergic reaction to bee stings can be obtained only by prescription. The drugs commonly prescribed are an antihistamine and epinephrine (adrenaline). Consult your doctor if there are any questions.

Here are some examples of treatments:

● Injected: EpiPen (sss.epipen.com/) or other insect sting kits are available with a prescription and include a syringe filled with epinephrine (adrenaline), with instructions for it to be injected under the skin (subcutaneously). Read instructions carefully (some kits may require refrigeration) and become familiar with how to administer the medication. Do **not** administer to others.

● Oral: Some over-the-counter antihistamines are helpful; your doctor may prescribe more powerful ones.
● Aerosol: An aerosol bronchial applicator, as for asthma sufferers, will offer quick relief of breathlessness as a result of a bee sting. Epinephrine inhalers are also effective. The dosage of two puffs should be repeated after 15 minutes.

Although the above information provides an outline of what might be done for systemic or general allergic reactions to bee stings, exact and precise medical information should be strictly adhered to. No one should attempt to self-diagnose a response to bee stings or to prescribe medications. **Seek the advice of a physician.**

If you develop an allergy to bee stings and wish to still keep bees, the only other alternative is to go through an immunotherapy session or venom allergy shots. It is normally expensive, complicated, and inconvenient, but it does usually eliminate future systemic reactions. Consult your doctor or allergist for more details (see Appendix C, and "Venom" in References).

Even if you are not allergic, tell your physician that you are a beekeeper in case of an emergency.

 **Notes**

 **Notes**

CHAPTER 6

Package Bees

A *package* of bees is a screened box containing several pounds of bees, a laying queen in a separate cage, and a feeder can of sugar syrup (see the illustration on this page). Packages are an excellent way to start beekeeping!

The package is prepared by package producers, usually from southern U.S. states, California, and Hawaii. A package of bees is produced by removing frames with clinging bees from a populous hive and then shaking the bees into a funnel that is inserted into a circular opening at the top of the package. Later this opening will be used to accept a can of sugar water. The bees tumble and fall down the funnel and into the package. After the desired number of bees, measured in pounds, are in the package, a newly mated queen taken from a queen bank is placed in a queen cage and added to the package. The cage is suspended from a slot adjacent to the circular opening. The feeder can containing sugar syrup is then inserted into the circular opening; a lid is placed over

the opening, and the package is ready to ship. You can also order a queenless package.

The bees in the package now have a foreign queen (not their own), but because she is caged, they are unable to harm her. While the package is in transit, the bees become "acquainted" with her. Each cage has a plug of "candy" to delay the queen's release; the placement of the candy is usually next to the opening through which the queen will exit (see Chapter 10, "Types of Queen Cages"); the candy hole is covered with a cork.

There are many methods used to install bee packages. The basic differences are in the manner in which the queen is released from her cage. In an *indirect-release method*, the queen remains caged and the bees are allowed access to a candy plug, which they must remove in order to release her. This method simply delays the queen from being freed among the other bees for a few more hours or days and increases the likelihood of their accepting her.

Package of Bees

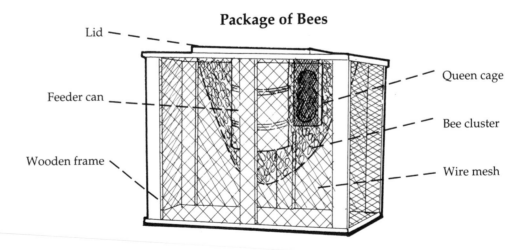

Lid

Feeder can

Wooden frame

Queen cage

Bee cluster

Wire mesh

In the *direct-release method*, the screen, cork (blocking outside bees access to the candy plug), or plastic cap is removed from the queen cage, allowing the queen to walk out onto the top bars of the hive among the other bees or into the entrance. When the queen is released directly, the bees still may not be fully acquainted with her, and as a consequence, they may form a tight ball around her and, by raising her body temperature, suffocate and kill her. This process, called *balling the queen*, may result in the queen's death or permanent injury; see "Unexpected Occurrences" in Chapter 5. Combinations and variations of the direct- and indirect-release methods are covered in "Installing Packages" in this chapter.

ORDERING PACKAGES

Packages can be ordered either directly from the producers or through a local bee supply house, a local dealer, or a bee club. Look for advertisements by local dealers, which may be found in publications from state beekeeping organizations and beekeeping journals (see Chapter 4, "Ordering Packages"). Sometimes a local bee supply dealer will drive down to pick up a load of packages; you could go along to help if you wanted to buy a large number of packages at a reduced price. Help is always appreciated, and it is an excellent learning experience. Some bee clubs often order together to get better prices.

Place your order early. November is not too soon to place an order for bees to arrive the next spring. A rule of thumb for selecting a date for when the bees should arrive is approximately one month prior to when the fruit trees bloom in your area.

If your bees are being mailed to you, call your local post office the week before the bees are expected and leave phone numbers where you can be reached (land line, work, and cell phone) so that the postal clerks can contact you when the bees arrive. When ordering, ask for guarantees about shipping in case of loss or damage. All such isues should be discussed before you pay for the order.

WHEN THE PACKAGES ARRIVE

When the packages finally arrive and everything seems to be in order, check for the certificate of inspection on each one. This certificate means that the bees were inspected for mites and diseases prior to their sale; see Chapter 14.

The bees may be buzzing loudly and wandering all over the package. They are not "mad" or ferocious. As soon as possible, the package should be placed in a cool (not cold), draft-free, quiet, and darkened area and fed heavily with sugar syrup. The bees will soon become calm. Do not place the package inside the house, as it may be too hot and bees will leave bits of wax and syrup on the floor. Feed the bees liberally by spraying sugar syrup from a spray bottle on the screen sides of the package. Do not soak them. Some beekeepers brush the syrup on the screen, but this can injure the bees—many of whom will have their tongues and feet protruding through the screen. Spray bottles work best!

Inspect the packages carefully before taking them home. Some dead but dry-appearing bees at the base of the package are normal. However, a package with bees that appear to be soaking wet, along with several layers of dead bees in a similar condition, is **not** normal. The surviving soaking-wet bees even after being released often will die; this package has overheated and will not recover. Do not accept such a package from a local dealer; if the package was received by mail, notify the sender immediately.

The cause of this condition is a result of bees becoming overheated due to high temperatures during shipment; or they are unable to settle quietly in the package and instead scurry about, becoming overheated. In an attempt to cool down, the bees regurgitate the contents of their honey stomachs. In a normal colony when the temperature exceeds 95°F (35°C), bees collect water, return to the hive, regurgitate it, and by fanning their wings induce evaporative cooling. Thus, the soaking-wet bees in the package may result from the regurgitation of the contents of their honey stomachs on one another when they are unable to cool themselves because of the lack of ventilation. When swarms are collected in makeshift cardboard or wooden boxes with inadequate ventilation, the same phenomenon is witnessed. Such a swarm has been referred to as a "cooked swarm." These bees will **not** recover.

There are a few things to note when examining your packages:

- If the package is guaranteed and the queen is found dead, usually discovered at the time of installation, and/or the package contains more than an inch of dead bees, the package seller should be notified immediately and asked when replacements can be expected.

● If the queen is dead, notify the seller immediately and ask for a replacement queen. Any replacements must not be delayed, or some workers will undergo ovary maturation and begin to lay eggs. Laying workers can produce only drone eggs; thus, the colony would be doomed; see below and "Laying Workers" in Chapter 11.

What to Do with a Queenless Package

If the queenless package bees can be provided with a frame or two of eggs and uncapped larvae from an established hive, laying workers will not develop and the bees will raise a new queen. But the colony would lag behind in growth because it takes three weeks for a new queen to develop, be mated, and lay eggs. Because few drones are available in the spring, and the weather may be too cold, the queen may not be mated, further delaying colony development. If the queen does not mate, you will have a drone-laying queen.

A short-term solution would be to add a queenless package to an existing but weak colony or combine the package with another package into a single hive until a new queen arrives. Then you could split that colony into two, requeening the queenless portion. These two solutions assume that you already have other colonies or that you as a beginner have ordered more than one package. New beekeepers would be wise to begin with three to five packages. This will provide you flexibility if there is a problem with one or more packages. Another solution would be to order an extra queen; she can be used later in splits or increases.

INSTALLING PACKAGES

Prepare a sugar syrup solution, which should be a mixture of one part white sugar to one part warm water, before the bees arrive; see "Sugar Syrup" in Chapter 7. If the package producer does not medicate the bees against nosema disease (ask the producer), add fumagillin to the syrup; see Chapter 14 for more information on nosema disease.

The common practice is to install packages in the late afternoon. If the weather is unusually cold, wait for the weather to improve (but do not wait more than a few days). Continue feeding the packages until they can be installed.

All equipment should be readied and in place well before the bees arrive. Equipment should include two to three deep hive bodies, 10 frames of foundation or drawn comb, a bottom board, inner cover, outer cover, and entrance reducer; see "Basic Hive Parts" in Chapter 3. Hive entrances on the empty equipment should be closed until the bees are installed, to prevent mice from entering the hive and damaging the comb. **Initially entrance reducers should be used for all methods.**

Determine the type of queen cage the producer used because there are a variety of cages now in use; see "Types of Queen Cages" in Chapter 10. Remember, there may be newer types of cages not illustrated here.

Indirect Queen-Release Method I

The bees should be fed with sugar syrup almost continuously for the last half hour before the package is installed, provided they continue to remove the syrup off the screens. Well-fed bees tend to remain calm. You can start packages on frames of foundation, empty drawn comb (from healthy colonies only, do not use *dead-outs,* equipment from colonies that died), or a combination of both. If using a combination, place the drawn combs in the center, because the queen can start laying eggs in the cells almost immediately. Make sure that any used frames are from hives that had no diseases; foulbrood and Nosema spores remain viable for a long time. If there is any question, do not use the comb.

Whether flowers are in bloom or not, since spring weather can turn from pleasant to inclement, causing a delay in foraging, supply the packages with a protein supplement/substitute patty or pollen patties, and fondant and/or sugar syrup. If the bees have no access to pollen, brood rearing will be delayed; see Chapter 7 "Feeding Bees" for more information.

To install a package, suit up, light your smoker, and follow these steps:

Step 1. Take the package to the preassembled hive site.
Step 2. With the sprayer containing sugar syrup, spray the bees to coat their wings, but do not soak them.

Step 3. Shake or jar the package so the bees drop to the bottom of the package. This will require you to bump the package firmly on the ground (do not drop) to dislodge the bees clustering around the syrup can, the queen cage, and the top of the package.

Step 4. Remove the lid, exposing the top of the feeder can and the slot from which the queen cage is suspended. Queen cages are suspended by various devices. In addition, the slot from which the cage is suspended may permit you to remove the cage without removing the syrup can. In other cases the slot is so narrow that the feeder must be removed and the queen cage guided out of its slot by way of the opening provided by the removal of the feeder can. The queen cage may be attached to a metal tab, or a wire, or a piece of screen, or, in the case of a plastic queen cage, a small disc. In any case, remove the queen cage carefully so it doesn't drop into the package.

Step 5. Once the cage is removed, replace the lid to keep the bees from escaping from the package.

Step 6. If the queen is alive, remove the cork or plastic "cap" from the end of the queen cage that contains the white candy plug; poke a hole in the candy plug with a nail, being careful not to go completely through the candy and kill the queen. The candy delays the queen's release, helping to ensure her acceptance by the bees.

Step 7. If no candy is present, after you remove the cork, plug the hole with a miniature marshmallow or homemade fondant or mock candy. If you don't have any on hand, leave the cork in place; you can proceed with the installation and return later to release the queen manually in a few days. However, it is much better to be prepared for the possibility that the cage does not have a candy plug; therefore, be sure to have a miniature marshmallow or some fondant candy available at the time of hiving packages.

Step 8. Place the queen cage in your pocket, with the screen side away from your body.

Step 9. Remove four or five frames from one side of the hive body.

Step 10. Remove the queen cage from your pocket and suspend the queen cage between the two frames closest to the oval hole of the inner cover. Once the queen cage is suspended, shake approximately a cupful of bees from the package onto the queen cage. The bees around the queen cage will begin scenting, and this will assist in drawing other bees eventually to that area. One of the main reasons why the queen cage should be located as stated is that in the event of cold weather, the bees are located within easy reach of the syrup; if not, the bees may not feed, protecting the queen instead, which could lead to the bees starving. Make sure the feeder is not leaking syrup.

Step 11. Place the entire package in the vacant space in the hive (where the frames have been removed), being sure that the open end of the package is up to allow the bees to escape (see the illustration of the indirect-release method I on p. 78).

Step 12. Place pollen patties (supplement/substitute) and/or fondant on the top bars of the frames, then place the inner cover on the hive, rim side down, to allow extra room above the top bars to accommodate the protein patties and fondant (if you are using it).

Step 13. Before placing the sugar syrup over the oblong hole of the inner cover, invert the syrup container outside of the hive so the initial drippings will fall into an empty bucket; otherwise, syrup dripped on the outside of the hive may invite robbing bees. If the feeder leaks, get another; see "Screw Top Jars or Feeder Pails" in Chapter 7.

Step 14. Now place an empty hive body on the inner cover to enclose the syrup container. Place the outer cover over this hive body. Partially block the entrance of the hive with a reducer cleat or with grass to discourage robbing and to slow the exiting of bees.

Step 15. After one week, check to see if the queen is released and all is normal. Light your smoker and gently blow smoke into the entrance. Lift off the empty hive body and feeder can (check to see if you need to refill it), then take off the inner cover.

Step 16. Remove the empty package and insert the necessary complement of frames. If bees are still in the package, shake them in front of the hive so they can walk in. Replace the frames you removed when you put in the package.

Step 17. Smoke the bees down just enough to retrieve the queen cage. If the cage is empty, the queen has been released; remove the empty cage, close the entrance, and adjust frames if needed. **Do not** otherwise continue with your inspection.

Step 18. If the queen is still in the cage, pull off the

Indirect-Release Method I

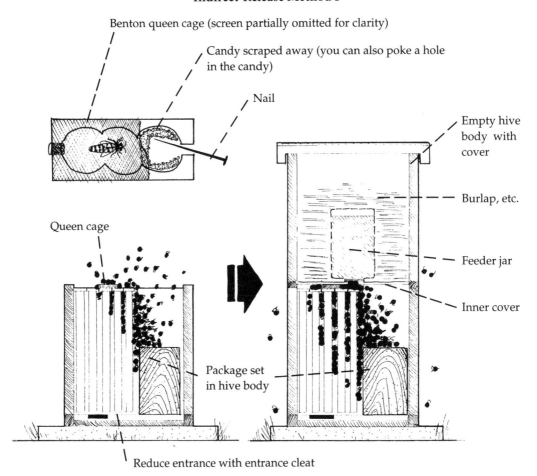

Benton queen cage (screen partially omitted for clarity)

Candy scraped away (you can also poke a hole in the candy)

Nail

Empty hive body with cover

Burlap, etc.

Feeder jar

Inner cover

Queen cage

Package set in hive body

Reduce entrance with entrance cleat

screen or cork opposite candy plug, or, in the case of a plastic queen cage, pull off the cap; be aware that this caged queen may be capable of flying off. Therefore, open the cage at the hive entrance and push it into the hive.

For the next 14 days, do not disturb the colony except to monitor the amount of sugar syrup in the feeder and refill as needed. When you are replacing the syrup, have a lit smoker ready and first blow smoke into the empty hive body at the top. Blow smoke around the feeder, then tilt up the empty feeder, direct smoke into the oblong hole of the inner cover to move the bees away, place a full feeder on top, and close the hive. If the pollen patty or fondant candy has been consumed, replace it.

Advantages

● Excellent chance of queen being accepted.
● Bees disturbed less than with method II.
● Less drifting.

● Easiest for beginners.
● Dead bees are left in the package and not on the hive floor.

Disadvantages

● Additional trips must be made to remove the queen cage, remove the shipping package, and replace frames.
● Egg laying is delayed since queen's release is delayed.
● Bees may build comb in the package, not in the hive proper, and the queen might start to lay eggs in these combs.
● Extra hive body is needed.

Indirect Queen-Release Method II— Bees Shaken Out

To install a package with this method, suit up and light a smoker just in case it is needed. Again, do this in the latter part of the day when the weather is warm enough for bees to fly. **Note**: This method is not the best alternative if temperatures are cold (below 65°F [18.3°C]). You will shake the bees out of the package, so have on hand your spray bottle full of sugar syrup and follow these steps:

Step 1. Take the package to the preassembled hive.

Step 2. Shake or jar the package so the bees drop to the bottom of it. This will require you to bump the package firmly on the ground (do not drop) to dislodge the clustering bees.

Step 3. With the sprayer containing sugar syrup, spray the bees to coat their wings, but do not soak them.

Step 4. Remove the lid, exposing the top of the feeder can and the queen cage.

Step 5. If the queen cage is attached to a metal tab adjacent to the feeder, remove the cage and replace the lid.

Step 6. If the queen cage is hung from a wire or piece of screen next to the feeder, first grasp the wire tab to keep the cage from falling into the package, then remove the feeder can followed by the queen cage. Now replace the lid or feeder to contain the bees.

Step 7. If the queen is alive, remove the cork or plastic cap from the end of the queen cage that contains the white candy plug; poke a hole in the candy plug with a nail, being careful not to go all the way into the cage and kill the queen. The candy will delay the queen's release, helping to ensure her acceptance by the other bees, again, see "Types of Queen Cages" in Chapter 10, if yours does not resemble this description.

Step 8. If no candy is present, after removing the cork or the plastic cap, plug the hole with a miniature marshmallow or homemade fondant candy. If you don't have any on hand, you can proceed with the installation by replacing the cork or plastic cap and return later to release the queen manually in a few days.

Step 9. Suspend the queen cage between the fifth and sixth (middle) frames of the hive, with the screen facing outward on a wooden cage; in the case of a plastic mesh cage, expose as much of the mesh as possible. In either case, the candy end should be up (see the illustration of the indirect-release method II on p. 80). Note: Avoid placing the cage

directly under the oblong hole of the inner cover because sugar syrup dripping or leaking on the cage could soak the queen, which can lead to her injury or death.

Step 10. Remove the package lid and shake approximately a cupful of bees onto the queen cage; replace the package lid.

Step 11. Place pollen patties and fondant on the top bars of the frames, then place the inner cover on the hive, rim side down, to allow extra room above the top bars.

Step 12. Before you invert the feeder can or jar over the oblong hole of the inner cover, first invert it away from the hive so that initial drippings will fall on the ground. Otherwise, syrup dripped on the hive or inner cover may invite robbing bees; if the feeder leaks, get another; see "Screw Top Jars or Feeder Pails" in Chapter 7.

Step 13. Now place an empty hive body on the inner cover with the feeder can, and cover with the outer cover. Some beekeepers fill the empty hive body with burlap, an old blanket, or other insulating material, but this is generally not necessary.

Step 14. Again spray the remaining bees in the package with syrup.

Step 15. Remove the package lid and shake a third of the bees out in front of the hive, allowing them to walk into the entrance. If it is cold, below 65°F, use the direct-release method instead.

Step 16. The freed group of bees will soon begin to *scent* (their heads will face the entrance, abdomens raised, wings fanning), releasing the "come hither" odor or pheromone from the Nasonov gland to attract loose bees to the hive.

Step 17. When the bees begin to enter the hive, slowly shake the rest of the bees from the package directly in front of the hive; this will help keep the bees from drifting. By shaking bees outside the hive, all the dead bees fall on the ground and not inside. Drifting may be a problem if the weather is warm, or if it is earlier in the afternoon, and especially when multiple packages are being installed.

Step 18. After most of the bees have entered, partially block the entrance of the hive with a reducer cleat or with grass; leave the entrance partially blocked for approximately two months (replacing grass when needed) to discourage robbing.

Indirect-Release Method II

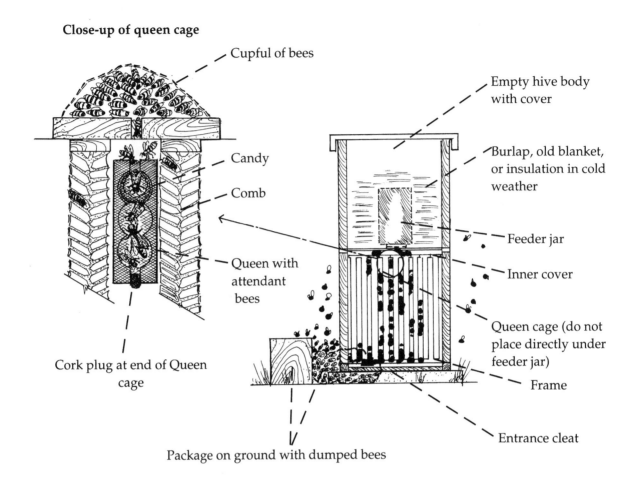

Close-up of queen cage

Cupful of bees

Candy

Comb

Queen with attendant bees

Cork plug at end of Queen cage

Empty hive body with cover

Burlap, old blanket, or insulation in cold weather

Feeder jar

Inner cover

Queen cage (do not place directly under feeder jar)

Frame

Entrance cleat

Package on ground with dumped bees

Step 19. Leave the package near the hive entrance overnight, open end up, to let any remaining bees escape into the hive.

After one week, check to see if the queen is released. Light your smoker and gently blow smoke into the entrance. Lift off the empty hive body and feeder can (check to see if you need to refill it), then take off the inner cover. Smoke the bees down just enough to retrieve the queen cage and close up the hive. If the queen is still in the cage, pull off the screen or remove the plastic cap and let her walk out onto the frames. If she is released, close up the hive without further disturbance.

For the next 14 days, do not disturb the colony except to replace syrup in the feeder can. When replacing the syrup, have a lit smoker ready and first blow smoke into the empty hive body at the top. Blow smoke around the feeder, then tilt up the empty feeder, direct smoke into the oblong hole of the in-

ner cover to move the bees away, replace with a full feeder on top, and close the hive.

The temptation to look at a new colony is more than most beginners can stand. To satisfy that urge, observe the entrance of the colony instead. For example look for:

- Undertaker bees removing the dead bees.
- Incoming forager bees with pollen on their legs (observe the color of pollen and try to determine from which flowers it was collected).
- Guard bees at the entrance, challenging incoming bees.
- Orientation flights, during which the young bees learn to locate their homesite.
- Signs of pests, such as skunks, ants, or mice.

If you do this every day, you can learn a lot about foraging activity, colony organization and structure, and the kinds of plants blooming in your area. If you don't see these activities, such as bees bringing back pollen, there may be some problems that require your attention; you should then inspect the colony.

Advantages

- Excellent chance of queen being accepted.
- Syrup located in vicinity of bees and queen, so likelihood of starvation is diminished.
- Easy way to feed medicated syrup.
- Bees will not abscond from the hive if queen is caged and unable to fly.
- Dead bees in bottom of the package will fall on the ground and not inside the hive.

Disadvantages

- Some drifting occurs, especially if the weather is warm.
- An extra hive body is needed.
- It may take a little more time than other methods.
- Queen cage must be removed at a later date.
- Egg laying is delayed since queen is not immediately released.

Open the hive on the 15th day after installing the package, weather permitting (see Chapter 5), using smoke as needed. If one or more of the frames shows a fairly compact brood pattern (capped cells and open cells full of eggs and larvae), all is well. Close the hive and leave it undisturbed for another week, replacing sugar syrup, pollen patties, and fondant as needed.

During the next visit to the hive, make sure the queen is present and the bees are drawing out the foundation. Refill the feeder and pollen patties and fondant as needed; continue feeding the colony until the first major honeyflow.

One to two months after installing the package, you should be able to add a second hive body with frames, if the weather conditions are good and bees are able to build up their population. By this time, the first hive body should be full of drawn comb and brood and the bees will now require room for expansion. If there is no major honeyflow and the new hive body contains frames with foundation only, feed the bees with sugar syrup to stimulate their wax glands. If the bees are not fed, they will chew the foundation. If the colony does not display conditions of normal development, your queen may be failing or the colony may be diseased or otherwise under stress, and you should get help from other beekeepers or the state apiary inspector.

Indirect-Release Method III— The Hansen Method

Follow steps 1 through 8 in the indirect-release method II, above. Then, follow these steps:

Step 9. Remove the queen cage from the package and suspend it between frames 6 and 7 of hive body 2 (this hive body will eventually sit on top of hive body 1 after step 13), so that the cage will be near but not under the oblong hole of the inner cover when it is replaced.

Step 10. Shake a cupful of bees onto the queen cage—these bees will begin scenting, which, as the process of installation continues, will help the remaining bees in the package orient to the queen cage's location; see the illustration of the indirect-release method III on p. 82.

Step 11. Remove the lid and feeder can from the package and place it in the center of hive body 1 (hive body 1 is sitting on the bottom board).

Step 12. Now place hive body 2 above hive body 1; this hive body has 10 frames and the queen cage plus a cupful of bees.

Step 13. Before placing the inner cover over hive body 2, add pollen patties onto the tops of the frames of hive body 2 and a feeder can or jar of syrup over the inner cover.

Step 14. Place an empty third hive body (3) over hive body 2. This empty hive body will enclose the sugar syrup container. Now add the outer cover and secure it with a heavy stone or cinder block.

Step 15. In the next day or two, remove hive body 3, remove the syrup feeder, puff some smoke into the oblong hole of the inner cover, and put a lid (you may use the lid that came with the package of bees) over the oblong hole.

Step 16. Set hive body 2 on top of hive body 3. Remove hive body 1, then remove the empty package from hive body 1. Now place hive body 2 on top of the bottom board and add hive body 3 on top of the inner cover. Then remove the lid over the oblong hole, puff some smoke into the oblong hole, and replace the sugar container over the oblong hole.

Step 17. Add the outer cover and secure it.

Step 18. After one week, check to see if the queen is released. Light your smoker and gently blow smoke into the entrance. Lift off the empty hive

Indirect-Release Method III
Hansen Method

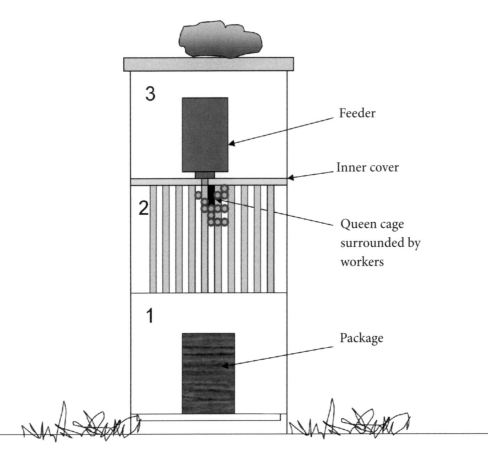

3

Feeder

Inner cover

2

Queen cage
surrounded by
workers

1

Package

body and feeder container (check to see if you need to refill it), then take off the inner cover. Smoke the bees down just enough to retrieve the queen cage and close up the hive. If the queen is still in the cage, pull off the screen or plastic cap and let her walk onto the frames. **Caution:** Since the caged queen has lost body weight, she is capable of flying if the weather is favorable. Remove the screen or plastic cap near the entrance and push the cage into the hive. If, on the other hand, the cage is empty, retrieve it and close up the gap created by the retrieved cage and close the hive without any further disturbance.

For the next 14 days do not disturb the colony except to add syrup to the feeder container. When replacing the syrup, have a lit smoker ready, and as you remove the feeder container to refill it, direct smoke into the oblong hole to prevent the bees from boiling out; return the filled container to its former position.

Advantages

- Excellent chance of queen being accepted.
- Minimal disruption of the bees in the package.
- Any dead bees remain in package.
- Initial drifting of bees is avoided; this is especially important if one is installing multiple packages in the same area on the same day.

Disadvantages

- Additional trips must be made to remove the queen cage, remove package, and rearrange hive bodies.
- Egg laying is delayed since queen's release is delayed.
- On occasion bees may build comb in the package, not in the hive proper, and the queen may start to lay in these combs.
- Initially two extra hive bodies are needed.

Direct Queen-Release Method

This method should be used only by experienced bee-keepers, as the queen could fly off if she is not clipped. Follow the first procedures as outlined for indirect queen-release method I as far as removing the queen cage (step 7), and then follow this sequence:

Step 8. If the weather is cool, place the queen cage in your pocket, screen side away from your body; if it is warm, put the cage in the shade.

Step 9. Remove four or five frames from the middle of the hive.

Step 10. Spray the bees in the package and remove the lid; shake all the bees onto the bottom board in the vacant space (or on ground at hive entrance).

Step 11. Spray the bees with sugar syrup again, to reduce their flying ability, but do not soak them.

Step 12. Dip the queen cage in sugar syrup or spray it, so the queen will not fly off when released. Do not dip the queen if the weather is cold; spray her instead.

Step 13. Carefully remove the screen from the queen cage (see illustration of the direct-release method on this page).

Step 14. Allow the queen to walk or drop gently on top of the bees.

Step 15. As soon as the pile of bees disperses, carefully replace the frames, taking care to avoid crushing the bees (unless you released the bees on the ground).

Step 16. Replace the inner cover; follow the rest of the steps (13–18) in indirect queen-release method I. Do not disturb this colony for at least 14 days except to monitor the syrup feeder.

Advantages

- Queen is released and can start to lay sooner.
- Easiest, fastest method; complete in one operation.
- No return trips to apiary needed except for feeding.

Disadvantages

- Extra hive body is needed.
- Queen could be killed or balled.
- Bees could leave (abscond) with free-flying queen.
- Queen could fly away (if she is not clipped) during installation, taking bees with her, or be otherwise lost.
- Queen could be superseded.
- Some bees and/or queen could be injured or killed when frames are replaced.
- Drifting often occurs if weather is warm or it is early in the day.
- Any dead bees in package are also shaken into the hive, making extra work for bees to remove them.
- Bees and queen could cluster some distance from the feeder container and be unable to access it during cold periods.

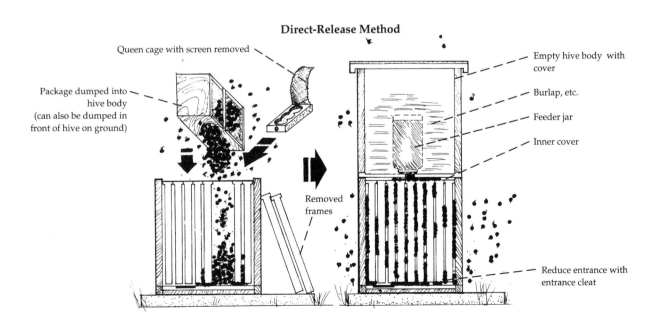

Direct-Release Method

Queen cage with screen removed

Package dumped into hive body (can also be dumped in front of hive on ground)

Removed frames

Empty hive body with cover

Burlap, etc.

Feeder jar

Inner cover

Reduce entrance with entrance cleat

Any combinations of the methods described can be done; just remember to keep the bees well fed and the queen protected.

Once established, the packaged bee colony may not build up enough stores or population to survive through their first year. If the population is sufficient, fall, late winter, and early spring feeding will be required. Note that a colony with a low population should be united to a strong one; if left alone, it is unlikely to survive the winter.

REASONS FOR PACKAGE FAILURES

Beehives started from packages sometimes fail after they are installed. Reasons for delay in development or failure include, but are not limited to the following:

- Queen has been superseded due to nosema disease or other reasons delaying normal colony development.
- Queen is unmated or poorly mated.
- Queen is balled as a result of too many disturbances of the hive by the beekeeper (especially during the first 10 days after installation).
- Weather has been too cold for bees to forage or obtain sugar syrup due to location or type of feeder. Note: By placing the queen cage near the oblong hole, access to syrup feeder is enhanced during cold weather in that the bees cluster near the queen, that is, in the vicinity of the oblong hole and hence the feeder.
- Bees have starved.
- Bees are infected with disease or mites and could perish.
- Bees have left the hive (absconded).
- Queens are stressed during shipment or are injured while being removed from a mating box or during placement into the queen cage or after her release, which may lead to diminished egg-laying capacity and/or pheromone production or premature death.

Any of these conditions will result in failure or substandard performance, which eventually will lead to supersedure (queen replacement by her colony), or upon her demise, the bees will be unable to replace her because there are no eggs or young larvae that can be converted to emergency queen cells. This will delay colony buildup (increase in colony population), or in the case of a queenless colony, laying workers will become established and the colony will decline and eventually die. Having extra queens on hand will mitigate these problems. No matter what method is used to install the package, there will always be a loss of bees due to drifting (bees failing to locate their hive during and after their installation). Drifting may be more severe when multiple packages are being installed. As each package is installed, some bees will expose their Nasonov glands, releasing a homing or orientation pheromone, which will attract drifting bees from other installations to their hive. Other bees will just drift away (Methods I and II will minimize drift).

Remember, feeding is also necessary to stimulate their wax glands; wax is essential for converting foundation into comb. Unless used equipment is available, packages will require all new equipment, so be prepared by ordering the necessary boxes, covers, and so forth, before the packages arrive. To be safe even though bees have been certified to be healthy, medicate all packages.

Installation is somewhat weather dependent. If it is a cold day, bees may cluster out of reach of the sugar syrup feeder, and if this is followed by a succession of cold days or nights, the cluster may starve. This situation usually can be prevented by placing the feeder close to, but preferably not over, the suspended queen cage. Bees will naturally cluster near the queen and as a result will be positioned near the syrup (this is accomplished by suspending the queen cage from frames close to the oval hole of the inner cover and by the placing the sugar syrup feeder over that hole). Here are some other reasons why packages fail to survive:

- By direct release of the queen, the cluster may form at some distance from the feeder.
- No eggs will be laid until the queen is released from her cage, and even if she is directly released (not recommended), comb cells need to be available for eggs to be deposited. Once egg laying is underway, the first new bees will emerge 21 days (3 weeks) later. During this hiatus, colony numbers are in decline, primarily due to natural attrition of the original adult population, death of older bees, bees lost by drifting, and foragers failing to return to their hive.

- Package bees installed on foundation have no food (honey, pollen, or bee bread) and thus must be provided with sugar syrup (to meet their carbohydrate requirements) and pollen or pollen substitutes (to meet their protein, fat, and vitamin requirements). In addition, since the timing for installing packages coincides with the scarcity of flowering plants, feeding is essential. Otherwise, the package could starve. So, if the bees are under all of these stresses they will not thrive. Make sure they can reach the food.

Other things to remember:

- Feeding is also necessary to stimulate their wax glands; wax is essential for converting foundation into comb.
- Unless used equipment is available, packages will require all new equipment.
- To be safe bees should be certified to be healthy.
- Package bees may be infested with pests such as mites and small hive beetles (see Chapter 14), so watch for these pests.

  Notes

CHAPTER 7

Feeding Bees

At certain times, it is necessary to feed bees sugar syrup, dry sugar, fondant, honey, pollen, or a substitute for pollen. Recognizing when bees are near starvation is an important step in learning beekeeping and could save the colony. Bees can exhaust their own food stores or for other reasons be unable to build up existing stores and eventually deplete them. In either case, the colony will be hard pressed to stay alive and must be fed. If this situation occurs during the flowering period, the colony may continue to exist on a day-to-day basis, but it will be weakened and, should an interval of inclement weather set in, may perish.

Bees should be fed under the following conditions:

- When no natural honey or pollen is available (in late winter or early spring) in order to stimulate brood rearing.
- When the colony is in danger of starving.
- When it is necessary to supply medication (chemotherapeutic agents).
- When installing a package or hiving a swarm for the above reason, as well as to stimulate wax glands when these bees or others are given foundation to draw.
- When requeening.
- When rearing queens and no natural honeyflow is on.

If stores are exhausted in the fall, winter, or early spring, the colony will die. Stores may have been depleted as a result of:

- The beekeeper removes too much honey, particularly in the fall.
- The bees consumed the remaining winter food by spring.
- The number of field bees becomes reduced due to spring dwindling, failing queen, or disease or parasites (see "Spring Dwindling" in Chapter 8).
- The bees' food consumption increases when egg laying resumes in midwinter, to provide heat and food for brood.
- An expected honeyflow fails to materialize; or inclement weather sets in at the time of the honeyflow, preventing bees from collecting fresh food; or a honey plant fails to yield expected food.

When colonies are in a condition where starvation is imminent, the bees must be fed to ensure their survival. The various methods of feeding bees with sugar syrup, dry sugar, honey, and pollen or pollen substitutes (or supplements) are discussed in this chapter.

SUGARS

Carbohydrates are organic compounds composed of carbon (C), hydrogen (H), and oxygen (O) atoms, with a general formula of $C_nH_{2m}O_m$; in other words, there are twice as many hydrogen atoms as oxygen atoms, and the number of carbon atoms can vary. Because carbohydrates are composed of many C—H bonds, which liberate a great deal of energy when they are broken, these molecules are an excellent candidate for energy storage. Sugars, starches, and cellulose are examples of carbohydrates.

The three different types of sugars that bees feed on

are glucose, fructose, and sucrose. *Glucose* ($C_6H_{12}O_6$, once called dextrose) is found in all living cells. It is produced by photosynthesis in green plants and is the primary energy-storage unit. Glucose is a simple sugar (a monosaccharide) that comes in two forms: a straight-chain molecule and, when mixed with water, a six-sided ring structure, a hexagon (see the illustration of sugars on this page).

Fructose (levulose or fruit sugar) is another monosaccharide and has the same formula as glucose but a different molecular structure. It is the sweetest of the simple sugars because the double-bonded oxygen is attached internally in the molecule, rather than on the end, as in glucose. Two different ring structures are possible in fructose molecules: a hexagon and a pentagon (five sides).

Sucrose ($C_{12}H_{22}O_{12}$, or cane sugar) is a disaccharide, or two sugars, made up of glucose and the pentagon form of fructose linked together. The major component of nectar, sucrose is broken down into the two monosaccharides by the action of the enzyme *invertase* (see Chapter 12). An enzyme is a protein that serves as a catalyst in chemical reactions, building up, breaking down, or rearranging the atoms.

Sugar Syrup

One gallon of sugar syrup (2:1 sugar: water, or 2 parts sugar to 1 part water) will increase the food reserves of a bee colony by about 7 pounds (3.2 kg). The following proportions (by volume) of sugar to water should be fed depending on the season and the purpose for the feeding:

● 1:1, for spring feeding and spraying onto bees in packages.

● 2:1, for fall feeding.
● 1:2, to stimulate brood rearing. Make only two to six holes in the lids of the gravity feeders, so the bees will be able to obtain only small amounts over an extended period of time; this effect will be similar to a light nectar flow.

Use white, granulated cane sugar only. Beet sugar may be from genetically modified (GMO) sugar beets. Never use brown or raw sugar, molasses, or sorghum, because these contain impurities and can cause dysentery in bees.

Mix the desired proportions of sugar and hot water and stir adequately until all the sugar is dissolved. Hot water from the faucet is hot enough to dissolve the sugar; you can also use water that has been heated over a stove but it must not be boiling. Never let the sugar-water solution boil over direct heat; syrup that is burned, or caramelized, will cause the formation of *hydroxymethylfurfural* or HMF, which is toxic to bees and will lead to high mortality. Honey and other syrups will also form HMF if overheated. Heating the mixture over steam or in a double boiler will prevent caramelization.

To prevent fall syrup from crystallizing, some beekeepers add cream of tartar (or tartaric acid) to the solution of sugar and water. Tartaric acid breaks down the sugars, but there is some concern that it is detrimental to bees; currently, its addition is not recommended. Other beekeepers add a little vinegar for the same reason, and to keep mold from forming in the syrup; a teaspoon of apple cider vinegar added to each gallon is safe for bees.

Other things added to syrup, including salt, vitamins, and bleach, are also being advocated but so far there is little research that these additives are advan-

Sugars

tageous. Check current journal articles for the latest information.

Bees should be fed early enough in the fall so that the sugar has time to cure; that is, the bees have time to reduce the water content near to that of honey (about 18 percent). If the syrup does not cure, it could ferment or freeze; both of these conditions are detrimental to overwintering bees.

High-Fructose Corn Syrup

High-fructose corn syrup (HFCS), now called corn syrup, is a sugar syrup produced by breaking down cornstarch, which is composed of two major chains of glucose (molecules linked to one another in long chains), into single units of glucose, fructose, and water. In 1969, researchers used enzymes, one that split the long chains of glucose into smaller ones, and another that reduced the smaller chains into individual glucose molecules. These enzymes are now produced by genetically engineered bacteria, replacing the process that used acid and heat. The resulting syrup contained 55 percent glucose, 42 percent fructose, and other substances.

The methods for feeding HFCS are similar to those described for other syrups. Unfortunately, HFCS may not be the best food to feed bees, and caution should be taken when feeding this to your colonies. Some products available today are mixtures of sucrose syrup and HFCS.

If you are going to feed HFCS, remember it is very similar to honey, so if it is stored in areas or storage containers subjected to extended periods of high temperatures, it could form HMF, which, as noted, is toxic to bees. And if this degraded syrup is mixed with water, formic and levulinic acids could form as well, which are also toxic to bees. When in doubt, get the questionable syrup tested or discard it.

TYPES OF SYRUP FEEDERS

Screw Top Jars or Feeder Pails

One of the best ways to feed bees at almost any time of year is with a 5- or 10-pound glass jar or a one-gallon plastic pail with a screened hole, turned upside down over the top bars of the hive or over the oblong hole of the inner cover. The jar lid can be perforated with small holes so the bees can insert their tongues into the holes to withdraw the sugary solution.

Plastic Feeder Pail

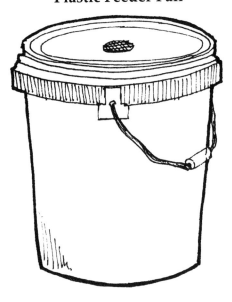

Don't use plastic jugs because they often collapse after being filled and inverted or leak. Metal pails are not recommended. If you use a plastic bucket, make sure it did not contain any toxic material. Although glass containers can break, they are easier to clean and inexpensive to replace, and you can readily see if they need to be refilled.

Use a shingle nail, or a three- or four-penny nail (1/16-in. [1.6 mm] diameter), to punch two or six holes in the jar lid; remember to remove the cardboard washer from the screw tip lid. A 1-inch hole can be burned out of the center of the top for the plastic pail, and a fine metal screen melted in place to cover the hole (see the illustration of the plastic feeder pail on this page).

At the hive, first invert the jar or pail so drippings will fall into some container or on the inner cover, rather than on the hive or the ground, so as not to encourage robbing. As soon as the dripping stops, place the feeder directly on the top bars near the cluster if the colony is weak; otherwise, place it over the oblong hole of the inner cover. Place an empty hive body rim around the feeder, and replace the outer cover. Make sure to put a weight on top of the outer cover to prevent it from blowing off.

The syrup will not leak as long as the holes are not too large and the feeder is level; if there is empty drawn comb in the hive, the bees will remove the syrup from the feeder and store it in the comb below. Make sure the feeders do not drip on the bees—it could kill them or cause robbing if the syrup leaks.

Miller or Top Feeder

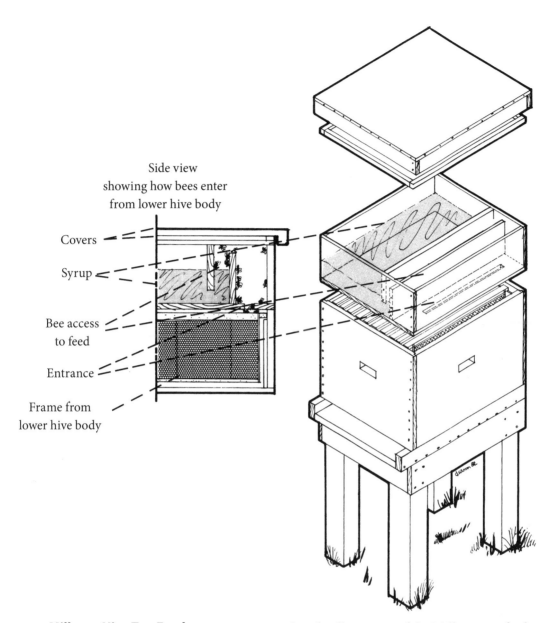

Side view
showing how bees enter
from lower hive body

Covers

Syrup

Bee access
to feed

Entrance

Frame from
lower hive body

Miller or Hive Top Feeder

One type of feeder, called the Miller feeder, or top feeder, originally was composed of either two aluminum or two plastic pans fastened together and hung within or otherwise attached to an empty shallow super. Some were made entirely of wood, but now most are a one-piece plastic super-shaped body, making it easier to fill and clean. An opening at one end (or in the middle) allows the bees to crawl up and over the sides of the feeder to get to the syrup. A control strip of wire mesh or roughened plastic allows the bees to cling to its sides without drowning in the syrup

(see the illustration of the Miller or top feeder on this page). This type of feeder is also available in nucleus sizes, making feeding nucs much easier.

The top feeder, as the name suggests, is placed on top of a hive, beneath the inner and outer cover; it can hold up to 4 gallons (15 liters) of syrup and can be rapidly filled or refilled. If this type of feeder is used, it must not leak and must be bee-tight on the outside, or else robbing may occur. There are many styles of top feeders available from dealers; check the types used in your area.

Division Board Feeder

The division board or *in-hive feeder* is a frame-sized container that can be inserted in place of a frame within a deep or medium super. Feeders of this type have some kind of flotation device or a strip of screen that allows the bees to reach the syrup without drowning. Older model division board feeders are made with Masonite, wood, or plastic and usually have Styrofoam, screen, or wooden floats. Newer plastic division boards have roughened inside walls to provide footing for bees. **Nevertheless, bees often drown in these feeders.**

Some beekeepers permanently keep one division board feeder in each deep hive body, usually on the outermost edge of the hive body. On cold days, however, a weak colony will be less able to obtain syrup when this type of feeder is used, unless it is located near the bee cluster. If the cold weather continues for a long duration, the bees may be unable to move to the feeder and could starve.

Conversely, if this type of feeder is left in place all the time, bees could start to construct comb in it, and the queen could lay eggs on that comb. Not only will that make it difficult to see what is on that comb but also it will be difficult to remove, and if you fill the feeder with syrup, the brood and maybe the queen could drown if they are inside. Some beekeepers fill this feeder with excelsior fiber, used for evaporative coolers. This will work, but the fiber should be replaced periodically because bacteria and decomposition fungi will eventually turn it into a smelly mess.

Empty Drawn Comb

Frames of empty drawn comb can be filled with syrup and placed in the hive. Slowly dip the frames into a tub of syrup or sprinkle them with syrup using a sprinkler can or other device. A steady stream of syrup poured directly from a container will not fill the cells because the air in the cells will act as a barrier to the liquid's entry (see the illustration on filling drawn comb on this page). This method can be used for emergency feeding, especially if the combs are located near or adjacent to the broodnest. As with other feeding methods, you have to look into the hive (and remove these frames) to determine whether or not they need refilling.

Filling Drawn Comb

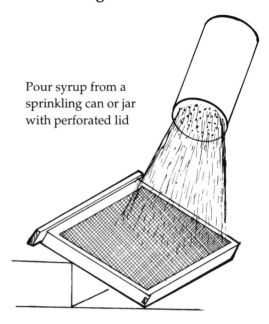

Pour syrup from a sprinkling can or jar with perforated lid

This method is often used when installing package bees in a hive with drawn comb. Newly hived swarms, however, should not be fed by this method. Bees in a swarm have full honey sacs, and if they are put on drawn comb, they may regurgitate the contents of their sacs into these cells; if this honey contains disease spores and is later fed to larvae, brood diseases such as American foulbrood could result (see "Collecting and Hiving a Swarm" in Chapter 11). It is very important to ensure that any combs used for feeding syrup have no history of brood diseases.

Boardman Feeder

A Boardman feeder is a wooden or plastic holder for a quart-size mason jar (see the illustration of the Boardman feeder on 91). The front portion of the holder's base is an entrance platform and is inserted into the hive entrance. The bees can obtain syrup by crawling into the holder's base to reach the jar.

The Boardman feeder is **not** recommended for feeding syrup because bees from other colonies can rob from it, which tends to encourage further robbing activity. If the weather is cold, the bees being fed will not break away from the cluster to reach the feeder and the cold syrup, and may starve. The feeder holds only a quart of liquid and would require frequent re-

Boardman Feeder

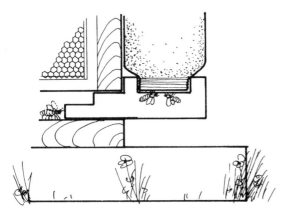

Side view
showing how the feeder fits
in the entrance

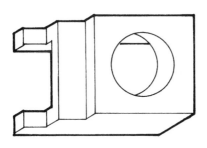

Top view

filling when the bees are actively feeding. Furthermore, in this highly exposed condition, the liquid could freeze, or the sun may decompose chemicals in the syrup that were added to medicate the bees. If the food is in there too long, the syrup could also ferment. Unless your hives are up on stands, Boardman feeders can be easily knocked off or taken by hungry raccoons or other clever mammals, who will scatter the jars throughout your beeyard.

The best use of the Boardman feeder is for *watering* bees during the summer months. If the weather is quite hot, bees can go through a quart of water in a day or two. This method of providing water also helps to keep bees from bothering your neighbors (see "Good Neighbor Policy" in Chapter 4).

Plastic Bag Feeders

Half-gallon plastic zippered bags, the kind used to store frozen food, are good emergency feeders if jars or pails are not available. Fill the bags one-half to three-quarters full, expel the air, and seal. Store them flat. When feeding your bees, lay one bag over the top bars or on the inner cover, and make a slit 1 to 2 inches (2.5–5 cm) long on the upper side of the bag as it is lying on the top bars. The bees will come up and feed as the syrup oozes out. Do not fill the bags too full, as you won't be able to put the outer cover on, and too much syrup may be squeezed out, which could drip on and chill a small cluster. Remove and discard the plastic bags when empty.

NON-SYRUP FOOD

Dry Sugar

Dry, white, granulated sugar can be used as an emergency food in late spring when outside temperatures are high enough to permit the bees to obtain water for dissolving the sugar; occasionally, water that has condensed in the hive is used for this purpose. If the bees are unable to store honey in the early spring, feeding dry sugar in late spring, prior to a honeyflow, may help prevent starvation.

The sugar should be located as close to the bees as possible. It can be spread around the oblong hole of the inner cover (rim side up), or on the back portion of the bottom board away from the entrance. Or the sugar can be spread over a single sheet of newspaper placed over the top bars of the hive body where the bees are located; the bees will chew through the newspaper to obtain the sugar.

Only strong colonies will benefit from the feeding of dry sugar; weaker colonies may not have sufficient numbers of bees to obtain the needed water.

"Mock" Candy

A quick candy can be made simply by mixing clean honey (not from diseased colonies or from store-bought honey) or thick sugar syrup with confectioners' sugar or Drivert sugar. Knead it like bread dough to form a stiff paste. Store it in the refrigerator or freezer and use it as emergency food, for queen cage plugs, or for feeding small queen-rearing nucs. As

Candy Board Feeder

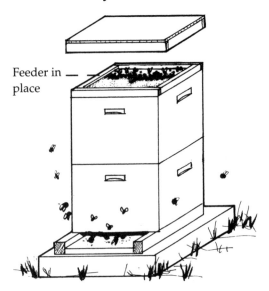

Feeder in place

an emergency food, place a thawed piece of flattened candy on the top bars or near the inner cover hole.

Fondant or Sugar Candy

Fondant candy can be made and fed to bees in small molds or with a special rim feeder called a *candy board* (see the illustration of the candy board feeder on this page). To make candy board, use an inner cover but add a rim that is 1 inch or deeper on the edges. When the board is filled with fondant, place a queen excluder over it (to prevent it from breaking up), then invert it over the cluster. This method is less sloppy than feeding syrup or dry sugar, but the preparations take much longer. The basic fondant or sugar candy recipe (to feed one colony) is:

- 2 cups white sugar
- 2 tablespoons corn syrup (light), or ⅛ teaspoon cream of tartar (tartaric acid)
- 1½ cups boiling water

Combine and heat ingredients, stirring until the sugar dissolves; heat without stirring to 238°F (115°C, or until medium ball on a candy thermometer); pour the syrup out onto a cold platter and cool it until it is warm to the touch. Beat the syrup until it turns white, and then pour it into molds or shallow dishes if you do not have a board feeder. **Be careful**, this is **hot**.

The molds can be made from metal or glass, but don't use plastic, as it may melt. Once the candy has

hardened, it can be inverted over the top bars near the cluster.

A recipe for making a larger batch of candy is from Penn State University, MAAREC (2004):

- 15 pounds sugar
- 3 pounds glucose or white syrup
- 4 cups water
- ½ teaspoon cream of tartar

Combine and heat the ingredients to dissolve sugar until the temperature reaches 242°F (117°C); cool slightly, to 180°F, and then beat with a rotary beater until thick. Pour into molds or candy boards to solidify. Feed bees by placing cake of candy above a cluster of bees. In feeding mock candy or fondant, make sure the bees have access to water so they can dissolve the sugars. Again, this is an emergency food and should not replace feeding syrup when available.

HONEY

The best food of all—when properly ripened, capped, and free of disease—is, of course, a super full of honey or several frames of honey placed next to the broodnest in colonies requiring feeding. Honey obtained from old combs and cappings, as well as crystallized honey, can be diluted and fed to the bees using any of the methods described for feeding syrup. Supers with frames that are "wet" or sticky after having been through an extractor can be placed above or below the broodnest or over the inner cover for the bees to clean.

Be careful when feeding bees with diluted honey or wet combs: the odor of honey will stimulate robbing. Therefore, feed honey or place wet combs on the hives in the early evening so that the bees will have sufficient time to remove and store it before morning. If this food is given to weak hives, reduce their entrances as a further precaution against robbers (see "Robbing" in Chapter 11). Make sure the equipment containing the wet comb is bee-tight to keep out robbers. Also, supers with wet combs should not be put on colonies in the late fall or winter. The entire colony of bees might move up into them and not return to stores below, and thus they may die eventually from starvation.

Store-bought honey should **not be fed** to bees. Honey from hives with foulbrood is sometimes extracted, bottled, and sold. There is an excellent chance that this honey contains foulbrood spores, which remain viable in the honey for up to 80 years, and its use could result in an outbreak of American foulbrood (AFB) in your apiary. (Fortunately for us, AFB disease spores contained in honey are not harmful to humans.) Be certain that any honey fed to the bees is free from spores that cause bee diseases.

Honey mixed with cappings, scrapings, or debris can be fed to bees if it is placed in a container or spread on a sheet of foil or on the inner cover; place an empty hive body and the outer cover over it to keep robbers out. Also, honey-filled combs that were broken, or broken combs that can be recycled, can be crushed in warm water and fed as syrup, or simply placed in a top feeder. The remaining wax can then be recovered and melted. As with any method of feeding sugar, make sure ants cannot get to the food!

POLLEN

Pollen, the male germplasm of plants, is the principal source of protein for honey bees and the building blocks of bee bread. The protein content of pollen ranges from 8 to 40 percent. An average-sized colony (approximately 20,000 bees) collects 125 pounds (57 kg) per year; other estimates are from 40 to 110 pounds (18–50 kg) a year for average-sized colonies (see the "Fun Facts" on the inside cover of this book).

When a bee collects pollen in her leg baskets, or corbiculae, she can pack in a pollen load weighing between 0.0035 and 0.0042 ounce (100 and 120 mg) or about one-half of her own body weight. These loads are called "pollen pellets." Foragers then drop them in cells near uncapped larvae. House bees pack down the many pellets, and one cell when full represents about 18 loads. The house bees mix the contents with a drop from the honey stomach (which contains beneficial bacteria) as well as other glandular secretions. This starts the fermentation process that will prevent the pollen grains from germinating and preserve and enhance the nutritional content of the pollen. The processed pollen (similar to silage for cattle) will turn into *bee bread*. Bee bread is a complex, nutritious food for bees that will not spoil. When bee bread is stored for several months, the bees will top off the cell with some honey and cap it with wax. The honey top not only serves as a physical barrier but also with its high acidity further deters spoilage and preserves the bee bread. In addition to protein, which bees can break down into simpler components called

amino acids, bee bread is practically the only source of fatty substances, minerals, and vitamins for larvae and adult bees. Bee bread and pollen are necessary parts of the diet of larval and young adult bees, which must have this protein within the first two weeks of their life in order to develop normally. If the pollen has been contaminated with pesticides, bees could perish or starve as some contaminants could kill the microbes that convert the pollen into bee bread. Make sure your bees are not in an area treated with these chemicals.

Foraging bees are apparently unable to discriminate between highly nutritive pollen and that which has low nutritional value. However, by collecting pollen from diverse plant species, they achieve a complementary balance of nutrition. This balance can be disrupted if bees are forced to collect pollen from only a few crops (such as when bees are moved into fields for pollination); in such cases, supplemental protein should be provided.

Bees increase their consumption of pollen and bee bread in the fall. This factor, coupled with a decrease in foraging and brood-rearing activities, seems to extend the longevity of worker bees beyond their usual summer life expectancy of six weeks. Many winter workers live longer than three months, less if they are nutritionally stressed.

It is also necessary for nurse bees to feed on bee bread, to activate their food glands (hypopharyngeal glands) to secrete a milky-white, protein-rich food that will feed the larvae and queen. This substance, called *royal jelly*, is fed in abundance to all larvae less than four days old and to queens during their larval stage and throughout their adult lives. Worker and drone larvae more than four days old are fed modified jellies and later a mixture of brood food, diluted honey, nectar, and pollen. Without pollen, bees could not manufacture royal jelly. About 0.0042 to 0.0051 ounce (120–145 mg) of pollen is needed to rear one bee, from egg to adult; this is around one and a half pollen pellets.

When flowers are available, bees will usually collect sufficient supplies of pollen in the hive. The demand for pollen increases in the winter when brood rearing resumes, and the remaining pollen stores can be consumed quickly. A good rule of thumb is for bees to have 500 to 600 square inches (1270–1524 cm²) of stored pollen reserves going into northern winters. Because bees cannot forage for pollen at this time of year, there must be ample bee bread stores to enable bees to feed the larvae that will replace them in the spring. If this is used up, or not available, bees must be fed some supplemental protein.

Beekeepers feed bees pollen, pollen supplements, and pollen substitutes to initiate, sustain, and increase brood rearing. By stimulating brood rearing, beekeepers profit from the increase in populations. Such colonies can send many foragers out to harvest nectar and pollen. Beekeepers utilize strong colonies to make divisions or splits, and to supply bees for pollination or to sell in packages.

Our knowledge of all the factors that return a honey bee colony from a broodless condition to one with brood is incomplete. Pollen seems to be one of the factors governing the initiation and maintenance of brood rearing; however, day length, pheromones, physical and chemical stimuli, and the commencement of nectar and pollen collecting may also play an important role in the brood-rearing activities of honey bee colonies.

The sequence of events that initiates brood rearing may begin with an increase in the amount of sugar intake by the queen, after which she begins to lay eggs. The presence of brood activates the special glands of nurse bees. These glands stimulate bees to feed on bee bread, which provides the glands with the nutrients necessary for their production of larval food. Certain ingredients in the bee bread may then initiate the elaboration by these glands of the larval food.

The best protein source for bees is pollen; in second place is pollen supplement, which is better than pollen substitute (see the section on pollen supplements and substitutes below). When feeding these substances, you should provide additional stimulus by simultaneously feeding sucrose syrup. Syrup feeding also stimulates the cleaning or hygienic behavior of bees, which in turn stimulates bees to forage.

Trapping Pollen

Various traps are available for collecting pollen. If the trap's design or the hive's position does not allow for the placement of the trap at the existing entrance, new entrances should first be established. Once the bees are familiar with their new entrance, close up the old entrance and set the trap at the new one. Some traps come with removable grids (the part the bees pass through); others do not (see the illustration of the pollen trap on p. 95). In any case, bees should be permitted frequent access to the free entrance of the hive so that they can replenish their own pollen stores. Check bee supply catalogs and other beekeepers for the styles of traps currently used. Bottom traps work very well, but they can collect a lot of debris, and if they are used in rainy or wet conditions, the pollen could be ruined.

Beekeepers may collect pollen for a variety of purposes: to study its characteristics (amino acid or protein content), to identify the flowers bees are visiting, to add to human and pet diets, and to feed bees during periods when fresh pollen is not available. Such feeding during the late winter and early spring may stimulate brood-rearing activities, which could result in a stronger colony.

An especially important reason for trapping pollen is to determine if it is contaminated with pesticides. For example, bees avidly collect sweet corn pollen; if the grower has used systemic-treated seed or sprayed a pesticide when the corn is in tassel, the pollen may be contaminated. If your bees are within range of these fields, pollen traps will collect most of the contaminated pollen, reducing the amount entering the hives. Since corn pollen is shed in the morning the traps can be removed each afternoon, giving the bees the opportunity to collect other, clean pollen (see "The Pesticide Problem" in Chapter 13).

There is always the danger that collected pollen may become contaminated with chalkbrood, American foulbrood spores, especially if the trap sits under the broodnest. Collect or purchase only pollen that has been obtained from disease-free colonies or has been sterilized by radiation. When in doubt, collect your own.

When the trap is in position on a hive, bees entering or leaving the hive must pass through the grid (a mesh of five wires per inch). The grid's dimensions will not permit a bee laden with pollen to pass through until its load (the pollen pellets) has been removed.

Ideally, a pollen trap is put on the hive during pollen flows and is kept on only for short periods of time. Some beekeepers keep the trap on throughout the summer but actually collect the pollen (set the trap) on alternate weeks or every three or four days. A colony deprived of pollen for an extended time period may quickly decline. To preserve its quality, trapped pollen must be removed regularly, every other day and especially if rain is imminent as it could wash out or ruin the pollen in bottom traps. Even in dryer

Auger-Hole or Exterior Pollen Trap

Front view showing how the trap fits in the hive

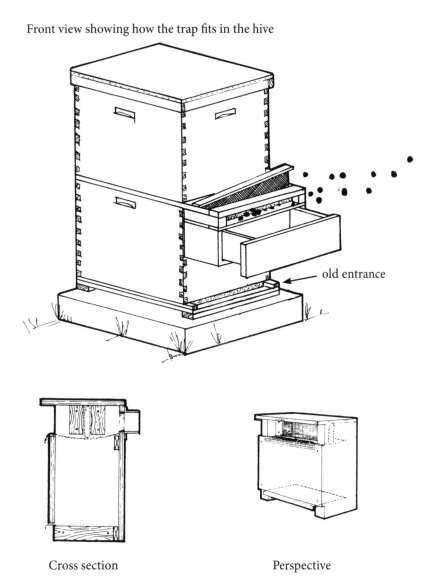

old entrance

Cross section   Perspective

Source: E.R. Harp. 1966. A simplified pollen trap for use on colonies of honey bees. Pub. 33-111. Beltsville, MD: USDA-ARS, Entomology Research Division.

climates, remove the trapped pollen regularly to keep ants and other pests from eating it. Clean the pollen of debris and insects by hand sorting, then quickly preserve and store the collected pollen to prevent spoilage; fresh pollen molds rapidly, especially in hot, humid weather. Pollen loses its nutrient qualities quickly.

Storing Pollen Pellets

Drying Fresh Pollen Pellets. Fresh pellets collected from a pollen trap can be dried in the sun for few days, in a warm oven, or with a lamp or food dryer. Heat to 120°F (49°C) for the first hour to kill yeast spores, then dry for 24 hours at 95° to 97°F (35–36°C). The pellets are ready when they do not crush when rolled between the fingers and do not stick to each other when squeezed. Store them in closed containers at room temperature.

Dry pollen may be fed directly to the bees, sprinkled into empty cells of a frame or mixed with other dry materials. If the dry pollen is to be added to wet mixes, it should first be soaked in water or syrup for an hour.

Advantage

● Inexpensive way of preserving pollen.

Disadvantage

● Less attractive to bees if no sugar is added.

Freezing Pellets. Place fresh, cleaned pollen pellets in containers and store them directly in a deep freezer at 0°F (–18°C) until ready to use; they will be moist when defrosted. Use immediately, or dry them as mentioned above.

Advantages

● Attractive to bees.
● Can be used separately or added to mixes.
● Best for human consumption, if selling pollen.

Disadvantage

● More costly to preserve.

Sugar Storage. Pollen pellets can be preserved with dry sugar. Fill a container alternately with layers of pollen and *white* sugar, topping it with several inches of sugar. Close the container tightly and store it in a cool place. Pollen should be mixed with twice its weight of sugar (1 part pollen to 2 parts sugar). Careful labeling of the container as to its amount of sugar and pollen will ensure that proper proportions are maintained when you are preparing mixes with other dry ingredients to make a pollen patty. You can also freeze this pollen and sugar mixture.

Advantage

● Attractive to bees.

Disadvantage

● Difficult to separate pollen and sugar if you want to feed straight pollen.

Methods of Feeding Pollen

One method of feeding pollen is to place dried pellets (whole or finely ground) on the top bars of frames beneath which most of the bees are located, or pour them around the oblong hole of the inner cover, especially if it is close to the active broodnest. Dry pollen without any sugar is not always attractive to bees and is not usually recommended. A better method is to pour the pellets into frames of empty drawn comb according to the procedure described below:

Step 1. Fill the comb on one side of a frame with pollen pellets. Insert the comb into the hive.
Step 2. If both sides of the comb are to be loaded, spray a thick sucrose syrup onto both sides before adding the pollen, to prevent the loose pellets from falling out.

Another method to feed pollen to bees, called *open feeding,* is to place ground pellets (grind them in a blender to break up the pellets) in a cardboard or plastic box or other container somewhere in the apiary. The container should be covered to prevent spoilage by rain or moisture. Naturally, it is necessary that the bees have access by means of a hole on the sides of the container. Because only strong colonies will benefit from open feeding, and inclement weather will prevent bees from foraging to the container, we recommend internal feeding over open feeding whenever possible. Open feeding also will attract other animals, such as raccoons and mice, which may disrupt colonies or otherwise foul the pollen. Therefore, secure a feeder away from these pests, such as by hanging it on a clothesline.

Making Patties to Feed Bees Pollen Patties

Pollen can be made into a dough by kneading it with clean honey or sugar syrup that has been heated before use (make sure the honey is not from diseased colonies). The dough should be stiff, not runny. Sandwich the patties between two pieces of waxed paper, not plastic wrap, to keep them moist and roll until about ¼ inch (3–5 mm) thick. Although the bees will eat holes in the waxed paper to get to the patties, cut a few slits into the paper to get them started; they will take over, discarding the waste paper. An easy formula is 4 parts hot water to 1 part pollen to 8 parts sugar. If you are making a lot of patties, investing in a commercial bread mixer or similar appliance will save a lot of time and effort. Make sure to store flattened, wrapped patties in the freezer.

Pollen Supplements and Substitutes

If you find it necessary to feed weak colonies or nucs, sugar syrup is preferred over honey, and trapped pollen or frames of bee bread over anything else. Only if pollen is not available or comes from a questionable source should you consider supplements or substitutes.

The terms *pollen supplement* and *pollen substitute* have been used in connection with feeding bees protein. These two terms have caused some confusion and are used here according to the following definitions: A pollen supplement consists of pollen and other substances of nutritional value to bees. A pollen substitute contains no pollen but consists of substances nutritional to bees.

Previously, soy flour made by the low-fat "expeller" or "screw press" method was a common ingredient in both, but now that the fat in soybeans is being expelled chemically, this flour is hard to obtain. Dairy products have also been used, but the milk sugars lactose and galactose are toxic to bees. Many different commercially made products are now available from bee suppliers or are advertised in the bee journals. Again, check with what others are using and experiment cautiously with new products. Brewer's yeast and soy flour are often the major ingredients in these products, with added vitamins and minerals. New products are available every year.

Pollen substitutes or other commercially purchased products available in bee supply houses may contain eggs or larvae of grain pests; these can be killed by freezing the material for several days at 0°F (–18°C); the material should then be placed in sealed containers to prevent subsequent contamination.

Feeding Pollen Supplements and Substitutes

If you find it necessary to feed colonies, it is important to start feeding in the early spring for 10-day intervals or as fast as the bees consume it. Do not stop feeding once a regimen has begun, but continue to feed until the bees will no longer take it. Once natural pollen is available, bees will not take the artificial feed.

Do not be tempted to mix vegetable oil in pollen patties, as the added lipids (fats) may be detrimental to bees. Keep up with current research articles on the subject.

Here are the different ways of making pollen patties, supplements, or substitutes:

- One pound (0.45 kg) of the dry material can be mixed with 4 cups of sugar syrup (2 parts sugar to 1 part water) to make several one-pound patties; make sure this mixture is thick enough so that it does not drip between the frames.
- Some beekeepers add anise oil, fennel oil, dark rum, or chamomile oil to make the material more attractive to bees.

When you are making patties, sandwich the mixture between two pieces of waxed paper (not plastic wrap); this way the they will remain moist. Tearing a few holes in the waxed paper on the underside of the patty will give the bees easy access to the food. Place the patty directly over active brood frames. Make sure it is not runny. If a patty is not consumed within a few weeks or if it starts to mold, remove it. Dry material can also be fed like pollen pellets.

  Notes

CHAPTER 8

Winter/Spring Management

The survival of colonies during the winter months (the most stressful period for honey bees) is dependent upon many factors, but most of these factors hinge on the degree and severity of any given winter. Efforts to bring bees out of the winter alive and in good health can be daunting. Successful overwintering of bee colonies may be achieved by following some standard practices

This chapter is organized to take you through the winter—if your bees are in a location that will experience several months of winter—to spring. If you live in southern climates, where the winter is much shorter, read the section "Wintering in Warm Climates." The next chapter, "Summer/Fall Management," is designed to help you anticipate the major tasks that must be attended to during the "busy" months.

WINTERING

General Rules

Colonies can survive very well without elaborate wintering techniques as long as the bees are protected from winter winds. Following the minimum procedures for wintering hives, however, can make the difference for overwintering success. These are the most common wintering practices:

- Invert the inner cover, to rim side down, to allow the warm, moist air to escape through the rim hole. If your inner cover does not have a rim hole, you may want to invert it and raise one side with a small piece of wood to allow for the escape of the moist air.

- Reduce the main entrance with a wooden cleat (or entrance reducer) or a piece of ½-inch wire mesh hardware cloth to keep mice from invading hives. Make sure that the opening of the cleat is turned up against the hive body, not against the bottom board; this will prevent the opening from becoming blocked by a layer of dead bees, which may accumulate on the bottom board during the winter.

- Provide top ventilation by propping up the inner cover slightly or by boring an auger hole, not more than 1 inch (2.5 cm) in diameter, in an upper corner of the top hive body; this opening lets moist air out of the hive and serves as another entrance. You may need to block this hole (with a cork) during periods of dearth to avoid robbing.

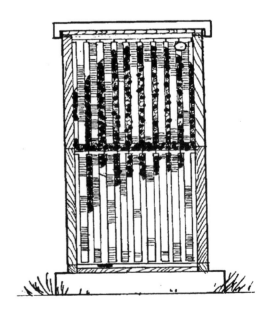

Winter Cluster

Temperatures at Which Different Bee Activities Take Place

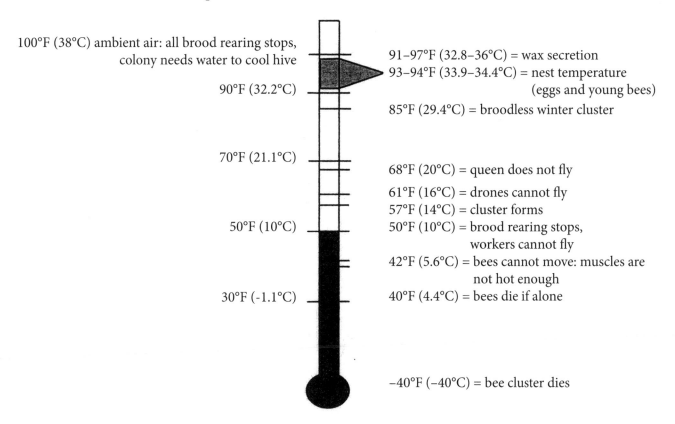

100°F (38°C) ambient air: all brood rearing stops, colony needs water to cool hive

90°F (32.2°C)

91–97°F (32.8–36°C) = wax secretion
93–94°F (33.9–34.4°C) = nest temperature
(eggs and young bees)
85°F (29.4°C) = broodless winter cluster

70°F (21.1°C)

68°F (20°C) = queen does not fly

61°F (16°C) = drones cannot fly
57°F (14°C) = cluster forms
50°F (10°C) = brood rearing stops,
workers cannot fly
42°F (5.6°C) = bees cannot move: muscles are
not hot enough
40°F (4.4°C) = bees die if alone

50°F (10°C)

30°F (-1.1°C)

−40°F (−40°C) = bee cluster dies

- Unite weak, but healthy hives. Kill all diseased colonies (or those heavily infested with mites); take your losses in the fall; note this in your hive diary.
- Place weights (rocks, bricks, or blocks) on top of hives so the covers will not blow off. Such weights are very important and could save your colonies from dying of exposure. You may also strap the tops onto the hive bodies with plastic webbing or metal straps.
- Remove all bee escapes, queen excluders, and escape boards.
- Leave ample honey (or cured sugar syrup) and pollen stores. Feed syrup early, while the days are still warm, to allow bees to cure such stores; allow about a month for the syrup to cure properly.

The Winter Cluster

In the late fall and winter, bees form a winter cluster. Honey bees do not "hibernate" but instead form a well-defined ball or cluster when the air temperatures are below 57°F (14°C). On days or weeks when the air temperatures are 43° to 46°F (6–8°C), most of the bees have joined the cluster. The winter cluster expands and contracts as the outside temperatures rise and fall. Bees remain relatively active in the cluster, eating, moving about, rearing brood, and generating heat by "shivering" (contracting their wing muscles). The by-product of these activities is metabolic water vapor, which must be allowed to escape. Ventilation during the cold months is as important as during the warmer months.

The amount of heat generated by the cluster depends, among other things, on whether brood is present. In the late fall, the colony is usually without brood, and therefore, the cluster will be producing only enough heat to keep the colony from freezing, maintaining a temperature of around 57° to 85°F (14–29°C). When the queen resumes her egg laying in midwinter, the cluster temperature in the vicinity of the eggs and brood will be maintained at around 93°F (34° C); see the illustration of the winter cluster on p. 98.

Connective clusters of bees form a bridge from the main cluster to the food stores. If these connectives are cut off, or if the winter is unusually long and very cold, the bees could starve even though there is honey elsewhere in the hive. The cluster must be able to move to the food continuously throughout the winter. In general, bee clusters will move upward throughout the winter, which explains why the bees are usually in the upper stories by spring.

Thus it is important to have stored honey above the winter cluster in the fall: when bees can move up, the winter stores are accessible. If during a late winter or early spring inspection you find that bees are not moving up into honey, tap the side of the hive to stir the bees into moving around. It might get them to move into reach of their winter food.

Bees retain their feces during the periods of confinement, a common situation during the winter months, even in southern states where rainy weather can keep bees confined. Periodically, air temperatures may reach over 57°F (14°C) during the winter, and on such days bees are able to break their confinement to take cleansing flights. If the bees are confined to the hive over long periods, the hive floor and frames can become littered with fecal material; and if excessive, this is *dysentery* and can weaken the bees further.

Winter survival depends on:

- A young, vigorous queen.
- Large population of bees (20,000–30,000 or 8–10 frames, covered on both sides with bees); see the information on estimating colony strength in the sidebar on p. 110.
- Adequate supply of honey (or cured sugar syrup) and pollen. This translates into one or two shallow or medium supers in addition to the one or two deep hive bodies, depending on your region.
- Some warm, sunny days during cold winters (to permit cleansing flights and cluster movement).
- Dry, cold winters so bees do not eat up their winter stores too fast and starve.
- Early springs with moderating temperatures.
- Disease- and mite-free conditions (medicate if these conditions do not exist).
- An upper entrance for cleansing flights.
- Top ventilation to release moist air (prop up or reverse inner cover).

- Protection from prevailing winds (use temporary windbreak or snow fencing).
- Reduced front entrance.
- Periodic inspections if warm enough to open hive.
- Maximum sunlight exposure.

The average loss of bees over winter, before mites became established, was around 10 to 20 percent. Now with mites, small hive beetles, and new pathogens in the picture, winter losses can be much higher, and some years can reach 60 to 80 percent. Even without these new problems, additional winter losses can be a result of:

- Wet, cool winters.
- Long, cold winters with few sunny, warm days, reducing or eliminating opportunities for cleansing flights.
- Unmated, injured, poorly mated, or diseased queens or drone-laying queens.
- Weak colonies in the fall, with improperly cured or poor-quality food stores.
- Queenless colony.
- Insufficient honey stores, especially if weather conditions were poor during the last honeyflow.
- Colonies placed in shaded, damp areas.

Wintering in Extreme Climates or at High Altitudes

In regions where average temperatures during the coldest months are below 20°F (–7°C), you should leave about 90 to 120 pounds (41–54 kg) of honey on each colony (or feed an amount equal of sugar syrup). This is about one deep body full of honey. If sugar syrup is to be fed to bring the food stores up to these numbers, remember it must be fed to bees while the weather is still warm so it can be properly cured. Make the syrup in a 2:1 ratio of sugar to water. You may also try cellar wintering if you live in extremely cold climates where winter lasts more than five months; see the discussion in this chapter.

The essential elements for a colony to overwinter successfully, as covered above, are similar in extreme cold conditions, with the addition of the need to provide adequate windbreaks, to wrap colonies, and to insulate hives.

Windbreaks

Apiaries should be located where they will be sheltered from prevailing winds to reduce the amount of cold drafts in winter and spring. Situate hives where barriers such as fences, evergreens, thick deciduous growth, walls, or buildings will take the brunt of the prevailing winds. When no windbreaks are present, construct temporary ones to lessen the velocity of the wind as it approaches the hives, while still permitting air drainage to take place.

A suitable windbreak would be a snow fence or slotted board fence 6 feet (1.8 m) high, set up at least on the north side of the apiary. The boards should be about 1 inch (2.5 cm) apart to slow the wind velocity and at the same time allow air to filter through. The first row of hives should be about 5 feet (1.5 m) from the windbreak.

Wrapping or Using Hive Covers

Hives can be wrapped with tar paper to protect the bees from chilling winds; also, the dark color will absorb the sun's heat. There are now several new hive covers made of materials ranging from corrugated plastic to treated paper, which are easier to install. Check what other beekeepers do in your area and investigate overwintering gear from bee supply catalogs.

There are several procedures for wrapping hives, and most of them incorporate these features:

- Top and bottom entrances.
- Top ventilation.
- Absorbent material enclosed in a super over an inner cover to draw off moisture (straw, shavings, porous pads, corrugated paper, fiberglass, foam insulating board, or other building insulation).
- Insulation under the tar paper covering in extreme weather.
- Dead air space underneath the hive.
- Use of mouse poison, such as treated grain, on the bottom board or the inner cover, or placing a mouse guard to prevent mice invasions.

Advantages

- Protects colonies from piercing winds.
- Allows colony to warm up when sun is out.

- Bees can move and recluster on honey if inside temperature is warm enough.

Disadvantages

- Time-consuming.
- Vapor barrier may form between hive and tar paper, resulting in excess moisture accumulation in the colony, which, if it freezes, will encase bees in an "icebox."
- Hive may warm up too much, and bees may begin premature cleansing flights before air temperatures are high enough to protect them from being chilled.
- Insulation may slow the bees' perception of suitable spring flight conditions, delaying foraging.

Insulation

Insulation will provide colonies with extra protection against cold winter temperatures. Before 1900, beekeepers would double wall their hives by adding another wider hive body and filling the intervening spaces with sawdust or straw (called *chaff* hives). Today, more modern insulating materials can be used, with caution. If you are experimenting, go slowly: use materials on only a few colonies at a time. Some insulating material may have harmful or toxic "gases."

Any one or a combination of these insulating materials, methods, and devices has been found to be of some aid (see the illustration on types of insulation on p. 102):

- Provide dead air space underneath the hive.
- Place *follower boards* against the inside walls. A follower is a solid piece of board or other durable insulation material, the size of a deep frame, of variable thickness, that hangs like a frame. It can be used to reduce the interior size of a deep hive body by substituting it for one or more frames.
- Insert an empty division board feeder filled with foam or other insulation (can substitute for follower board) on the outer frames of the colony.
- Insulate the top with moisture-absorbing material (such as newspaper, straw, leaves, sawdust, old blankets, or shavings, or a combination of these), placed between the outer cover and the inner cover (rim side down); make sure the oblong hole is open.
- Place 1-inch foam board on top of the inner

Types of Insulation

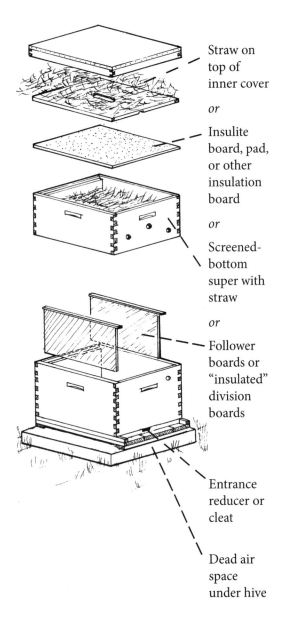

Straw on top of inner cover

or

Insulite board, pad, or other insulation board

or

Screened-bottom super with straw

or

Follower boards or "insulated" division boards

Entrance reducer or cleat

Dead air space under hive

cover; remember to cut a hole in the center to let out moist air.

● Place an empty hive body with a screened or cloth-covered bottom on top of the inner cover and filled with insulating material.

Some beekeepers paint hive bodies with insulating paint or put a super of dry, drawn comb on top of the inner cover, although these combs can collect excess moisture and be damaged. Supers of honey are also good insulators, so leaving extra honey on never hurts.

Check how other beekeepers in your area winter their colonies. If you do nothing else, protect them from piercing winter winds.

PREPARING FOR THE NEXT SEASON

The following tasks should be attended to during the winter months in preparation for the next spring season:

● Order packages to arrive before fruit trees bloom in your location.
● Clean stored supers and frames of burr comb and propolis; replace old combs.
● Build new equipment for the coming year: frames, hive bodies, tops, and bottoms.
● Paint and repair equipment; replace with new equipment if necessary.
● Sort and cut out sagging, diseased, damaged, or drone combs and replace with foundation. Old wax combs need to be replaced every 3 to 5 years.
● If you use plastic-coated foundation, you may want to scrape off the older cells and let the bees draw out fresh wax.
● Melt down old wax and prepare to sell or make candles.
● Order other equipment you will need for the coming year (gloves, extractors, bottles, new veil).
● Check the apiary periodically for damage from downed trees or limbs, wind, or vandals or other predators.
● Attend bee meetings, read or write for your local bee club's newsletters, and review your hive diary notes.
● Browse new catalogs, bee journals, and web pages; buy a new bee book and read it.

LATE-WINTER MAINTENANCE

There are important tasks that need to be done in the late winter and early spring, weather permitting, before the fruit trees begin to bloom. Some of this work may have to be repeated in late spring and thus may be referred to again in the next section on spring maintenance. But this will give you a heads-up of what you will be doing.

The tasks include:

● Checking for dead colonies; removing or closing

up any dead hives to prevent them from being robbed.

- Determining if the dead colonies succumbed to disease (nosema, foulbrood), mites, or starvation (see next point). New research suggests that mite-killed colonies may harbor virus and may not be suitable to rehouse bees. In addition, old comb and pollen may also contain virus. **Never feed honey from diseased colonies to any other colony!**

- Determining if the bees died with their heads inside the cell, which means they probably starved.

- If there is a sour or foul smell in a colony, checking for scales or dead brood tongues that are sticking perpendicular to the cell wall (this could be foulbrood, in which case the hive equipment and bees should be isolated; see "American Foulbrood Disease" in Chapter 13). Before disinfecting any equipment, check with the local bee inspector.

- Checking to see if a large amount of fecal matter (yellow/brown spots) is on the frames, covers, or bottoms, which could indicate a *Nosema* sp. infection or dysentery. Clean off fecal material and fumigate (or dispose of the dirty equipment). If you feed nosema-contaminated honey frames to hungry bees, you are not helping them unless you also medicate for it (see "Nosema Disease" in Chapter 14). If dysentery, the honey stores may have fermented or were not properly cured. If *Nosema*, mark these colonies and treat with fumagillin.

- Determining if a colony is queenless or weak but healthy (i.e., no mites or diseases), in which case it can be united with another colony (see "Spring Dwindling" in this chapter and "Uniting Weak Colonies" in Chapter 11). Again, unless you know your equipment is disease free (you've had it radiated or otherwise treated), don't mix equipment.

- To determine if bees had a disease, collect adult bees and freeze or store them in alcohol to have them checked for disease or mites. Tracheal mite populations are highest this time of year; see Chapter 14.

- If the weather is very cold, determining the amount of stores by lifting or tilting the hive; if the hive seems light, feed the colony. Emergency food during cold weather includes fondant candy, "mock" candy, frames of honey, or dry sugar sprinkled on top of the frames or inner cover; see Chapter 7.

- If warm enough, beginning to feed sugar syrup and pollen to stimulate brood rearing. Feed pollen, supplements, or substitutes in moist patties (see Chapter 7). Provide frames of sealed honey if the weather is too cold to feed syrup.

- If the air temperature gets above 75°F (24°C), checking the colony for the queen's condition by examining the brood pattern: a compact pattern of worker brood indicates that a healthy queen is present. This examination should be brief; otherwise the brood can become chilled.

- If a healthy colony is near starvation, provide clean frames of honey on either side of the cluster as well as above it. Feed it depending on the present and future weeks' weather conditions.

- Preparing for the arrival of package bees or nucs by making certain you have enough equipment.

- Making new equipment for the coming year.

If you are preparing hives to move colonies for pollination services, stronger colonies are needed. First, select strains of bees that will overwinter with good bee populations. Feed these colonies with a good amount of syrup and pollen to keep them strong over winter. In the early spring, feed a pollen patty and some light sugar syrup (1 part sugar to 1 part warm water) to stimulate brood rearing. Later, equalize these colonies so they are all about the same strength. Follow the overwintering procedures discussed above, but mark the hives destined to be moved for pollination for special attention. Check to make sure these hives have new equipment or that all hive parts are in good shape, with no cracks or weak spots that could break while being trucked.

SPRING MAINTENANCE

The following apiary tasks should be started when the dandelion and fruit trees bloom in your region and when all danger of frost is past. Do the following:

- Clean the apiary of any winter debris.
- Unwrap or take down winter protection.
- Remove temporary winter windbreaks.
- Remove any insulating materials.
- Remove entrance reducers from strong colonies. Keep them in weaker colonies, but switch the entrance hole to the larger opening. Make note of these weaker colonies.

- Mark colonies to be requeened, to replace poor performers; see Chapter 10.
- Clean off bottom boards. Note which colonies have few dead bees and debris on the bottom board, and overwintered in healthy conditions; breed queens from these colonies.
- Feed hives that require additional food.

When air temperatures reach 75° to 80°F (24–27°C), inspect the hives for diseases, mites, brood pattern, and amount of remaining stores. You may need to:

- Reverse brood chambers if the cluster has broken and bees are flying. If the lower hive body is empty, most of the bees, brood, and queen will be in the upper bodies. Put this body with bees, brood, and the queen on the bottom, and at least one empty super and some frames of honey on top. Since the queen and bees move upward, this reversal relieves congestion and provides room for expansion.
- Replace dark, old, sagging, and uneven combs, or broken frames. It is a good time to start recycling old, dark comb for newer ones or frames of foundation. Provide foundation frames when the spring bloom of fruit trees and dandelions is underway. Replace all your frames every two to three years to reduce the buildup of toxins and disease spores that are in the wax.
- Provide additional space (hive bodies or supers) as needed.
- Make increases in number of colonies only when the weather is warm enough so the brood will not become chilled; see "Relieving Congestion" in Chapter 11.
- Medicate or treat the colonies for mites, brood diseases, and/or nosema; do not give medication once the bees have begun to store nectar, or the nectar and its end product, honey, will be contaminated. Read labels carefully for proper dosage and timing.
- Look for signs of swarming and, if necessary, initiate swarm prevention/control techniques; see "Swarming" in Chapter 11.
- Investigate clean water sources (to determine that bees are not getting into contaminated water) or provide fresh water.
- Register hives and apiaries with the state agriculture department.

Spring Dwindling

In some colonies, older bees may begin to die faster than young bees emerge, such that the number of bees is reduced to a point at which the process cannot reverse itself and the colony dwindles to nothing. This is called *spring dwindling* because it usually happens or is noticeable in the spring. This condition may have several causes, such as pathogens (spiroplasmas, protozoans, bacteria, or viruses), mite predation, poor queen genetics or race of bees, condition of colony going into the winter, poor winter stores, or current weather conditions.

While weather plays an important role in overwintering success of colonies, spring dwindling may be prevented or checked by:

- Making sure you have strong populations of young bees and a new, young queen in the fall.
- Wintering only strong colonies with ample stores of honey and pollen; combining, requeening, or destroying weaker hives in the fall, if necessary.
- Providing high-quality winter stores such as low-moisture honey and extra pollen supplements.
- In the fall, having a young queen of a race that is known to overwinter well in your area.
- Using windbreaks to protect your hives from winter and spring drafts and dampness.
- Having ample colony strength in spring; if there are only three or four frames of bees, unite it with a stronger colony or destroy the weakened hive.
- Take steps to protect bees from pesticides during the previous summer. Toxins are found in older comb, so replace dark, old comb with new foundation in time for the bees to draw it out and store honey.
- Preventing drifting, to avoid making some colonies weaker in the fall.
- Medicating the bees against nosema disease in the fall or spring, if your colonies tested positive for nosema spores.

OTHER WINTERING OPTIONS

This section on wintering honey bees was written by R.W. Currie of the Department of Entomology at the University of Manitoba at Winnipeg.

Indoor Wintering of Honey Bees

History of Indoor Wintering

In geographic regions with cold and long winters, many beekeepers choose to winter their colonies indoors. Indoor wintering can be done on a large or small scale and probably has its origins in the late 1800s when beekeepers would place small numbers of colonies in an unheated "root cellar" in order to help them survive the winter. Wintering large numbers of colonies in specially constructed underground chambers was also attempted, but both small- and large-scale methods fell out of favor because of difficulties controlling temperature and air quality and the labor associated with carrying colonies into and out of the wintering chamber. However, some advantages of indoor wintering, including the ability to monitor and feed colonies during the cold period, and innovations that eliminated some of the disadvantages, have made this an economically viable alternative in northern climates.

In the early 1960s Agriculture Canada in Brandon, Manitoba, initiated studies into the use of high-capacity air conditioners in combination with temperature- and humidity-control equipment to winter bees, but beekeepers in Nebraska probably first adopted this idea later in that decade, when up to 700 hives were wintered in an insulated wood frame building. Success by those beekeepers led to the widespread adoption of the technology by beekeepers much farther north, in several regions of Canada.

Modern building techniques, insulation methods, and ventilation controls, as well as a wide range of equipment that has been developed for moving and manipulating hives, have made critical contributions to the success of this method. Today indoor wintering is practiced in northern regions by beekeepers ranging from large commercial producers who winter several thousand colonies in specialized buildings to hobby-level beekeepers who winter a few colonies in a garage or shed.

Building Design Requirements

Whether a small- or large-scale operation, the requirements for an effective wintering room are similar. Essentially the space for hives should be well insulated, temperature controlled, completely dark, well ventilated, and with adequate volume to contain the hive and provide good air recirculation and air exchange. To create a wintering chamber, some producers modify an existing space (e.g., a hot room or garage), while others construct buildings specifically for the purpose of wintering colonies.

The building should be constructed with a polyethylene vapor barrier (on the inside of the wall), with walls insulated to a minimum value of R20 (RSI 3.5) and ceilings insulated to a minimum value of R30 (RSI 5.0). Where cement foundations are used, they should be insulated to at least 2 feet (0.6 m) below grade with polystyrene (2 inches [5 cm] thick). All doors should be insulated and well sealed (or covered with black plastic) to prevent any light from entering the chamber. Windows should not be present, or they should be well covered to exclude all light.

The shape and height of the building are not important, but it should be large enough to accommodate the appropriate number of hives and have dimensions and access points (e.g., loading dock) compatible with whatever type of equipment is going to be used to move the colonies (e.g., a hand cart or forklift). Since hives are normally stacked on top of each other, the floor space required is highly dependent on how colonies will be arranged within the building. Typical recommendations for floor space vary from 2.7 to 3.2 square feet (0.25–0.3 m²) per hive. The room volume should allow for space of 24 to 30 cubic feet (0.7–0.9 m³) per hive. At least enough space should be left between rows of hives for the beekeeper to move and periodically sweep dead bees that fall from the hives. Higher colony densities than those recommended above are possible, but in some cases this can increase problems with air temperature regulation and air circulation.

External (white) light is highly disruptive to the bees and must be prevented. Any ventilation inlets or outlets should be shrouded with a "light trap" consisting of a black box that surrounds the fan housing and prevents the direct entry of light. The room should also be fitted with light sockets that can accommodate red light bulbs (lights are normally left off) to facilitate work within the room when necessary. Alternatively, many beekeepers carry a red-light flashlight with them when working in the wintering room. Red lights are much less disruptive to the bees.

Ventilation and Temperature Regulation Requirements

Although bees from colonies held within the wintering building do not forage or leave the hive to defecate during the entire storage period (typically late October to mid-April), they do consume honey, produce heat, and respire. Thus, ventilation and temperature controls along with good air exchange and air circulation are required to maintain stable temperatures and good air quality.

The wintering building should be maintained at a constant temperature of about 38° to 45°F (4–7°C). In most commercial facilities in northern climates, this is usually done through the use of ventilation fans that bring cold air from outside into the building. Normally a series of exhaust fans sequentially activate as the temperature increases, creating a negative pressure in the building to draw in the air from an outlet going to the outside. However, some buildings are based on a positive-pressure system (fans force cold external air into the building and it exits through outlets), and some utilize external air-conditioning units to help maintain the desired temperature. Although colonies produce a considerable amount of heat (each colony can produce the equivalent of 8–28 watts), it may also be necessary to provide supplemental heat on exceptionally cold days to make up for heat lost from building ventilation and by conduction. If recommended colony stocking rates are used, supplementary electric heat should be sized to provide about 10 watts per hive.

In some facilities, producers reduce heat loss in midwinter and reduce heat gain in spring by drawing fresh air from an attic or honey super storage area, rather than from outside the building, to moderate the effects of the extreme temperatures outdoors. This is needed because outside temperatures in winter can reach −40°F (−40°C), but in late winter or early spring outside temperatures often rise to the point where building temperatures cannot be maintained at the optimal rates. Although bees in the building can handle temperatures of over 100°F (38°C) without any problem, higher temperatures can result in increased loss of bees. Short periods of high temperatures are usually dealt with effectively by maintaining high rates of airflow. However, if temperatures remain too high for too long or an air-conditioning system is not available, beekeepers often choose to remove some or all of their hives from the building.

As bees within colonies respire during winter, they also produce excess moisture and carbon dioxide. To prevent the buildup of this moisture and carbon dioxide in room air, buildings must be continually ventilated at a rate of at least 0.53 cubic foot per minute (0.25 liter per second) per hive through the use of the ventilation fans (described above). As temperatures increase, ventilation rates should correspondingly increase through the activation of more fans, to increase levels to a maximum of 10.6 cubic feet per minute per hive (5.0 liters per second per hive). Similar ventilation approaches are used for hobby wintering facilities, except commercial livestock ventilation fans are usually replaced with something much smaller, such as a bathroom ventilation fan on a timer. Maintaining an appropriate exhaust rate for building air is critical as carbon dioxide levels can quickly increase to a point where they will be fatal to colonies. Humidity levels are typically not regulated but will usually fall within the range of 50 to 75 percent relative humidity if the ventilation system is working properly.

Appropriate recirculation of air is also important as stale air can pool in some areas of a building. Plastic air recirculation tubes, such as those found in a greenhouse or livestock barn, are often utilized for this purpose and are sized at a rate of 10.6 cubic feet per minute per hive (5.0 liters per second per hive). Other commercial producers favor slowly rotating ceiling fans, and hobbyists often use an oscillating fan such as what you would use in a home during a hot summer day. The important point is to provide good recirculation of air throughout the entire room.

Colony Management

Colony preparation for indoor wintering is similar to that for outdoor wintering. Colonies should have a young productive queen, be free from diseases and parasites, and have adequate food stores in preparation for the wintering period. Sucrose syrup or high-fructose corn syrup is typically the preferred type of food. Beekeepers should provide about 40 to 50 pounds (18–23 kg) of syrup for a single-hive body and 60 to 70 pounds (27–32 kg) for two bodies. Syrup is preferable to honey as the latter can granulate during winter, resulting in increased stress to the bees. Colonies can be wintered indoors in a single-brood chamber or double-brood chamber with equal success, and small five-frame nucleus colonies are also often wintered using this method.

Colonies are typically moved into the wintering buildings in late October to mid-November, where the hives are stacked adjacent to each other in rows (usually about five colonies high) with adjacent rows separated by at least 3 feet (1 m). Hive entrances are usually removed. Many producers do not feed colonies during winter storage; however, if colonies are light going into the building, this is an option. Feeding of syrup (or water) can be done through entrance feeders (e.g., Boardman feeder). Pollen or pollen supplements are not fed during winter as they are believed to put additional stress on the bees (in indoor wintering, the bees do not have the opportunity to defecate until they are removed from the building).

Colonies are typically removed from the building in late March to mid-April, when the first pollen flow occurs or when temperatures are too hot to maintain the bees within the building. In order to minimize losses, bees should be removed from the building in early morning or late evening and placed in a location with no snow. Doing otherwise causes excessive disorientation of the bees and high rates of loss.

Wintering in Warm Climates

In contrast to wintering bees in the middle and higher latitudes, wintering bees in southern areas does not require the same attention or effort. Here, the daylight is longer and the nights are shorter than in northern regions. In the south, winter temperatures fluctuate between 45° and 68°F (7° and 20°C). Although temperatures can fall below freezing for short periods, it is not necessary to provide the same winter protection, such as wrapping and insulation, for your colonies.

In addition, plants that yield nectar and pollen are available almost year-round, so brood rearing is interrupted for only a brief time. It is in these climes where the package bee industry is located and where many commercial beekeepers overwinter and rear new queens for their colonies. It is also now an area where many of the pesticide-resistant mites and other resistant pathogens are passed around from apiary to apiary.

Most of the published information concerning wintering honey bees is based on apiculture in temperate climates. As a consequence, these sources frequently concentrate on wintering because that is usually when there is the most colony loss. It is often recommended that you overwinter colonies with two full-deep chambers as a standard configuration: a brood chamber below (with a large population of bees) and a food chamber above, filled with honey and pollen stores.

In more southern regions, different conditions prevail. The cold weather is not as severe, and therefore, it is unnecessary to pack colonies for protection. Brood rearing continues longer into the year; there may, in fact, be no broodless period in some southern states. The two full-depth brood chambers are often reduced to one deep brood chamber with a medium-depth honey super as the standard hive configuration.

Dr. Malcolm Sanford, retired Extension Service entomologist from Florida State University, offers some observations on wintering in southern regions. The length of daylight, or photoperiod, in warmer zones increases as you approach the equator, and warmer zones have a greater diversity of plants, which produce relatively less nectar per individual species. These two factors appear to be the most important reasons why honey production is not as pronounced or intense as in more temperate regions. In the true tropics, moisture availability, with a more pronounced summer wet and winter dry season, takes precedence over temperature.

The relatively long season for brood rearing, the warmer winters, and the more sparse nectar production are the cornerstones of southern beekeeping. Thus colony management must be spread out across the year more evenly, and timing of specific tasks is radically different from that in northern areas. As examples, two management techniques extremely important in successful wintering stand out: requeening and controlling varroa mite populations.

Requeening in southern regions is done in the fall and with queen cells rather than mated queens, as is common in the northern areas. This may change, however, with incursions of Africanized honey bees infiltrating southern states. Mite treatments are scheduled differently during the longer growing season in southern regions, since varroa and tracheal mites can be present in the colony year-round. Because of the later honeyflows, a summer/fall varroa treatment must be timed carefully to keep bees free of mites but not contaminate honey. In addition, another varroa treatment may be necessary in the early spring before the colonies build up.

Here are some points to remember when wintering in southern areas:

- Winter in one brood box, with a minimum of a medium-depth (6 ⅝ in.) super containing honey.
- Install entrance reducers.
- Keep bee populations between 15,000 and 20,000 (5–8 deep frames covered with bees).
- Treat for diseases and mites.
- Provide upper entrance.
- Protect colonies from cold winter winds.
- Periodically check to see how food stores are holding up; replenish as needed by supplemental feeding.
- Check water sources and provide if weather is too dry.

Timing is as critical to beekeeping as it is to most endeavors. To time your beekeeping activities properly, it is necessary to keep accurate records of past seasons, to observe carefully the flowering periods, and to recognize the needs of your colonies. Even then, there are capricious fluctuations from year to year or season to season that will make beekeeping a continual challenge. Be prepared—no two seasons are alike.

 **Notes**

CHAPTER 9

Summer/Fall Management

SUMMER MANAGEMENT

Each colony should be examined periodically and especially before the advent of a major honeyflow in your area. Check the colony strength to determine whether it is populous enough; a colony should reach a population of over 40,000 by the time of major honeyflow. Weak colonies should be united or requeened; inspect these more often in case you have to feed or medicate them. This is where having a colony on a hive scale is invaluable. Keeping one hive on a scale per apiary can be helpful in deciding some critical hive manipulations (such as when to put on honey supers).

Here are some methods for estimating colony size:

- Try to count the number of bees coming and going at the entrance; if they can be easily counted, the colony is weak; a number between 30 and 90 bees per minute indicates a strong colony.
- One deep frame covered with adult bees equals about 1 pound (0.5 kg) of bees (3500 bees).

For additional methods, see the sidebar "Estimating Colony Strength" on p. 110. Other tasks during this time should include the following:

- Requeen as necessary (i.e., when the colony is weak or otherwise not up to standard).
- Monitor colony strength, uniting weaker ones with healthy, stronger colonies.
- Check for diseases and mites.
- Rear queens for fall requeening.
- Check the colony's food stores.
- Reverse the brood chambers again if necessary.

If the colony is unusually strong and ready to swarm, practice swarm prevention procedures, or split strong colonies, or use the colony to rear queens for fall requeening.

- Add honey supers as needed; when a super is two-thirds full (six or seven full frames), add another super; see "Rules for Supering" in this chapter.
- Add frames of foundation in supers only if there is a good honeyflow; otherwise, bees will chew holes in the wax, or the foundation will warp and bend.
- Keep burr comb and propolis scraped off the frames and hive walls; collect and process.

Cooling the Hives

When the temperatures are above 90°F (32°C) for an extended period of time, you may need to help the bees keep from overheating. Here's how:

- If the hives are in full sun, you can provide temporary shade in an emergency with fencing, boards, or shrubs, or break some branches and place them over the hive cover.
- Stagger supers slightly to increase the airflow throughout the hive. Some beekeepers raise the inner cover or the front of the bottom super with a small block; others bore a 3/4-inch (18.8 mm) auger hole in an upper corner of the top deep super. Do **not** bore a hole above or in the handholds! Trying to lift these supers off is a stinging adventure you want to avoid. Plug these holes later with a cork to keep robbers from entering.
- Make sure fresh water is available; this can be done with a metal or plastic cattle water tank; to keep

Colony Estimations

Estimating Colony Strength (all European)

- A shallow frame fully covered with bees will hold about 0.25 pound (0.1134 kg) of bees or about 875 individuals.
- A deep frame fully covered holds 0.5 pound (0.2268 kg) of bees or about 1750 individuals.
- There are about 3500 bees per pound (0.4536 kg).
- A sheet of deep wax or plastic foundation 8.5 × 16.75 inches (42.5 × 21.3 cm) equals 3350 cells per side. That is 83.75 cells linearly and 40 cells vertically; the number of cells on both sides is 6700.
- One square inch of comb has 25 cells (5 worker cells per linear inch, 4 drone cells per linear inch).
- Count the number of bees returning to the hive for 1 minute. To estimate the number of bees in the deep frames of a colony, use the following equation: number of bees returning per minute × 30 min × 0.0005. The factor 0.0005 assumes that one deep frame, both sides, contains 2000 bees, or 1 divided by 2000 = 0.0005; the time of 30 minutes assumes the amount of time needed for any one bee to make a return trip.

Estimating Weight of the Colony

- Multiply by the number of frames in the colony.
- A deep frame fully filled with honey weighs about 10 pounds (4.536 kg).
- A full medium frame (U.S.) weighs 7 pounds (3.2 kg).
- A full shallow frame (U.S.) weighs 5 pounds (1.8 kg).

bees from drowning, fill the tank with sand or pebbles. You can also use a dripping hose or install a Boardman feeder filled with water in each hive.
- Use screened bottom boards to give more ventilation.
- Paint metal covers or wooden tops with white paint to reflect the maximum amount of sun.

Signs of Honeyflow

Prepare for a honeyflow in the winter or spring months; do no wait until the last moment to make extra frames or supers or you could lose a honey crop! Start by repairing frames or by making frames with foundation for the honey supers. Keep fresh wax foundation sheets in plastic bags to protect them against wax moth infestation and to keep them from drying, because dry foundation becomes brittle and breaks easily.

Honeyflow is the time of year when bees are able to collect ample supplies of nectar. A honeyflow may last a few days or a few weeks. Bees are natural hoarders, and the presence of empty combs stimulates this hoarding behavior, so if you provide the bees with more storage space than they need, you will be inducing more bees to collect nectar. This behavior of storing more food than is immediately needed is a sound evolutionary trait, for the amount of food any one colony will need or produce is unpredictable. By understanding this aspect of bee behavior, you increase the opportunity to obtain a surplus (more than the colony needs) of honey. This surplus is stored by the bees as honey in supers (located above the brood chamber). Good beekeepers harvest only this surplus and leave ample food reserves.

A honeyflow is indicated by one or a combination of the following signs:

- Fresh, white wax evident on edges of drawn comb and on top bars.
- Dramatic weight gains (as indicated by the hive scale) over several days or weeks.
- Wax foundation drawn out quickly.
- Large amounts of nectar ripening in cells.
- Bees fanning at hive entrance.
- Much foraging activity.
- Odor of nectar (ripening honey) often pervading the apiary.
- During a honeyflow, bees are extremely docile and easy to work.

During the main honeyflow, you should avoid opening the hive to look at the brood area unless you are doing some major management operations (requeening, treating for diseases, etc.), nor should you place pollen traps on hives. Check the colonies before a major honeyflow, because if you disrupt the bees by tearing down the hive during a flow, it will disrupt their gathering activities and may even reduce the amount of honey being brought in for several days. You will also kill many more bees (the population is much higher at this time) and even the queen if you work bees intensively during this time. So, restrict your curiosity to smaller colonies, or set up an observation hive in your house.

If you must look at your colony, try to minimize your time inside the hive to reduce the disruption.

Apiary Tasks in Summer

In general, the tasks at the apiary just before and during a major honeyflow should include the following:

- Super hives as needed, placing honey supers above the broodnest (see next section).
- Provide adequate ventilation so bees can cool the hive and cure the honey adequately.
- Reverse honey supers occasionally so the fuller one is always on top.
- Keep supers on the hive until honey is capped, as unripened honey will ferment when extracted.
- Avoid adding too many supers at once, as bees may partially fill all of them instead of filling one completely (called the *chimney effect*). Put on only enough frames for the bees to fill in a few weeks; this is a judgment call, one gained by experience.
- Never super a weak colony.
- **Never medicate** colonies that you plan to extract honey from. It is illegal, and you can contaminate the honey. Honey collected by diseased colonies that required medication must not be used for human consumption.
- **Do not feed** syrup during this time because it will probably end up in the honey supers.
- Requeen weak colonies or those heavily infested with parasitic mites so as to break the brood cycle and disrupt the mites' life cycle (see Chapter 14).
- This is a perfect time to raise queens, so take advantage of a natural flow to raise high-quality queens.

Super Sizes

When a honeyflow is in progress, the bees will deposit nectar in supers placed above the brood chamber. These supers may vary in size from full-depth supers

Super Sizes

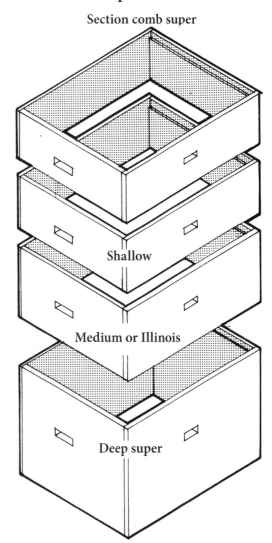

Section comb super

Shallow

Medium or Illinois

Deep super

(deep hive bodies) to the shallow or section comb supers (see the illustration of super sizes on this page). There is no hard and fast rule about which super size to use; personal preference, one's physical strength, and the quantity of the expected surplus should be your guide. Some beekeepers keep all hive furniture the same size so they don't have to mix and match different sizes of supers and frames or foundation. There are pros and cons to mixing or to keeping the super sizes the same, so chat with some beekeepers about which super sizes work best in your region.

Rules for Supering

The most important rule for supering is to keep the queen out of the honey supers. The presence of brood interspersed in honey frames will make the honey

Beehive Components

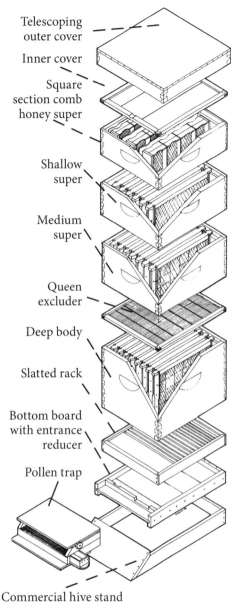

Telescoping outer cover

Inner cover

Square section comb honey super

Shallow super

Medium super

Queen excluder

Deep body

Slatted rack

Bottom board with entrance reducer

Pollen trap

Commercial hive stand

(when extracted) unsanitary because of the presence of larvae and pupae. The extracting tank will have honey and brood mixed together and brood would be wasted. In addition, it is difficult to cut away the cappings of brood combs that have been darkened with propolis. Cull any dark and old frames and save them for other uses, such as bait for swarm traps; otherwise, melt them down for wax.

Use one of the following methods to restrict the queen from laying in the honey supers:

● Place a queen excluder above the broodnest (see the illustration of beehive components on this page).

● Place a super of light-colored comb or foundation above the broodnest; as long as the queen is not crowded for space, she will prefer to lay her eggs in the darker comb.

● Keep a hive body filled with honey directly above the broodnest. Such a honey barrier often keeps the queen from moving upward.

● Place a section comb honey super above the broodnest; the queen generally will not lay in the section boxes or other types of comb honey sections.

Some general guidelines for supering bees during a honeyflow are listed below:

● Stagger the honey supers to hasten the ripening of honey, leaving a 1/4-inch gap, especially in hot, humid areas (but do this only on very strong colonies to avoid robbing by robber bees); you can also use a screened bottom board instead.

● Use only eight or nine frames in the supers destined to be extracted, so the bees will draw cells out wider than normal; this makes it easier to cut the cappings off when extracting honey.

● Bait an empty honey super with a frame or two of capped or uncapped honey if the bees seem reluctant to move up; this will attract bees to move into the super.

● Some beekeepers use drone comb foundation in their honey supers; the cells are larger and honey seems to extract readily from them. Drone foundation can be obtained from bee supply houses.

● Rearrange frames in supers periodically, so the full ones are at the ends and the empty ones are in the middle (bees fill the middle ones first). If you are low on supers, this exchange will give you some time to ready more supers.

Methods of Supering

There are two basic ways to super for honey; these are reverse supering and top supering (see the diagram on supering on p. 113).

Reverse or Bottom Supering. This method generally needs a queen excluder to keep the queen from laying in the honey supers (see the diagram on supering on p. 113 and the one on the reverse supering sequence on p. 114) and can also be used for comb honey production. A super with foundation or dry combs (S2) is always placed below a super at least half full of honey (S1). Because the emptier supers are on top of the broodnest, the queen excluder is necessary. As the supers are filled, they can be removed, or stored above the emptier ones. With the empty super closest to the broodnest, and the full super above the empty one, the bees will be drawn up into the empty super above them.

Top Supering. This method does not require a queen excluder because the queen rarely will go into a super full of honey. Put honey supers with dry comb or foundation (S2) **above** honey supers that are at least half filled with honey (S1); see the diagram on top supering on this page. Keep adding supers as the bees fill the ones below, until you remove the honey.

There are many methods of supering using these two themes; talking with local beekeepers may be helpful in determining how to super in your particular area. Success of either method often depends on the type of honeyflow in your location—short in duration and intense or long but less intense. Short and intense flows will enable you to make comb honey; less intense flows may be for extracted honey. See also the diagram on the top supering sequence on p. 114.

Comb Honey

Harvested honey can be left in the comb or extracted from it. Honey in the comb is referred to by various names. Normally found at fairs and honey shows, *bulk comb honey* is an entire frame of capped honey

packaged without cutting. If the honey-filled comb is cut into chunks and packaged, it is referred to as *cut comb honey*. Cut combs placed in a bottle and then filled with extracted honey are called *chunk comb honey*. Comb honey contained in small wooden frames or plastic rings (Ross Rounds) or plastic boxes that is **not** cut out of the frames is referred to as *section comb honey*; see Chapter 12.

Thin, unwired foundation is used to produce bulk, cut, chunk, or section comb honey. As soon as the combs are sealed, they should be removed from the hive to prevent the white cappings from becoming darkened with propolis, soiled by travel stains, or damaged by wax moth, small hive beetle, or the bee louse, which lays its eggs in honey cappings. (This last insect is rarely seen these days.)

The supers containing frames for comb honey production should be placed on only the strongest colonies, either those consisting of two brood chambers or colonies reduced to one brood chamber (as described for the production of section comb honey in the diagram on p. 115). Place an excluder above the broodnest and super the hive using the same rotation method as that illustrated for reverse supering. It is a good idea not to mix the comb honey supers and the extracted honey supers in any one hive. Some colonies will fill sections quickly and with very white cappings—such colonies should be reserved for section comb honey only. The color of wax cappings (snow white or "wet") is a genetic trait.

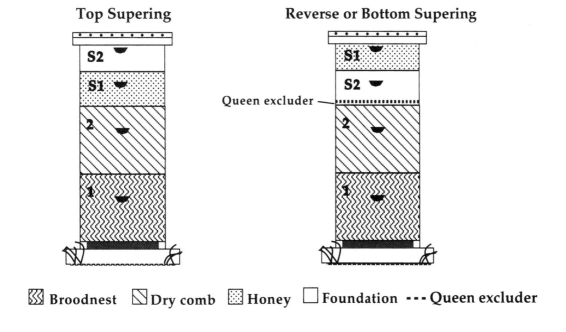

Top Supering   Reverse or Bottom Supering

Queen excluder

🔲 Broodnest   ◻ Dry comb   ⊞ Honey   ☐ Foundation   --- Queen excluder

Reverse Supering Sequence for Extracted Honey

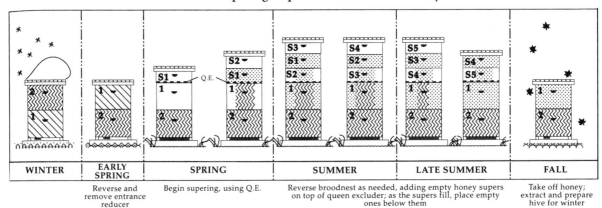

WINTER	EARLY SPRING	SPRING	SUMMER	LATE SUMMER	FALL
	Reverse and remove entrance reducer	Begin supering, using Q.E.	Reverse broodnest as needed, adding empty honey supers on top of queen excluder; as the supers fill, place empty ones below them		Take off honey; extract and prepare hive for winter

⊠ Broodnest   ◻ Dry comb   ▨ Honey   ☐ Foundation   - - - Queen excluder (Q.E.)

Top Supering Sequence for Extracted Honey

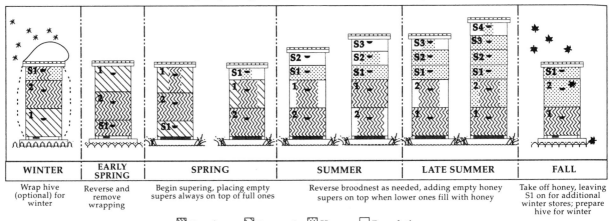

WINTER	EARLY SPRING	SPRING	SUMMER	LATE SUMMER	FALL
Wrap hive (optional) for winter	Reverse and remove wrapping	Begin supering, placing empty supers always on top of full ones	Reverse broodnest as needed, adding empty honey supers on top when lower ones fill with honey		Take off honey, leaving S1 on for additional winter stores; prepare hive for winter

⊠ Broodnest   ◻ Dry comb   ▨ Honey   ☐ Foundation

Supering for Section Comb Honey

Comb honey, especially section comb honey, is difficult to produce because success depends on a heavy honeyflow, exceptionally strong colonies, and time-consuming hive manipulations at the correct intervals. The Miller method of supering is one that is used for section comb honey; this is described below and is illustrated as method A in the diagram on p. 115.

A colony used for section comb honey production is generally wintered in two deep hive bodies (1 and 2). In the spring, this colony must be built up to full strength prior to the major honeyflow, and the brood chambers should be reversed to provide ample room for the queen to lay. This may need to be done several times to maintain enough empty comb for the queen to fill with eggs.

As soon as the honeyflow begins, reduce the strong two-story colony to one deep (2). Set up this colony so it contains two empty brood frames (in the middle) and as many frames of capped brood as possible on either side, with accompanying queen and worker bees.

The following is the procedure for method A, without using queen excluders:

Step 1. Reduce colony to one deep (2); frames of honey and any remaining brood frames should be given to other, weaker colonies.

Step 2. Over the reduced hive (2), place the first section super (ss1), with thin foundation in the section boxes or rounds (or whatever system you use).

Step 3. When ss1 is half full of honey, place a second section super (ss2) below it.

Comb Honey Using Section Supers (ss) Supering Sequence

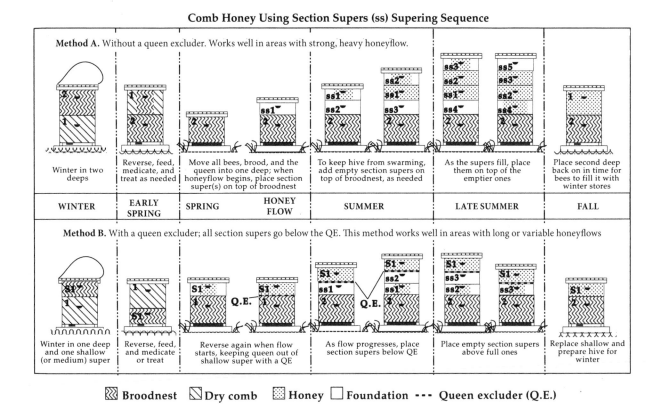

Method A. Without a queen excluder. Works well in areas with strong, heavy honeyflow.

WINTER	EARLY SPRING	SPRING	HONEY FLOW	SUMMER	LATE SUMMER	FALL
Winter in two deeps	Reverse, feed, medicate, and treat as needed	Move all bees, brood, and the queen into one deep; when honeyflow begins, place section super(s) on top of broodnest		To keep hive from swarming, add empty section supers on top of broodnest, as needed	As the supers fill, place them on top of the emptier ones	Place second deep back on in time for bees to fill it with winter stores

Method B. With a queen excluder; all section supers go below the QE. This method works well in areas with long or variable honeyflows

Winter in one deep and one shallow (or medium) super | Reverse, feed, and medicate or treat | Reverse again when flow starts, keeping queen out of shallow super with a QE | As flow progresses, place section supers below QE | Place empty section supers above full ones | Replace shallow and prepare hive for winter

⬚ **Broodnest** ◹ **Dry comb** ⬚ **Honey** ☐ **Foundation** - - - **Queen excluder (Q.E.)**

Step 4. When ss1 is almost filled, place ss2 below it (so the full super is above the empty one).

Step 5. If the honeyflow is strong, add a third (or subsequent supers, ss3) above the broodnest until ss2 is half filled, then reverse ss1 and ss2 so the full supers are above the emptier ones. Before placing another empty super on top, make sure the section supers are full from end to end, or the bees may funnel up the center, ignoring the end frames. You can correct this chimney effect by removing the full frames, or rearranging the super so the full frames are on the ends and the emptier ones in the center. If more supers are needed (ss4, ss5) place them over the broodnest.

Step 6. Remove the completely filled section supers either as they fill up or all at once. Use bee escapes or bee blowers to clear the bees out of the supers. **Never** use a fume board to remove the bees, because the honey might be adversely flavored.

Method B is slightly different; it uses a queen excluder and a single deep brood chamber plus a shallow super. This method uses the idea to always have a full honey super above the section boxes, to encourage the bees to move upward.

Comb honey should be marketed as soon as possible to reduce the danger of its granulating or being damaged by wax moth, the bee louse, or small hive beetle (see Chapter 14 for more information about these pests). Storing comb honey in the freezer will help eliminate these problems (see Chapter 12). After the honeyflow is over and the section comb honey production ceases, take off all section supers and unite the reduced colony with another hive or otherwise allow it to build up enough stores to overwinter in two deep hive bodies.

Harvesting the Honey

In some regions, two crops of surplus honey can be expected, one in the summer and another in the fall. Some beekeepers harvest the summer and fall crops separately; others harvest both at the end of the fall honeyflow. Average yields of honey depend on the amount of open land filled with honey plants within flying distance from the apiary (2-mile [3.2 km] radius). Yields vary from 25 pounds of surplus per colony up to over 100 pounds (9.3–37.3 kg). For hives located in temperate climates, 90 pounds (33.6 kg) or more of honey should be left as overwintering stores for each colony (see "Wintering" in Chapter 8).

For late-summer honeyflows, consider taking off the honey supers as soon as possible, unless they are to be kept on as winter stores. If varroa mite populations are high, it may be necessary to treat the bees in the fall with approved acaricides and allow enough time for the treatments to work before winter. To do this, you **must** take off the honey supers. For more information, see "Varroa Mite [Varroosis]" in Chapter 14.

Removing Bees from Honey Supers

Honey supers are often free of bees when the temperatures get very cold (in the early fall), because the bees leave the honey supers to join the warm cluster below. But remember, once you commit yourself to taking off honey, be prepared to extract or otherwise process your honey in a few days. You cannot store supers of capped honey for very long, unless you place them in a cooler or freeze them, because of the danger that wax moth, small hive beetles, or other pests will damage or destroy them. Five methods, listed below, describe ways of removing bees from honey supers.

Shaking or Brushing. Remove a frame with sealed honey from the super and shake the bees off in front of the hive entrance, or gently brush off the bees with a soft, flexible bee brush or a handful of grass. Allow the bees to fall at the hive entrance. Then place the frame, free of bees, into an empty super and cover it with burlap or a thick, wet cotton sack (robbing cloth) to keep out robbing bees. If robbing is particularly intense, an additional cloth might be needed to cover the super you are working. If robbing becomes unmanageable, put the honey frame(s) into a vehicle and close all doors and windows; stop taking honey from colonies in that apiary for the day.

Advantages
- Able to select frames containing *capped honey* (honey covered by thin layer of wax).
- Relatively easy if bees remain calm and only a few colonies are involved.
- Inexpensive.

Disadvantages
- Timing and good judgment are critical to avoid robbing.

- Method is time-consuming.
- Brushing may excite bees to sting.
- Cappings over the honey cells may be perforated, especially if handled roughly.
- Honey could be tainted with smoky flavor, from excessive use of the smoker.

Bee Escape or Escape Boards. Bee escapes are used primarily to remove bees from hive bodies containing honey so that the honey can be harvested free of bees. When an inner cover or modified cover contains at least one bee escape, it is referred to as an escape board. The inner cover can be made into an escape board by using a small device called a *Porter Bee Escape* (see the illustration on this page). The Porter Bee Escape is an inexpensive metal or plastic gadget that allows bees to pass through it in only one direction. The escape fits into the oblong hole of the inner cover—or any cover that has been modified to hold bee escapes—to facilitate the passage of bees.

Other styles of escape boards and bee escapes are available from bee suppliers, and all provide a means for clearing supers of bees; check bee supply catalogs and talk to other beekeepers. Some of these boards are made with cones, screens, or other escape devices. All provide passage of bees in one direction. An escape board made with several bee escapes is more efficient and will clear supers of bees faster.

Escape boards can also be used to move a hive that consists of more than two hive bodies. First, place the escape board below the honey supers. After the bees exit, the extra bodies are removed and the remaining

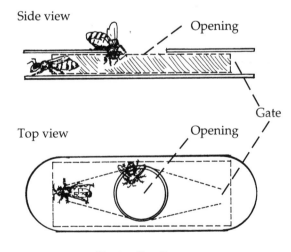

Porter Bee Escape

Escape Board
(at arrow)

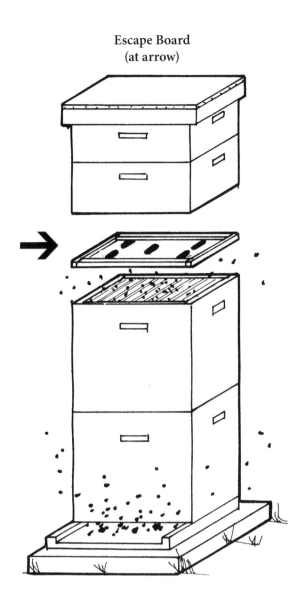

supers can easily be lifted and moved (see "Moving an Established Colony" in Chapter 11).

The escape board is placed directly below the honey supers you wish to remove (see the illustration of the escape board on this page). Usually within 48 hours after the escape board is in place, the bees move down to seek the warmth of the broodnest or the bee cluster, and because many of these bees are field bees, they want to leave the honey supers to resume foraging activities. Do not leave the board on for more than 48 hours, for the bees will figure out how to reenter the supers, especially if a cone escape board is used.

There must be no cracks or holes in the honey supers placed above an escape board, because bees from the same hive, robbing bees, and other insects (such as yellowjackets) will invade and remove the honey.

Tape or otherwise close off these inadvertent entrances to the unprotected honey supers. If the outer cover is warped and you are using the inner cover as an escape board, put an extra inner cover above the topmost super to close off the top and to keep out all robbers.

In extremely warm weather, install the escape board during the late afternoon and remove the supers as soon as they are free of bees—the next day if possible. A word of caution: In warm weather, comb may melt if bees are not available to fan it.

If the supers contain brood, the bees will be less likely to abandon them. If this is the case, pull out the frames of brood and place them in the broodnest or into a colony that will benefit from additional brood to allow the young bees to emerge. If the frames are of different sizes, place the brood frames into the correctly sized super; once filled, you can either put it back on the same colony or give it to a weaker one. Meanwhile, you can select honey-only frames to place above the escape board.

Advantages

- Does not excite bees.
- Easy.
- Inexpensive except for the price of the escape board.
- Usually effective.

Disadvantages

- Honey could be removed by bees from the same hive or by robbers if supers are not bee-tight.
- Not always effective.
- Drones or dead bees may block escape, keeping bees in supers to be vacated.
- Involves extra trips to apiary to insert board, remove supers, and so forth.

Note: If you find brood in your honey supers, try another supering technique; it is very time-consuming and wastes brood to have to sort through your honey supers. If you used a queen excluder, check for the leak that allowed the queen to move through. Whether your excluder leaked or another supering technique failed, when you place another queen excluder below the brood-filled honey supers, in 25 days the supers will be cleared of brood as the young bees emerge, providing the queen is not there. Repair, if possible, or discard the "leaky" excluders.

Fume Board (Fume Pad). Many commercial or side-liner beekeepers drive bees out of the honey supers using fume boards with a chemical odor offensive to bees. Premade fume boards consist of a wooden frame with a metal cover and flannel lining to hold the repellent. The cloth is then saturated with a chemical that repels bees (sold as Bee Go, Honey Robber, and Fischer's Bee Quick, to name a few). Some fume boards have a black metal top that will absorb solar heat and make the chemicals work better. These boards work best when the bees are in shallower supers. Make sure you have removed the queen excluder before starting so bees will move down quickly.

To use a fume board, pick a warm, sunny day; then:

Step 1. Sprinkle repellent onto the pad, following the directions on the label for the repellent.
Step 2. Remove the outer and inner covers, using smoke as needed.
Step 3. Scrape any burr comb off the top bars.
Step 4. Use smoke to drive the bees downward between the frames.
Step 5. Place the board over the frames (see the illustration of the fume board on this page).
Step 6. After no more than five minutes, the bees will have left the topmost super.
Step 7. Remove the first super and repeat the process for subsequent supers below.
Step 8. Air the supers thoroughly and store them in a covered place to prevent robbing and invasion by the small hive beetle (see Chapter 17).

Use the fume board only long enough to get the bees out of the supers. Do not leave it on the hive for more than a few minutes. Do **not** use on comb honey supers, as it will impart an adverse flavor to the honey.

These chemicals work best on a hot sunny day, but do not overdo this, as you can drive all the bees from the hive.

Federal and state laws may restrict the use of some or all of the chemicals used for bee repellents, so comply with all regulations in this regard. These chemicals are hazardous and must be treated with respect. You are responsible for following directions on the label with regard to their use, storage, and disposal. The use of chemicals is further complicated by the

Fume Board or Fume Pad

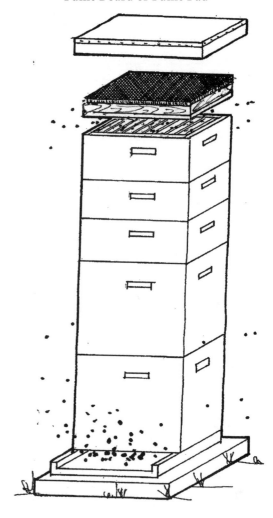

fact that their efficiency is governed by the air temperature; therefore, the desired result is not always certain. Never store or transport the fume boards or the supers treated with these chemicals in a closed vehicle.

Advantages
- One trip to remove honey.
- Easy.
- Relatively inexpensive.

Disadvantages
- Excites bees.
- Dependent on temperature.
- Could adversely flavor honey.
- Rapidly crowds bees to the lower hive bodies and may force them outside.
- Need to remove queen excluder to facilitate bees exiting from honey supers.

The Bee Blower. A bee blower is a portable gas or electrically powered device that produces a blast of air strong enough to blow the bees off the frames and from the supers. The air blast can reach a speed of 200 mph (322 km/h). Set the super on its end, not on its side, near the hive entrance and blow air between the frames from the bottom bars to the top bars. You may have to move the frames from side to side as you blow the air up through them, to clear out all the bees. Some beekeepers have a special stand that holds each super as it is blown free of bees. In this case, the stand is placed near the hive entrance and the bees are blown down from the top. If the weather is cool, blow the bees down through a super that is still on the hive, to keep from chilling bees.

Sometimes leaf blowers can be modified to become a bee blower; they are cheap, commonly available, and adequate for most jobs. Add more hose (15–20 feet of 2½-inch hose) to give you room to maneuver. If you are too close to the parent colony, loose bees may get into the blower's intakes and clog them up; you can screen the intakes and motors. A blower is very useful if you misjudge how quickly bees will leave the supers when you are using a fume board (i.e., supers are still full of bees when you are ready to load) or if there is some brood and the bees have not left the super. Blowers are a good backup system, especially if you are harvesting more than 50 colonies. Many beekeepers use both, when they have help in the yard, to remove supers efficiently.

Advantages

- Fast.
- Effective.

Disadvantages

- Expensive.
- During cold weather, bees blown out may be unable to return to hive.
- Queen could be blown out and lost.
- Requires two people—one to load supers, the other to work the blower—to work efficiently.
- Requires more equipment to haul, blower, gasoline, and so on.

Abandonment or Tipping Supers. This is a good method, but it should be used only by experienced beekeepers because it requires considerable expertise and know-how. W. Allen Dick of Alberta, Canada,

outlined his methods of tipping. To use tipping successfully, you must be able to recognize the difference between bees leaving a hive (or super) and those robbing. Be sure you understand the conditions, both seasonal and weather related, that influence how well this methods works; remember, conditions could change quickly.

Here's how tipping works:

Step 1. Choose a day when outside temperatures are warm enough for free bee flight and a good honeyflow has been on for several days. Late afternoon works well, as the bees are beginning to come home for the evening.

Step 2. Remove full or partially filled honey supers; try to get supers that are brood free, preferably those from above a queen excluder.

Step 3. Place each super you remove on end, either on the ground to one side of the hive's entrance or on top of a hive nearby or on another empty rim. Do not block the flight path of bees returning to their colonies.

Step 4. In a short while, the bees in the tipped supers should finish their tasks, clean up any drips from burr comb, and fly back to the hive from which they came. This may take minutes or hours, depending on the temperature and the intensity of bee flight activity. You may even have to leave them overnight.

Step 5. Pick up the supers when all the bees have left, and take them out of the apiary. You are done.

The reason this works is that all the bees are too busy gathering honey elsewhere to be looking for a source nearby. But be sure there is a good honeyflow on, or the bees could start to rob.

Advantages

- Fast and easy.
- No extra equipment or chemicals are needed.

Disadvantages

- Weather can change fast, as can the temperament of the bees. Bees that were happy and gathering nectar at one moment may turn to frantic robbers, resulting in (total) loss of the honey.
- Queens, if excluders are not used, may be in the honey super(s) you removed. Careful blowing,

brushing, or shaking toward the correct hive is then required, as the bees may not leave the queen. Brood in the supers has the same effect; that is, bees will stay with the brood.

- Requires a second trip to the apiary to collect supers.
- Requires an experienced beekeeper to know the temper of bees in the hives and appropriate weather conditions.
- Beekeeper should be nearby; that is, use this method on the home apiary so you can monitor the bees frequently.

Getting Along with Your Back

Lifting full supers of honey might be the reward of a productive year, but it can also be a literal pain in the back. Unless you are careful in lifting these heavy boxes, you can do serious damage to your back. Proper lifting and strengthening exercises might be needed if chronic back pain is a problem. In any case, medical advice should be sought. For some general information on back care, see "Honey and Honey Products" in the References.

Extracting the Honey

Now that you have removed the honey, place the supers in your honey-processing area (a basement, garage, trailer, or special honey house); you are now ready to process it. You have already prepared this space before you took your honey supers off, right?

Decide what type of honey you are going to process—extracted, cut comb, section comb, or a combination of all of these. Try to do all your extracting at one time to avoid cleaning up the floor, the extractor, and other equipment more than once or twice. This is a messy job and will try the patience of all involved, but the reward is worth the effort.

The usual process for getting honey out of the wax cells is to remove the cappings with a hot knife called an *uncapping knife* (see illustration on cutting wax cappings on this page) or uncapping plane and to put the uncapped frames into a centrifuge, called an *extractor*. As the extractor spins, the honey is forced out of the cells and against the cylindrical wall of the extractor, leaving the frames of wax combs empty of most of the honey. A small gate at the bottom of the extractor can be opened to let the honey flow out into other containers. **Keep your eye on the honey** drain-ing into these containers—overflows are costly and messy.

Honey left in supers before extraction can be ruined if it is stored in a humid or wet area, because even capped honey can absorb water vapor. To prevent this, stack the supers in a staggered arrangement to allow for ventilation, and use either a dehumidifier or a fan to blow warm, dry air over the frames before extraction. You can also place extracting supers in a warm, well-vented room for a day to "dry down" the honey and make extracting easier. This will further reduce or maintain the existing low moisture content of the honey. Warm your supers at least one day before extraction if the frames are cold; warm honey will move out of the cells faster and will pass through a filter in less time.

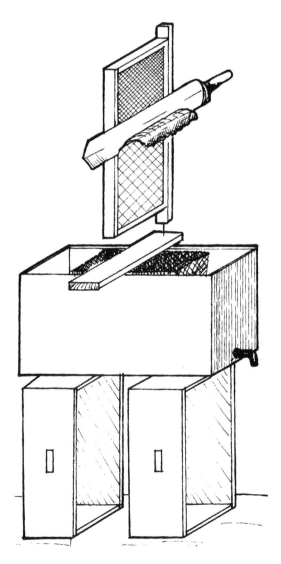

Cutting Wax Cappings from Honey

If you are in a very hot, dry location, such as the southwestern United States or parts of California, the honey will be too dry and the humidity too low for easy extraction. You will therefore need to increase the humidity of the honey house by hosing down the floor. This is especially true for extracting mesquite (*Prosopis* spp.) honey in Arizona. Because the honey crystallizes quickly, beekeepers there must add humidity in their honey house to get the honey out of the frames.

Frames with honey to be extracted should be completely or almost completely capped. Uncapped cells will contain honey with higher moisture content. Extracting honey from too many partially capped frames will increase the moisture of the honey and invite spoilage by fermenting yeasts (see "Extracting Honey" in Chapter 12 for more information on this topic).

Remember: **Honey is a food product—someone will be eating it. Keep it as clean as possible to have a quality product.**

Laws on the books require certain minimal sanitary practices, especially if you are selling your product. Honey is a quality hive product and it is up to all beekeepers to keep it pure, uncontaminated, and clean. Check with local or state authorities for food-processing rules and legislation; also the National Honey Board website is a good source of information.

In addition to collecting nectar from floral or extrafloral nectaries (e.g., cotton), bees collect another sugary liquid called *honeydew*. Honeydew is the sticky, sugary liquid excreted by certain plant-sucking insects that feed on plant sap. One of the principle producers of honeydew are aphids (order Hemiptera). Other insects that exude honeydew include whiteflies, gall insects, scale insects, mealybugs and leaf hoppers; for more information see Chapter 12: Products of the Hive. Bees will sometimes collect honeydew in large quantities and store it as honey. Honeydew honey is usually dark in color, contains less of the two principal sugars derived from nectar honey, and has more protein and less acidity (higher pH) than honey made from nectar. It is considered to be a low-quality honey in this country, but is quite popular in Europe and other areas. If you find you have lots of it, seek out special markets for this particular honey rather than blending it in with your regular crop.

Note: In areas where the small hive beetle lives, extract honey quickly without storing.

Harvesting Comb Honey

When you are harvesting comb honey, break apart the comb honey supers the day before harvest and scrape off any burr comb the bees may have built up between supers. Overnight, the bees will clean up any honey drippings, and the combs will be clean of leaking honey or drips.

Take off these section supers carefully, using as little smoke as possible and trying not to crack or bend the bodies or the frames, because the sections can leak if broken. Do not jar or drop such supers, and keep them covered when trucking them to the honey house to exclude dirt and debris. Once in the honey house, process them quickly, to keep small hive beetles and other pests from getting into these supers (or store them in a freezer).

Take out the sections from the frame, then cover and store them in the freezer (to kill any wax moth eggs or larvae) until they are ready for sale. Allow the comb honey sections to come up to room temperature before you put on the labels; this allows the sections to dry from water condensation (see "Selling Your Hive Products" in Chapter 12).

As with the wet extracted frames and supers, you can place any partially filled comb honey and supers back on colonies for the bees to clean up before storing them for winter; or use as emergency food for spring feeding.

FALL MANAGEMENT

After the fall crop has been removed and the supers have been cleaned and stored, each colony should be checked and attended to as follows:

● If possible, pick a day when there may still be a light honeyflow and forager bees are out. Bees are more prone to sting when a hive is being manipulated and after honey has been removed, because of the dislocation of bees. Also, in cool weather, most of the bees are inside the colony.

● Check for brood diseases and mites.

● Do not attempt to overwinter a colony found to have American foulbrood or excessive mite infestation, deformed wing virus, or other problems; destroy the colony and disinfect the equipment.

- Medicate for nosema as a preventive measure; see Chapter 14.
- Remove the queen excluder and any section comb honey supers.
- Remove any honey supers you don't want bees to move into over the winter.
- If requeening, check colony after seven days to see if the queen has been accepted.
- Check winter stores; about 90 pounds (40 kg) or more of surplus honey should be left for each colony in areas where winters are severe; see "Wintering in Extreme Climates" in Chapter 8.
- Feed the colonies whose stores are low; feed early enough in the fall so the syrup can be properly cured by bees (about a month).
- Treat for mites and place grease patties for tracheal mites (see "Tracheal Mites or Acarine Disease" in Chapter 14) if necessary.

FALL MANAGEMENT OF SOUTHERN COLONIES

Many beekeepers live in the southern states where beekeeping practices are a little different. Dean Breaux of Florida had the following suggestions for managing colonies in the South. The best way to overwinter colonies in the South is to start with young vigorous queens. Requeening in the fall is necessary because honeyflows occur so early in the year (February) that you could miss your first crop unless your bees are strong. The bees do not have a chance to utilize a queen raised in March, so begin fall management in October by requeening your colonies in the late fall. If populations drop below five frames, the colony will in all likelihood produce a meager crop in the spring or be lost in the winter. If possible, requeen with queen cells; use mated queens for the colonies that were not successfully requeened with a cell. In areas where the Africanized bees have colonized, requeening with a queen cell is not advised, because the virgin queens may mate with African drones, giving the offspring some African temperament.

The fall requeening procedure is:

- Remove all surplus honey.
- Treat colonies for varroa mites.
- Remove old queen.

- Install queen cells (or caged queen). You can add grease patties for tracheal mites at this time.
- If you have many colonies, it may be faster to not kill the old queen but to put a queen cell in the hive with a cell protector. While you don't need to find the old queen, the results are variable.
- Return in two to three weeks to check and see that there are mated queens in the hives; if the hive is queenless, install a young mated queen in a cage. Add another grease patty to all the hives if the other patty is gone to control tracheal mites.

By the middle of November, requeening should be completed.

Winter preparations for these southern colonies can now proceed. Here are some tips to remember:

- Overwinter in a deep brood box with a minimum of one 6 ⅝-inch super of stores.
- Install entrance reducers and mouse guard.
- Provide upper entrances if you have several months of cold weather.
- Make sure the bee population is between 15,000 and 20,000.
- Test for mite levels periodically to ensure mite- and disease-free colonies; treat if necessary.
- Protect colonies from cold, wintry winds in marginal regions.
- Periodically check to see if honey stores are holding up; feed if necessary.

Once most of the cold weather is over, prepare for the first spring honeyflows by doing the following:

- From the first of December to the end of January you can move hives from fall locations to spring yards (primary citrus groves) to prepare for the spring flows.
- Try to locate spring yards in areas that have red maple (*Acer rubrum*) trees; they bloom in late December through January and provide pollen for the bees to start building, in preparation for the early citrus flow in the beginning of March.
- Begin stimulatory feeding with light syrup the beginning of February to ensure that the queens are laying well. This feeding will ensure a strong population of young bees in March for the early honeyflows.

● Add supers to colonies in March in preparation for the flows.

BEEKEEPING IN THE SOUTHWEST

Drs. Gordon D. Waller and Gerald Loper, both now retired, worked at the USDA Carl Hayden Bee Lab in Tucson, Arizona, and offer some insight into keeping bees in the desert Southwest. Beekeeping in this region is much different from any other type of beekeeping. It presents certain challenges that are not seen in other regions, including sparse bee forage, little water, and extreme temperatures. Because there is such a diversity of landscapes and temperatures, from low plains and valleys to high mountain ranges, it is difficult to generalize about Southwest beekeeping techniques.

Rainfall can vary from 3 inches per year (in Yuma, Arizona) to over 30 inches in the center of the state, and comes twice a year, the slower rains in winter and the fast and spectacular thunderstorms during the summer monsoon season. Although most beekeepers locate their apiaries in irrigated agricultural areas, they often move them to take advantage of different cultivated crops and natural vegetation at varying elevations. Like most commerical beekeepers throughout the United States, many move their colonies to southern California for almond pollination.

Forty or 50 years ago, the bees in this region were mostly kept in permanent yards year-round, using only standard Langstroth deep supers with either two or three on each colony. Frames were removed and honey was extracted whenever it was available; some beekeepers had mobile extracting plants to facilitate honey removal. Such methods have been replaced with more normal colonies producing honey in shallow supers that are taken into a permanent building for extraction.

Colonies overwintered in riparian areas are likely to get their first pollen from the cottonwood trees (*Populus* spp.). African sumac, planted as an urban tree (*Rhus lancia*), as well as *Eucalyptus* species, bloom in January and February. Annuals, such as London rocket (*Sisymbrium irio*), bladderpod (*Lesquerella gordoni*), filaree (*Erodium* spp.), and *Phacelia* species also bloom at this time. March is when the citrus start to bloom, and some areas produce good honey crops from this. The main desert honey, however, is from the mesquite (*Prosopis* spp.) and catclaw aca-

cia (*Acacia greggii*), which bloom from April into June, depending on rainfall amounts. Other plants blooming then provide additional food and include palo verde (*Cercidium* spp.), creosote bush (*Larrea tridentata*), fairy duster (*Calliandra eriophylla*), ironwood trees (*Olneya tesota*), and of course the iconic saguaro cactus (*Carnegiea gigantea*).

A good desert honeyflow will depend on the winter rainfalls, which could extend or shorten the blooming times. Mesquite can bloom a second time as well, but will yield no appreciable nectar if strong winds or unseasonal rains occur during bloom. Honey is usually extracted at this time, in the hottest and driest part of the year. For this reason, extracting houses are usually hosed down to increase the humidity that is necessary to allow the honey to flow more freely from the combs. Another characteristic of honey derived from mesquite, catclaw acacia, and fairy duster is that it crystallizes very quickly, so it must be removed and extracted at once. The honey, however, has an exceptional consistency with very fine, smooth crystals that stay soft and creamy.

Following the spring mesquite bloom is a maintenance time, when not much is blooming. If beekeepers can find alfalfa or cotton fields, they can get another honey crop, but this is becoming more difficult since some areas that were used for cotton have switched to other crops or are using genetically modified cotton, which some beekeepers say bees do not like as much. Bees do not normally pollinate cactus, such as prickly pear (*Opuntia* spp.). Mostly the colonies are maintaining until the winter rains. If bees are moved for pollination, the most commonly grown crops include melons and other cucurbits, alfalfa, onion and carrot (grown for seed), and citrus.

By the late summer and fall, desert areas having a heavy, brush growth of salt cedar (*Tamarix pentandra*), or the related wild tree known as athel (*T. aphylla*), may bloom again and will provide bee forage throughout the summer if moisture is adequate. The plants that are good for bees include burroweed (*Haplopappus* spp.), and at elevations of 4000 to 5000 feet, sandpaper plant (*Mortonia scabrella*) will also be blooming, as will some of the agaves. "Sugar-bush," which is a local name for buckbrush (*Symphoricarpos oreophilus*), is found at elevations over 5000 feet north of Phoenix.

Overwintering bees in the Southwest need good stores of honey, just like colonies in the northern states. Brood rearing may stop for intermittent periods when pollen becomes scarce, but otherwise, brood may be found throughout the winter. If the colonies are located near large numbers of *Eucalyptus* species, trees, and other ornamentals, they often provide sufficient pollen and nectar throughout the winter months to support continued brood rearing. An excellent reference on the bee plants in this area is "On the Pollen Harvest by the Honey Bee (*Apis mellifera* L.) near Tucson, AZ (1976–1981)" by R.J. O'Neal and G.D. Waller, available from the USDA Carl Hayden Honey Bee Research Center, Tucson, AZ.

Probably nowhere else in the United States are beekeepers more concerned about providing shade and water for their bees than in the desert Southwest. Summer temperatures can often exceed 110°F (43.3°C) in the shade at lower elevations, and at these temperatures, bees spend a lot of energy carrying water back to their colonies to air condition the broodnest. Therefore, beekeepers must either supply water or locate their apiaries near cattle watering tanks, irrigation ditches, or other places that have a continuous source of water, within a ½-mile (0.8 km) flight range.

Some beekeepers will build shade structures called "ramadas," a framework covered with shade cloth or brush. In addition, trees such as mesquite, tamarix, and cottonwood provide excellent shade. Not only is this necessary for the bees, it is also good for the beekeepers.

Other challenges to desert beekeeping include hazards such as rattlesnakes and scorpions, poisonous spiders that nest under the outer covers, fire ants that love to invade bee colonies, not to mention the numerous plants covered with spines and thorns. Additionally, most of the feral bee colonies are now Africanized, and beekeepers have had to learn how to manage these bees. Many are doing so successfully, however, and good honey and pollen crops can be obtained in these regions. Pests and diseases common to bees elsewhere also occur in the Southwest, such as wax moth, which can be a year-round problem, as well as the two foulbroods and chalkbrood. Varroa mites are now causing problems as well. One mechanism the Africanized honey bees use to control these pests is to swarm frequently and abscond from their nest. For more information on Africanized honey bees, see Appendix E.

 **Notes**

CHAPTER 10

Queens and Queen Rearing

Although queens may live for four years or more, the most productive queens are usually between one and two years old. There are several reasons why they have such short productive periods. Queens lay up to 2000 eggs per day (60,000 eggs per month) for four to five months in northern latitudes, which takes its toll on them! In addition, their pheromone levels decline with age, which could result in swarming or supersedure.

Added to these natural factors shortening the productive period is the presence of varroa mites in colonies. These parasites feed on the blood of honey bees. Older larvae, pupae, and adult bees fall prey to the mites, which, while feeding, pass viruses to their hosts. In order to save colonies from varroa mites, beekeepers have come to rely on miticides. The mites, the viruses they transmit, and the insertion of miticides into bee colonies are having debilitating consequences for the bees. Evidence indicates that some of the chemicals contained in some miticides cause a decrease in the viability of drone sperm and the sperm stored in the queen's spermatheca. In addition, the growing queen larvae may not be fed adequately, or with food of poor nutritional value; this may also make the resulting queens weaker and have a shorter life span. This is one of the reasons why many beekeepers are replacing queens on an annual or biannual basis (see "Marking or Clipping the Queen" in this chapter).

In addition to the problems mites create, studies have shown that a colony with an older queen is more likely to swarm than a colony with a young queen. Given this information, annual requeening may provide a threefold benefit: (1) an old queen will be replaced with a more productive young queen; (2) a younger queen will release adequate or more than adequate pheromone levels, making the colony less likely to cast a swarm or supersede the queen; and (3) in the process of requeening, a natural break in the colony's brood cycle will occur, with the consequence of interrupting the mites' life cycle. Given the problems concerning the mites, their viruses, the side effects of miticides, and the need for adequate pheromone levels to keep the colony intact, annual requeening may become the norm. Naturally, at any time of the year, queens whose performance is questionable should be replaced.

The effects of mites and miticides, and whether the correct time to requeen should be in the spring, summer, or early fall, continue to be debated. Keep current with the latest information on this and related topics in the bee magazines and on the Internet.

Bees make new queens for two basic reasons: to divide the colony (swarm) or to replace an inferior or missing queen (supersedure or replacement). Bees preparing to swarm or supersede their queen are requeening their colony. This is natural requeening, replacing a mother with a daughter. Colonies that swarm (queen replacement) may issue a single swarm or multiple ones (afterswarms), if the colony is exceptionally robust. The departing swarm reduces the colony's population, thus reducing colony congestion and at the same time, reproducing more colonies. But from the beekeeper's perspective, whose goal is honey production, a colony diminished in numbers equals a reduction in honey production.

If a primary swarm (or *prime swarm*) is captured, then it usually contains the old queen. This queen should be replaced with a young queen for reasons noted (see "Swarming" in Chapter 11). Returning to the colony that cast the swarm, the young queen that emerges is a virgin and will soon take a nuptial flight where she will mate with from 1 to 24 drones (average of 14). The drones within her flight range may not be genetically diverse enough, and although her mother may have had many attributes (which include her own genes and the genes found in the sperm donated by the drones), the new queen may not be able to sustain the same vibrant colony that her mother once headed.

If some of your colonies frequently cast swarms, requeen them with stock from several different queen breeders. If you choose to do otherwise (i.e., requeen colonies with swarm cells taken from colonies that frequently swarm), you may be promoting the tendency to swarm since swarming behavior has a strong genetic component.

By diversifying your queen stock, you are also inadvertently diversifying the drone population that congregate in *drone-congregating areas* (DCAs). DCAs that have greater genetic diversity provide a powerful advantage to queens undertaking mating flights. Research has shown that queens that are multiply mated with genetically diverse drones are better able to resist disease, overwinter better, and produce more honey. Queens with low sperm viability and amount can decline in the fall or go into the winter with low bee populations.

Queen supersedure, on the other hand, takes place only after a colony has been declining because it is headed by a failing queen (see "Queen Supersedure" in Chapter 11). Her replacement may also be inferior, especially if colony numbers (of young nurse bees) and food stores are inadequate for rearing quality queens. She may also mate with drones whose genetic stock has poor attributes. In areas where Africanized honey bees (AHBs) are present, mating may include African drones, and this will turn your colony or colonies into an Africanized one within a year. Africanized bees raise more drones than European queens, and therefore are more likely to mate with virgin queens (see Appendix E on Africanized bees).

You should give serious consideration to requeening colonies that exhibit the following:

- Low bee populations for no other reason than a failing queen.
- Bees that are prone to diseases and manifest high mite populations.
- A queen laying more drone than worker eggs. Once the cells of larvae are capped, the large domed cappings over drone cells (shaped like the tip of a bullet) are easy to identify. If the drone cells far outnumber worker cells, this indicates that the queen may not be properly mated and is producing an excess of drone eggs.
- An injured queen, laying only drone eggs, or having drone and worker larvae scattered over the comb, rather than in concentrated areas. A good queen will lay mostly worker eggs.
- Diseased queen, brood, or workers or a colony that never seems to expand and is always weak.
- Poor brood patterns, which may be a symptom of disease or poorly mated queens.
- Defensiveness.
- Excessive propolizing, unless you are collecting and selling propolis.
- Excessive debris on hive floor (nonhygienic).
- Poor wintering success (colony reemerges from the winter in a weakened condition with a low population).
- High honey consumption, which is measurable with a hive scale). This is difficult to discern because a large bee population may be one of many factors that result in high honey consumption.
- Poor honey production. If all other colonies have equal populations, only then can a true assessment of this be made.
- High tendency to swarm.
- Workers failing to form a retinue around queen.
- Low sperm viability; colonies peter out in the fall or go into the winter with low bee populations.
- Spotty brood pattern with lots of open cells containing no brood.

WAYS TO OBTAIN QUEENS

Types of Queens to Purchase

Queens can be obtained by purchasing them, raising your own, or obtaining them from colonies preparing to swarm or supersede their queen.

There are five categories of purchased queens:

1. *Virgin queens* (or queens emerging from queen cells) are unmated, and introducing them into normal colonies requires extra work. These queens must be less than four to eight days old when purchased; if they are older than that, they will not fly to mate and will become drone layers.

2. *Untested queens* are shipped once they start to lay eggs, but the qualities for which they were bred have not been tested; most queens sold to bee-keepers today are untested.

3. *Tested queens* are not shipped until the first worker brood emerges, to determine the purity of mating. Other variables may not be observed.

4. *Select-tested queens* are tested not only for purity of mating but also for other characteristics such as tolerance to disease or mites, gentleness, and productivity.

5. *Breeder queens* are tested for one to two years, evaluated, and used to raise daughter queens with similar genetic traits. Commercial beekeepers and queen breeders often buy some breeder queens to serve as the mothers of the queens in their opera-tion. These queens are usually instrumentally in-seminated with semen from selected drone stock and can be very expensive.

Whenever you purchase queens, make sure they come "certified mite-free" so you are not introducing more mites to your apiary. It is also recommended that all new queens should already have been medi-cated with fumagillin, to protect them from nosema disease. Mark all new queens. By marking queens, you will be able to tell whether or not the queen you find in a colony is the one you originally purchased. If marked queens are absent, then you will know that she was either superseded, died, or departed with a swarm. If your queens are unmarked, you have no reference to their history.

There are reports that commercially reared queens have been showing problems with acceptance and longevity. The causes may be related to mites, mite treatments, nutrition, or the lack of adequate and vi-able drones in their mating yards. If you see a colony you recently requeened with some of these character-istics, requeen it with a queen from a different source. If you are purchasing queens from a local or un-known breeder, start slowly and keep good records of how that queen performed (or didn't perform) so you can decide which breeders to use in the future.

Here are some things to look for:

- Queens in the cage are not attracting workers, or retinue is not formed around queen.
- Bees are unusually loud, as if queenless (decibel level will be above 65).
- Colony suddenly becomes queenless.
- Supersedure or emergency cells are not built de-spite a failing queen.
- Many larval and pupal deaths are evident (be sure, however, this is not the result of foul-brood).
- Eggs and brood scattered about the broodnest in-stead of being in compact, concentric patterns; we use brood pattern as our major means of evaluat-ing our queens.
- Pollen cells mixed in the honey supers or other-wise is not near brood-rearing areas.
- Honey is stored below the broodnest or scattered throughout the colony.
- Colonies are abandoned but still contain lots of honey; such colonies may contain toxins or virus and should not be reused.
- Hybrid brood cappings—more domed than worker cappings but flatter than drone caps, a cross between the two—are seen.
- Supersedure rates are high.
- Bees fail to form a cluster during cold periods.
- Bees are seen scent fanning in large numbers when you open the hive.
- Bees swarm (or abscond) at unusual times (dur-ing the fall or winter).

Purchasing and Ordering Queens

Queens can be purchased alone or in packages from reputable queen breeders throughout the United States. Purchased queens may arrive in a queen cage (with or without attendants) or in separate cages in a larger package that includes many worker bees. If the purpose of ordering queens is to requeen an existing apiary, order queens so they arrive during the season when there is a good honeyflow in your area (again, by keeping records annually you will have a reason-ably good idea when this is). Requeening hives just before the peak in varroa mite population breaks the mite cycle (as long as the colony is otherwise healthy) and serves as a method for controlling mite popula-tion buildup. If you plan to requeen in the fall, you can still have your queens delivered in the summer; plan to install them in nucs to get them laying in preparation for fall requeening (by uniting the nuc to the colony; see "Uniting Weak Colonies" in Chapter 11). Order queens as you would packages, as early in the year as possible, because most breeders start making queens when snow is still on the ground in northern states. You can have your queens arrive either in the spring or in the summer. Some breeders will ship queens with or without attendants; others will place the cages in a screened bulk shipping container (especially if you order 10 or more queens at a time) that includes loose worker bees and some kind of food.

Treat your queens carefully and follow the installation instructions given below. It is advised that you **not** treat new queens with miticides until they have been laying eggs for about 30 days.

Many journal advertisements market "mite-resistant" queens. Be sure to inquire which mites they are resistant to and what the basis for such claims is. Do your homework, research the claims, and talk to other beekeepers who have used queens from the same providers. To date, the only known bees with some resistance to mites are the hygienic line and hybrids with Russian and SMR (VHS) stock.

Types of Queen Cages

Purchased queens are packaged in queen cages and mailed from breeders to all parts of this country and abroad. Some of the procedures for notifying the postal service of the arrival of bee packages should also be followed when queens have been ordered (e.g., call to alert the post office of your order, as suggested in Chapter 6). Again, all queens must be certified to be free of mites.

In the United States, queen bees are shipped in several kinds of cages. One style is the *Benton* (also called two- or three-hole) mailing cage; it is a small wooden block, approximately 3 × 1 × ¾ inches (7.5 × 2.5 × 2 cm) with wire screening stapled over the length of it to cover the two or three holes to the compartments (see the illustration of queen cages on this page). Generally, two compartments serve as the living quarters for the queen plus several attendants, and a third compartment contains candy. At either end of the cage is a small, bee-sized hole. The hole at one end is adjacent to the compartment

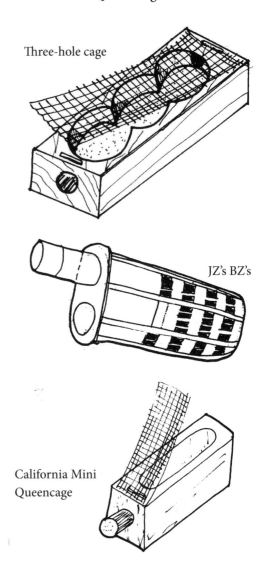

Queen Cages

Three-hole cage

JZ's BZ's

California Mini Queencage

that is filled with candy, while the hole at the opposite end is used to place the queen in the cage; both openings are closed with cork plugs upon shipment. The candy serves two purposes: it provides food for the bees inside and delays release of the new queen from the cage into the colony. The hive bees must eat the candy plug before the new queen can be freed. This usually takes about two to three days, by which time the colony will have adjusted to the new queen's scents, improving her acceptance.

Other cages used by bee breeders (see the illustrations) include JZ's BZ's plastic cage and the California Mini Queencage, developed by C.F. Koehnen and Sons. The latter is a narrow wooden block, about 2 ½ × ¾ × ¾ inches (6.5 × 2 × 2 cm), with one long compartment. It has a large candy-filled plastic tube inserted in one end, and a corked-plugged hole at the other end. The smaller size of this cage allows more queens to be *banked* (or stored in queenless colonies until needed), and when it is inserted between frames, it does add extra space that bees will fill with comb.

The plastic shipping cage called JZ's BZ's is in the shape of the lowercase letter "d." It is much smaller than the wooden cage, its overall length, including the stem, being 2 ½ × 1 × ½ inches (6.5 × 2.5 × 1.5 cm). The stem of the "d" holds the candy plug. These cages are sometimes shipped in battery boxes.

Increasingly common is the cardboard/screened "battery box" for shipping queens. Caged queens are secured in a small screened box, and young worker bees are then added to attend all the queens in the cages. This method is preferable because the loose bees can protect the caged queens from temperature variations during shipping. Other types of cages are becoming available so check the bee supply catalogs or talk to or visit individual queen breeders.

REMOVING ATTENDANTS

Before the queen is introduced to a new colony, many beekeepers remove the accompanying attendants (if there are any) because they are foreign bees and the workers in the colony may become aggressive toward them and even the new queen, biting them through the wire screening of the cage. This activity may release alarm odors, alerting other bees to surround the cage, and in the process, they could injure or kill the queen. In addition, these attendant workers may be infested with tracheal or varroa mites, so they should be eliminated.

Three Ways of Removing Attendants

An easy method of removing attendants from the cage is to hold the cage between your thumb and middle finger; remove the cork from the non-candied end and place your index finger to cover the uncorked hole. Hold the cage vertically and, as attendants move up the cage, remove your index finger, permitting the bees to exit. If the queen moves toward the opening and attempts to leave, cover the hole. This method, like the others, should be done within an enclosed area; in the event the queen slips out, she can be easily recaptured.

Another method is to replace the screen with a piece of plastic queen excluder that has been cut to fit on top of the cage. First, remove the staples that are keeping the screen in place while you hold the screen. Put the plastic queen excluder on top of the screen and carefully slide the screen out from underneath, taking care not to injure the queen if she is on the wire screen. Then tack the piece of excluder to the cage. Now place the cage near a light source (natural or artificial) in a darkened room, and the workers will exit through the excluder. You can also do this in a closed vehicle or room. Once the attendants have left, replace the screen but don't forget to restaple it.

A third way is to remove the cork opposite the candy end of the cage (or pry up one of the staples and lift up a corner of the wire screen) in a darkened room next to a closed window. Hold the cage in a vertical position, with the opening at the top, and the bees, like most insects, will climb upward. When the bees exit, they will continue to move toward the light, and you can easily recapture the queen from the window. Pick her up only by the wings or thorax and return her to the cage, and replace the corks or screen. This is a good time to mark and/or clip the queen (see "Marking and Clipping Queens" below).

CARE OF CAGED QUEENS

When caged queens arrive, they should be properly cared for and placed into a colony as soon as possible. If the cages contain candy and attendant bees, they can be kept in a warm (about 60°F [15.5°C]), dark place free of drafts for up to a week provided that they are not unmated queens. The attendants help keep the queen warm, but you should cover the cages with a light cloth to keep the warmth in. A very, very small drop of water should be placed on the screen twice a day. Do **not** get the candy or bees wet, as the bees could become chilled. Remember you are not providing water for horses! If you live in a very dry region, you may want to place a wet sponge in or on the queen cage to increase the humidity and prevent the bees from dehydrating.

If the queens are to be introduced by the indirect method, the attendant bees should be removed first, as we already described. If you use another method where the attendants are needed, check the condition of the queen and accompanying bees to see if there are any dead attendants in the cage. If there are some,

Queen Cage Holder or Banking Frame

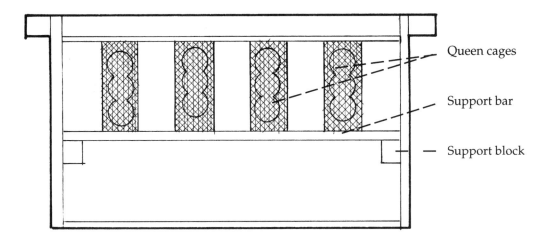

Queen cages

Support bar

Support block

remove the queen and place her in a new cage with attendants of newly emerged workers from one of your colonies; older bees will in all likelihood attack the queen. If the cage becomes fouled for any reason, remove the queen and place her in a new cage, then fill one compartment with candy (made of clean honey and powdered sugar) or a marshmallow (or several miniature marshmallows).

If you are not planning to install the queen for more than a week (providing she is not a virgin), you can bank the queen (with no attendants) in a queenless colony for up to three weeks. You can bank many queens this way. First, remove the attendant bees, then place the cage, screen side down, between the top bars in a queenless nuc that contains lots of young workers. Make sure there is plenty of food in the nuc, or you can feed the bees medicated (fumigillin) sugar syrup and a pollen substitute. (You can also use a full-sized queenless colony.)

You can make a special queen cage–holding frame or banking frame (see the illustration of the queen cage holder on this page). One end of the cage goes against the underside of the top bar of an empty frame (without comb or foundation), and the other end rests on a bar of wood that has been nailed in to run the length of the frame. The bar is nailed to the side bars of the frame; this way 10 to 20 cages can be banked in several frames.

Insert the banking frame into a strong, queenless or into a queenright colony above a queen excluder. If you surround the cages with frames of emerging bees, pollen, and honey, the caged queens will be well cared for until they are needed. If you are storing them in a queenless colony, add some frames of capped brood once a week and feed the colony. A free queen must **not be allowed** in the queenless colony or above the excluder in a queenright colony; otherwise, the caged queens may be killed. As an added precaution, place all frames of open brood in the bottom hive body of a queenless colony with an excluder above it, in case the queenless colony rears a new queen.

Marking or Clipping the Queen

It is advisable to mark the queen with a spot of paint or a color disc (with or without numbers) on her thorax, or clip her fore and hind wings on one side (see the illustration of the queen with a numbered tag on p. 131). Marking or clipping the queen allows you to keep a record of the age of the queen, and the use of color will make it easier to find her, especially if she is dark in color. Clipping the wings of a queen will not control swarming, contrary to what the literature erroneously states. In some instances, clipping may cause the bees to supersede or replace her. Therefore, our advice is not to clip the queen's wings. If you must keep the queen from leaving (e.g., if she is an expensive breeder queen), place a strip of queen excluder in the hive entrance instead. But remember, this will also keep the drones from leaving, so use this only on nucs. Keep records in your apiary diary on where and when you purchased the queen; this will allow you to determine her worth, her age, and whether or not you would consider rearing daughter queens from her.

When you pick up the queen to mark her, never hold her by the abdomen; see the illustration on this page for the proper method of holding a queen. Picking her up by the abdomen can cause injury or compromise her egg-laying ability. As a result of this or any injury to the queen, the colony may supersede or replace her. To avoid injuring a queen, first practice picking up a few drones. Remember, the queen will not sting you unless you handle her roughly.

To mark the queen:

Step 1. Grasp the queen by the wings with one hand, and then with the other hand, hold the sides of her thorax or a few of her legs, releasing her wings.

Step 2. If you are marking with paint, use a fast-drying paint, like nail polish or model paint, and mark only the thorax.

Step 3. If you are gluing on a numbered disc, first apply adhesive on the thorax, then firmly attach the disc.

Step 4. Allow the paint or adhesive to dry before you gently return the queen to the frame on which you took her, or place her on the top bar of a frame and let her walk down the comb. Sometimes the paint and adhesive odors will cause the hive bees to attack the queen; cover her with a few drops of honey to mask the odor of the paint or adhesive so the bees will clean her off instead.

The queen can be removed from the colony without alarming the other bees; in fact, for short periods (5–10 minutes) they will not be aware she is absent.

**Clipped-Wing Queen
with Numbered Tag**

Proper Method of Holding Queen to Clip or Mark

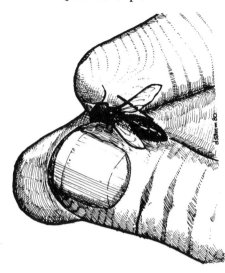

It is becoming more common to requeen colonies every year to maintain quality stock, reduce swarming, or keep diseases and mites under control. Make sure you note down in your hive diary the date you purchased the queen, the source of the queen, and the color with which you marked her. This is becoming especially important in areas with Africanized honey bees.

You can use any color to mark the queen, but make sure they are bright, primary colors so she will stand out. Some bee suppliers sell special queen-marking kits and color sticks, which make it easier for inexperienced beekeepers to use. If you wish, use the International Color Chart system that is on a five-year sequence for marking the queens:

- White or gray—year ending 1 or 6.
- Yellow—year ending 2 or 7.
- Red—year ending 3 or 8.
- Green—year ending 4 or 9.
- Blue—year ending 5 or 0.

Advantages
- Queen can be easily found.
- Queen's age can be determined.
- The marked queen's absence indicates a change in the colony (e.g., queen swarmed or was superseded), or the colony was usurped by Africanized bees.

Disadvantages

- Bees might supersede "maimed" queen.
- Clipping does not prevent or control swarming.
- Queen could be injured while being handled.
- Queen that was clipped may be a virgin and thus would be unable to fly and mate.
- Virgin queen might sting when handled (but this is not likely).
- Color spot or disc may come off.

Failure to Find a Marked Queen

In the absence of finding a marked queen, you can assume that either the colony swarmed with her, she was superseded (replaced), she was accidentally lost and subsequently replaced, or an Africanized swarm (or another swarm with an unmarked or virgin queen) has moved in. Another reason could be that the paint or disc came off. Clipped queens are permanently "marked." If a queen is not found, but eggs and brood are present, you can be assured that there is a queen in the hive (marked or not).

If you do not find a queen after a thorough search, and no eggs or brood are present, look for queen cells about to hatch, or suspect that a new young queen or virgin is present but has not yet begun to lay eggs. Check the colony for eggs in another week; if you still find none, requeen.

A virgin queen is more difficult to locate because she tends to move more quickly on the comb than a laying queen. She might also be a little smaller than a laying queen (more the size of a worker) because she has not gained body weight after the mating flight and needs more time to gain it before she starts to lay eggs.

The absence of eggs or brood in a colony, therefore, could mean that a virgin queen is present but has not yet mated or has not begun to lay eggs. Before you requeen such a hive, be sure that the colony is queenless, since any attempt to introduce a new queen into a hive with a virgin queen or newly mated queen is likely to fail.

In areas of Africanized honey bees, your colony may have been taken over by an invasion or usurpation swarm. This is a difficult situation to change, and if you live in such an area, you may have to requeen your colonies frequently to keep your hives European.

If the colony is definitely queenless, you will eventually have laying workers present. If you find scattered drone brood, scattered capped drone cells, and several eggs in each cell that are attached to the cell wall instead of the cell bottom, you have laying workers (see "Laying Workers" in Chapter 11).

SPRING REQUEENING

Requeening can be done in the spring, summer, or fall. It is always preferable to requeen during a honeyflow, since a colony is almost certain to accept a new queen when food is incoming. It is easiest to requeen in the spring or early summer, but fall requeening is becoming more popular, as it helps in lowering varroa mite populations. In the absence of a honeyflow, you can mimic one by feeding bees sugar syrup.

Spring is the time of year that purchased queens are readily available from bee breeders. You may also wish to raise queens from your own colonies during the spring and early summer, when bees are naturally inclined to rear queens. Remove these swarm cells and place each in queenless nucs (you can also use these larvae to prime queen cups). All of these can be used as a source for making your own queens. In addition, the bee population is more manageable now so you can locate queens (or the lack thereof) easily. If you do make a mistake (queen introduction doesn't work) you have all summer to manipulate the colony you are requeening into a queenright state. Before mites, beekeepers often requeened weaker colonies coming out of the winter with new queens they purchased or raised. Now, requeening takes place all season long; however, it is still easier to requeen colonies in the spring.

Advantages

- Colony is less likely to swarm if it is requeened before the swarming season has started in your area.
- Vigorous egg layer will produce large bee populations, ready for subsequent honeyflows.
- Colony will enter the winter with a large population.
- Old queen is easier to find when colony numbers are low, as is the case in the spring.
- Bees are calm and less prone to sting, run, or rob, especially during a honeyflow.
- There is less chance of swarming the following

year, because queen will be only one year old.

- There is time to assess queen's performance and to replace her, if necessary.

Disadvantages

- Dependent on weather; if you hit a rainy or cold spell, it could be many days before you can re-queen the colony.
- Queen could be superseded or killed if inclement weather sets in and bees go hungry. This may be overcome by providing sugar water as needed.

SUMMER/FALL REQUEENING

With mite infestations, many beekeepers find it better to requeen later in the year. By doing so, they break the bees' brood cycle and consequently the mites' cycle (see Chapter 14), especially if they install queen cells. Queens have to emerge from their cells, take mating flights, and then begin laying, which creates a longer hiatus between brood cycles. This allows bees to clean out diseased or varroa-damaged brood (if hygienic), and for the colony to be treated to reduce mite populations.

Advantages

- There is less chance of swarming the next year.
- Colony enters winter with a strong population and a young queen.
- Young queens will lay more eggs in the late winter and spring than older queens.
- Colony emerges in spring with a high bee population ready for the honeyflow.
- Timing breaks the brood cycle, thus reducing disease and pest problems.

Disadvantages

- Hive is populous, making it difficult to find the resident queen.
- If no honeyflow is on, bees are prone to sting, rob, and run when hive is opened; robbing could be serious.
- Time-consuming.
- In fall, fewer opportunities to check if queen was accepted.
- Less time to assess queen's performance.
- Could end up with queenless colony and laying workers overwinter if the queen is not accepted.

QUEEN INTRODUCTION

Although many methods, including some ingenious ones, have been devised for introducing queens into colonies for the purpose of requeening, none can guarantee absolute success. Often the more time-consuming ones are the most likely to succeed.

It is generally agreed that no matter what method is employed, the most opportune time to requeen is during a honeyflow. If no honeyflow is evident, feed the colony to be requeened at least one week before and during the introduction, as well as after the new queen is installed, to simulate a honeyflow. All the methods listed here, except the division-screen method, require that the colony be dequeened (queen taken out to make hive queenless) for a period up to 24 hours prior to the introduction of the new queen.

During the time you are introducing the new queen, observe how the bees react to her. If the bees in the colony tightly cluster over the queen cage and appear to be trying to bite or otherwise injure her, there already may be a queen in the colony that you are trying to requeen. On the other hand, if you observe the bees forming a loose cluster and try to feed or lick the new queen, they are probably ready to be requeened. If the bees completely ignore the queen, you may have to try another one. It may be difficult to truly discern that the bees are ignoring the queen. Check back in one week, and if the queen is present and young brood is evident, then all is well.

The methods used for requeening can be divided into two categories: the *indirect release*, in which there is a delay before the bees have direct access to the queen, and the *direct release*, in which the queen is immediately released among the bees. Some of these methods can be combined with swarm control or making increases in the apiary (see "Prevention and Control of Swarming" in Chapter 11).

After you requeen a colony, check after a week or so to see if the new queen has been accepted. Look for eggs and young larvae; if they are present, you have a laying queen. If the colony has a queen other than the one you introduced (because it has a different mark or is unmarked and the newly introduced queen was marked), you can either assess the qualities of the queen that is present for a month or so, or requeen the colony again later in the year.

Indirect Release

Push-in Cage

The push-in cage is a tried and true method of introducing the queen (see the illustration of the push-in cage on this page), first used in the 1900s. This method is the best one to use, especially if you are introducing an expensive breeder or inseminated queen.

The cage is made by folding all four edges of a square or rectangular piece of hardware cloth at right angles, to form a box with a top and sides but no bottom. The top of the formed box should be at least 3 × 4 inches (8 × 10 cm) and the sides at least ½ inch (1.3 cm) deep (see the illustration). You can also make a bigger one, 5 × 8 inches (12.7 × 20.3 cm). Pinch the corners of the box into a triangle so they will fold cleanly on the edges. Use 1/8-inch (3.2 mm) hardware cloth that has been soaked in hydrogen peroxide and rinsed with water; brand new hardware cloth has toxic substances that will kill bees and queens. Plastic push-in cages are now commercially available, and ads for these cages can be found in the bee journals and online.

Dequeen the colony at least one day (12–24 hours) before installing the new queen. Feed the colony to be requeened with sugar syrup for a few days before you introduce the new queen, and then go in and find and remove or kill the old queen. When you are looking for the queen, use smoke sparingly, as too much smoke will cause the queen to run, making it more difficult to locate her. When you find her, kill or remove her.

After the colony has been dequeened, remove a frame that has capped brood and some honey, and

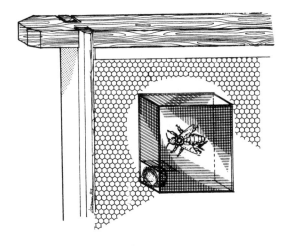

Push-in Cage

remove all bees from the area to be occupied by the push-in cage by shaking or brushing off the bees. Place the new queen on a patch of capped brood and honey and push the bottom of the cage into the comb over the newly obtained queen.

The comb under the cage should contain capped brood, a few cells of honey, and no adult bees. As bees under the cage emerge, they will care for the queen. After seven days, release the queen by removing the push-in cage. Some beekeepers cut a hole in one of the corners of this cage and plug it with candy made from a small amount of honey mixed with confectioners' (powdered) sugar or a piece of marshmallow (see the illustration); the bees will release the queen after eating through the candy. The disadvantage to this method is that the bees may eat under the cage, by scraping away the comb, and kill the enclosed queen.

Indirect Release with Queen Cage

This method of requeening employs a queen cage; here are the steps to follow:

Step 1. Dequeen the colony 12 to 24 hours prior to replacing with new queen.

Step 2. Select a caged queen with no attendants.

Step 3. Remove the cork in the candy end, and if the candy is hard, use a nail to make a small hole through it to make it easier for the hive bees to free the queen by eating out the candy. The hole should not be too large; one of the purposes of the candy plug is to delay the queen's release and thus enhance her acceptance. Do not make a hole if the candy is soft. Be careful not to impale the queen with the nail!

Step 4. Find a frame or two containing uncapped larvae and cut out some of the comb to allow the queen cage to fit vertically from the top bar, screen side out. Push in the cage, candy side up, and place the frame back in the hive. Now place a frame of young brood next to the queen and push the frames together so the bee space is maintained. (If you leave a gap, the bees will fill it with comb because they do not tolerate additional space.) Make sure the hive bees have access to the screened or open side (if plastic) of the cage and to the candy plug and queen. Alternatively, place the cage horizontally between the top bars of two frames of brood, screen side down, so the bees have plenty of access to the queen inside and to the candy.

Step 5. Remove the cage after a few days to prevent excess comb from being constructed in the additional space this technique creates.

Step 6. Observe how the bees react to the caged queen. If they are defensive, you may have a loose queen in the hive. If they cluster on the cage, trying to feed her and scent fanning, the hive will probably accept her. This observation can take place only by removing the cage after five minutes and looking and observing the activity taking place on the screened portion. This is time-consuming, and interpretation can be problematic but may be a good exercise for beginners.

Step 7. Examine the colony after one week; if the queen is still in her cage because the bees have not eaten through the candy plug, enlarge the hole in the candy to let the bees finish releasing her; or release her directly by pulling out the cork in the other end, or by pulling off the screen. When you are using a plastic queen cage, if the queen has not been released, poke a bigger hole in the candy plug and lift the lid that was originally used to insert the queen into the cage. This delay in releasing the queen due to a hardened candy plug has provided the bees more than sufficient time to "familiarize" themselves with the queen, unless during this period the bees began to raise a replacement queen. If no replacement is being raised, the caged queen will more than likely be accepted whether the release is further delayed or the queen is directly released.

Direct Release

Nucleus Method

This method does not require the attendant bees to be removed from the queen cage. Remove two or three frames of capped and emerging brood from a strong colony and shake or brush off all the bees back into the parent colony. Place these frames, now free of bees, in a three- to five-frame nucleus hive; fill the rest of this nuc with honey and pollen frames. Directly release the queen (and her attendants) onto a frame of the soon-to-be-emerging brood; any young workers will immediately accept the new queen. Use this method during the hot summer months so the brood (and the queen) will not be chilled on cool

nights. On the other hand, make sure the temperatures do not get so high that they overheat the nuc. This method should be tried on a limited basis until you have a sense of its effectiveness. You can save this nuc and use it later to requeen an established colony.

To requeen an established colony with the now populous nuc, follow these steps:

Step 1. Dequeen the colony to be requeened at least one day before.

Step 2. Place the nuc containing the new queen next to the dequeened hive.

Step 3. Apply a small amount of smoke into the nuc entrance, being careful not to disturb the bees too much.

Step 4. Remove two or three frames from one side of the dequeened hive and replace them with the frames from the nuc box containing the bees and the laying queen; the new queen should be between two of the inserted frames. Any extra frames can be given to a weak colony or to start a new nucleus colony.

Step 5. Close the hive and check after one week to see if the new queen was accepted. The only way to confirm acceptance of the nucleus queen is if she is laying eggs (and is marked).

Another nuc method is similar to the one just described, except first the attendant bees are removed from the queen cage. Remove frames of capped brood from a hive and shake off any attached bees. Place these frames in a nuc and directly release the queen onto these frames, then replace the cover. Now remove two or three frames of young, uncapped brood, making sure the resident queen is **not** on these frames, and shake these frames containing young nurse bees in **front** of the nuc so the bees can enter it. Move this new nuc to another location and check it in a week. Once the queen is accepted, you can requeen another colony later, as outlined above.

Honey Method

Dequeen the colony at least a day in advance, and then proceed as follows:

Step 1. Open the hive and remove the nearest frame; check each frame until you find one with young larvae and honey; remove it, shaking off all the adult bees.

Step 2. Break the wax seal over some honey and, without injuring the new queen, coat her with honey.

Step 3. Release the queen on the frame with young larvae and then gently return the frame to the hive; replace the remaining frames and close the hive.

Step 4. Check in one week to see if the queen was accepted.

Scent Method

The scent method employs a scented syrup (a few drops of peppermint, lemon, vanilla, wintergreen, onion, anise oil, or grated nutmeg per cup of syrup) that temporarily masks the odor of the introduced queen. As the scented odor gradually diminishes, the queen acquires the odor of the hive and the bees will accept her. Dequeen the colony (or nuc) at least a day in advance and feed it with the scented syrup. Then proceed as follows:

Step 1. Remove two or three frames containing bees and spray them with the scented syrup, but do not soak the bees. Try not to get the syrup inside uncapped larvae, as it may kill them.

Step 2. Spray the queen and release her onto the top bar of a sprayed frame. Guide her down between two sprayed frames.

Step 3. After she has crawled down, close the hive and then feed the bees with more scented syrup.

Step 4. Check after one week to see if the queen was accepted.

Smoke Method

Dequeen the colony at least a day in advance, and then proceed as follows:

Step 1. Blow four or five strong puffs of smoke into the entrance. You can also scent the smoker fuel with one of the scented oils listed above.

Step 2. Reduce the entrance to at least 1 inch (2.5 cm) with loosely packed grass, rags, or newspaper.

Step 3. Close the entrance for one to two minutes.

Step 4. Open the entrance slightly, release the queen, and guide her into the entrance; smoke a few puffs after she enters.

Step 5. Close the entrance for three to five minutes.

Step 6. Reopen the entrance to about 1 inch wide again and walk away; the bees will remove the remaining material within a few days if it is loosely packed.

Step 7. Check after one week to see if the queen was accepted.

Caution should be used with this method if the weather is extremely warm, and the colony especially populous, because the reduced entrance could make it difficult for bees to ventilate and cool the hive.

Shook Swarm Method

This method requires a nuc box that has a screened bottom, called a *swarm box*. You can make this by screening the bottom of a nuc box and attaching some legs to keep the bottom off the ground. A swarm box is normally used to collect swarms and does not contain an entrance, as it is a collecting box.

Dequeen the colony at least one day in advance, and then proceed as follows:

Step 1. Remove four or five frames with bees and spray the bees and frames with syrup (can be scented syrup).

Step 2. Shake these queenless bees into the swarm box or another container (the container should be large enough so bees are not overcrowded and should be screened for ventilation). After the bees have been "shook" off the frames into the screened box, remove any frames with brood and give to other colonies (or use in nucs for new queens, as described previously). Add enough frames of dry comb, honey, or foundation to the hive to fill the box; cover. You can make a feeder lid by drilling a hole large enough to hold a quart jar, and nail or staple hardware cloth over the opening. To feed this "shooked" swarm, place the feeder jar over the screen hole.

Step 3. Put the shook bees in a cool, dark place and feed them with 1:1 sugar : water syrup for at least eight hours.

Step 4. The next morning, dump the bees directly in front of their old colony (like it was a swarm) and release the queen on top of the bees; she can be sprayed, along with bees, with a scented syrup. Alternatively, you can introduce a caged queen into the shook bees for eight hours before shaking the bees back into their original hive.

Snelgrove, Division Screen, or Screen Board Method

A Snelgrove or screen board, sometimes called a division board, is a double-screened, rimmed board the size of the inner cover, which has a small entrance on one side of the rim (see the illustration of the Snelgrove screen on this page). It is used to make an increase or a split, or to start a two-queen colony. The screen separates the queen and bees in the lower part of a hive from the queen cells or new queen and bees in a hive body placed above. The smaller colony above can take advantage of the heat generated from the colony below.

To requeen by this method, follow these steps (see the illustration on the division screen method on this page):

Step 1. Reverse the colony, placing the queen below.
Step 2. From any strong hive, remove three or more frames of capped brood from which bees are emerging; replace these frames with foundation or drawn comb.
Step 3. Shake the frames so all adult bees fall back into the parent hive.
Step 4. Place these brood frames in the center of an empty hive body (number 3 on the illustration of division screen method of requeening).
Step 5. Place frames of sealed honey with some pollen on each side of the brood frames.

Snelgrove Screen or Division Screen

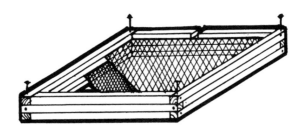

Step 6. Fill the remaining space with frames of drawn comb. The new hive should be composed of the following parts:
● One deep hive body with original queen (2).
● One shallow, medium, or deep body with new queen (3), and the inner and outer covers.
● Division screen placed between (1) and (3).
Step 7. Requeen the top hive body with a queen cell, ready to emerge, or a marked queen (by any method discussed previously). As the young bees emerge, they will accept the queen as their own.
Step 8. The entrance of the division screen should be small so only a few bees can pass through at one time (this entrance should be pointed in the opposite direction to the main entrance). Close this new entrance with loosely packed grass for a week until the brood emerges.
Step 9. Check colony after one week.

Division Screen Method of Requeening Existing Colony without Making an Increase

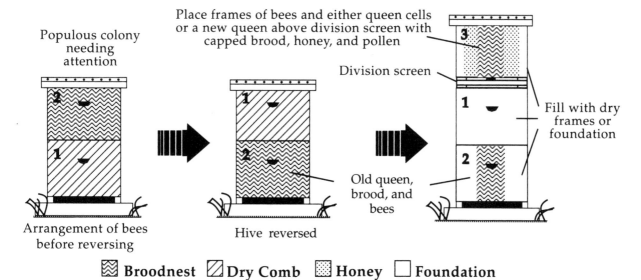

Three weeks later, replace the division screen with a queen excluder. The hive can now be run on a two-queen system (see below) until after the honeyflow, or the hive (3) can be united (with 1 and 2) by removing the excluder just before a major honeyflow. Find and remove the older queen or just unite the colonies by removing the excluder; usually the older queen will be killed. Alternatively, you can make 3 a separate colony by placing it onto a new bottom board when the bee population is high.

TWO-QUEEN SYSTEM

Some beekeepers use two-queen systems of colony management to improve honey yields, especially in marginal areas. Two separate colonies, one above the other and each with its own queen, are joined, but a queen excluder is placed between them to protect each queen from the other.

There are various methods of managing hives with two queens; see diagram p. 139. Here is one method:

Step 1. Split a very strong colony and requeen the upper queenless portion by the division screen method, placing only capped brood above the screen (see the illustration on the two-queen colony manipulation on p. 139).

Step 2. Replace the division screen with a queen excluder when honeyflow is underway, supering both colonies for honey. You can also put an excluder above and below a honey super to further separate the queens and keep them from killing each other through a single excluder.

Step 3. At the end of the summer, remove the excluder and either kill the less desirable or older queen (or let the two fight it out) or split the colony into two separate ones, each with its own queen.

Advantages
- Strong hive for wintering if queen excluder is removed and colony enters the winter with one resident queen.
- Better yields, since one large colony will produce more than two separate colonies, each having half the strength (and fewer than half the foragers) of the large one.
- More bees, which can be used later for making an increase or for strengthening weak hives.
- If one queen fails, you have a backup queen.

- Method can be combined with swarm control and requeening techniques.

Disadvantages
- Even if all manipulations are completed successfully, they will not be effective if local honeyflows and colony manipulations are not synchronized.
- Time-consuming.
- Difficult to operate.
- Hard to make work successfully.

REQUEENING DEFENSIVE COLONIES

Occasionally, a colony exhibits extreme defensive behavior, such as flying at and stinging with little provocation. These colonies, at peak population, are very difficult to work, and if they release excessive amounts of alarm pheromone when worked, the defensive stance of the other colonies in the yard could be elevated as well. If this happens, you may have to quickly abandon beekeeping activities and wait for a better day.

If this colony is close to homes, schools, or high-use areas (such as parks or sports fields), it could become not only a public nuisance but also a liability. Any stinging incident results in unhappy neighbors and could lead to zoning regulations restricting beehives in residential areas.

If you experience defensive bees, first check for signs of skunk or other pest activity to see if their actions are the cause of the heightened level of defensive behavior (see "Animal Pests" in Chapter 14). If you live in regions of the United States where Africanized bees have become established (Texas, New Mexico, Arizona, southern Utah, southern California, Oklahoma, Arkansas, and parts of the Gulf states including parts of Florida as of this writing), extreme defensive behavior may indicate your hives are occupied by Africanized bees; these bees exhibit a highly defensive posture with a minimum of provocation and could be very difficult to requeen. Check the USDA website for the map of the current locations of this bee.

The Tucson USDA-ARS bee lab and the Department of Agriculture in Florida have the necessary equipment to differentiate Africanized bees from European races; see the References for addresses. However, there are many signs that can help you tell the two apart.

Africanized bees:

Two-Queen Colony Manipulation

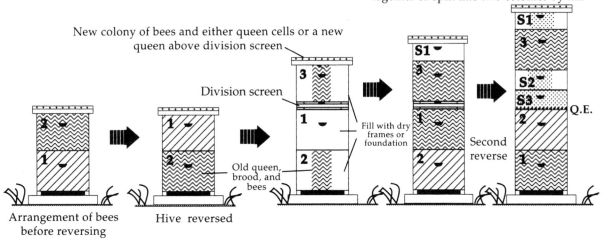

Super hives and replace division screen with queen excluder during honeyflow; join together or split into two colonies by fall

New colony of bees and either queen cells or a new queen above division screen

Division screen

Old queen, brood, and bees

Fill with dry frames or foundation

Second reverse

Q.E.

Arrangement of bees before reversing

Hive reversed

⌗ **Broodnest** ⌗ **Dry comb** ⌗ **Honey** ☐ **Foundation** --- **Queen excluder (Q.E.)**

- Show extraordinary defensive behavior, many bees running on the combs and lots of stinging bees in the air and on your veil.
- Sting without provocation such as when entering the vicinity of the apiary.
- Pursue you far from the apiary; see Appendix E.
- Sting animals, sometimes fatally (usually tethered or otherwise constrained and unable to run away).

Even European bees may have genetic traits for a heightened defensive response. In each case, the remedy requires that such colonies be requeened immediately with a purchased queen of known gentle stock, or that the colony be destroyed outright.

After you successfully requeen, the change in the behavior of the colony will be apparent as soon as the new queen's offspring populate the hive. Thus it will take almost two months for all the existing workers and larvae to be replaced; this may be unacceptable if a more immediate solution is needed. You may have to remove the colony to an isolated location until such time as the new workers replace all the defensive ones.

Regardless of which method you use for requeening, the hive first needs to be dequeened. Finding the queen of a volatile hive is not easy, especially if the colony is populous.

Here is one method of requeening called the non-shook swarm method:

Step 1. During a favorable day, when most bees are out foraging, move the colony to a new location.

Step 2. If you suspect the bees are Africanized, place an empty deep hive body on a bottom board with a cover, in the old location. The foragers that return will collect in this equipment. In the evening, kill these bees; pest removers in Africanized bee areas often spray bees with soapy water to do this.

Step 3. If the bees are European but highly volatile, place in the old location a deep hive body with all its complements, plus one frame of eggs and young larvae from a gentle colony. They should begin to raise a new queen on this frame. An alternative is to place a ripe queen cell or a caged queen from gentler stock into the colony. Add a frame of capped brood and fill the remaining space with dry, drawn comb. Field bees from the defensive colony will return to the old location and the new hive. Since they will live for only another three weeks, they will be replaced by the emerging brood.

Step 4. Back at the hive you moved, with its population reduced, the bees will be more manageable. Find the old queen and kill her.

Step 5. The next day, introduce the new queen by any method previously described.

Step 6. Wait seven days; then, if a caged queen was introduced, check to see if she has been accepted.

Step 7. At the end of two months, all of the undesirable bees should be replaced by the new stock.

The non-shook swarm method, while effective, is very time-consuming, so if possible, try to requeen the problem colony at its original location if the queen can be found quickly. Or you can separate the brood chambers with one or more queen excluders. Return in four days: the queen will be in the super with eggs. If the hive is too populous, split it (separating the hive bodies and supplying each its own bottoms and tops), wait four days, then dequeen the split containing the queen. Requeen one or both splits or unite them after one has been successfully requeened.

QUEEN REARING

Natural Queen Rearing

The simplest way to rear queens is to let the bees make their own queens. Kill off the older queen so the bees make emergency cells. This method has risks, as the bees may fail to rear a new queen or the daughter queen may be inferior to her mother. A superb queen can probably be found in any apiary with five or more colonies. Obviously, if queens could be raised from the larvae of such a colony and later be introduced successfully to other colonies, the entire apiary could be upgraded, but this could also lead to inbreeding unless you have 40 or more colonies (see "Queens" in the References).

Good queens are reared by bees in a strong colony when there is an abundance of food (honey or sugar syrup and pollen) available to the nurse bees. Queen-rearing operations can coincide with swarm control manipulations (see "Prevention and Control of Swarming" in Chapter 11).

Once you have killed the old queen, provide the queenless colony with a frame of open brood with eggs and young larvae taken from a colony with desirable qualities. You can also take extra queen cells from colonies preparing to swarm.

For more control, split a populous colony into two or more smaller ones or in nuc boxes, each with its own top and bottom and with entrances facing a direction different from that of the parent hive. Let the bees raise their own queens, making sure either some queen cells (swarm cells) or the right-aged larvae or eggs are available in each nuc you start. The chances are high that one of these splits will take and you can always choose one daughter to replace the mother.

Advantages

- Can be easy.
- Inexpensive.
- Will usually succeed in obtaining queens.
- Few manipulations needed.
- Can coincide with mite control, as it breaks the brood cycle.

Disadvantages

- Queen could be inferior or may mate with inferior drones.
- Cycle of the dequeened colony will be disrupted; no new brood will emerge for up to 43 days.

Conditions Needed to Rear Queens

Some beekeepers prefer to raise their own queens rather than purchase them from a commercial breeder. While educational and exciting, rearing queens can be tricky, time-consuming, and often unsuccessful, but with practice it can be accomplished.

Before you begin to rear queens on a larger scale, read books on the subject, talk to beekeepers, and take a class (see "Queens" in the References). Hands-on practice is essential for success. Then you should choose whether you want to graft larvae into artificial queen cups (Doolittle method) or cut out comb containing such larvae and let the bees make their own queen cells (Miller method).

Next you should ask yourself three questions: Why? What? and When? The answer to why was already covered at the beginning of this section—to replace inferior queens or to increase your colony numbers.

The next question: What is a good queen? Pick your breeder queens carefully, from stock that you can trust and that has the characteristics you seek. Decide if you want to instrumentally inseminate the queens or open-mate them, and what apiary has the best drones. Then make sure your queen larvae are fed copious amounts of food, because the number of ovaries is directly proportional to the size of the abdomen—bigger queens are generally the better queens.

Finally, the last question you need to ask: When do you want to rear them? Prepare and organize your bees and yourself so you can have abundant workers, food, and drones at the right time.

Conditions needed for successful queen rearing include:

- Abundance of workers.
- At least 20 pounds (7.5 kg) of honey.
- Plenty of pollen or pollen substitute.
- Intense honeyflow, or feed syrup (1 water : 2 sugar) three days before you start rearing queens.
- Plenty of mature drones (14 days old) available one week after the queens emerge.
- Superior queen mothers.
- Large populations of young "nurse" bees.
- Correct-aged larvae from which queens can be reared.
- DCAs with genetically diverse drone population.

While there are numerous methods of queen rearing (see "Queens" in the References), it falls outside the scope of this book to go into great detail on this subject; however, some simple methods are given below from which you can start experimenting. You will need breeder queens; starter, cell-builder, and finisher colonies; mating nucs; mating yards; and drone mother colonies.

Breeder Queens

First, you must select your best queens from which to raise new ones; these are your *breeder* queens. Carefully choose these breeders by testing 3 to 20 colonies (depending on how many colonies you have) that possess the characteristics you wish to perpetuate in your bees.

To choose which queens are best, give some sort of test to potential breeder queens. The weather conditions, nectar/pollen flow, and temperature should be similar. Grade the bees, with a letter or number grade for each option, and select at least the top five colonies with the highest grades (see the sample preselection data sheet on p. 142). Here are some characteristics on which to score your colonies; there may be others, so add or subtract them from your list:

- Brood production. A compact, solid brood pattern means good brood viability; a spotty brood pattern indicates larvae were removed because of poor viability, poor mating, or disease.
- Disease and pest tolerance. Hygienic behavior means bees uncap and remove dead or infested larvae in less than 24 hours. Grooming behavior means some bees remove varroa mites from each other; also important are tracheal mite resistance, wax moth resistance, and chalkbrood removal.
- Overall population. How populous is the colony compared with others?
- Propolis. Does the colony use excessive or low amounts?
- Temperament. Test without smoke; the size of the colony is not important. Grade from 1 (bees do not react) to 5 (bees sting, smoke needed).
- Composure on comb. How quiet or runny are the bees when you examine frames?
- Pollen arrangement and hoarding. Is there a clearly defined ring of pollen around the broodnest or is pollen scattered? Gauge amount of pollen stored in frames, especially in the fall.
- Honey production. Also note the ability of the bees to move up into the supers.
- Beeswax. How fast are the bees drawing out foundation? Also note the color or consistency of the cappings—white or wet?
- Swarming tendency. Note whether the colony swarmed and, if so, how many swarms it cast.
- Robbing tendency. Score the number of times robbers were seen at other colonies by dusting robbers with flour and following them back to their home colony. Conversely, is the colony constantly being robbed or nondefensive?
- Wintering ability and spring buildup. How many frames of bees are found in spring and the rate of buildup under early spring conditions?
- Flight time. How late and at what temperature do the bees begin and end foraging?
- Body color. Are the bees light or dark? The evenness of the color of workers and drones indicates purity of mating.

Remember, records should be kept on **each** queen you choose to be a breeder. It would be helpful if you had similar records for your drone mothers too, as the worker offspring of your new queens will carry the genetic traits of both the breeder queen and the drones she mated with. For more information, check "Queens" in the References; you can also

Preselection Data Sheet for Breeder Queens

Date_____	Queen no._____		Queen origin _____		Date mated _____	

	Score					
	5 *Exceptional*	4 *Excellent*	3 *Average*	2 *Fair*	1 *Poor*	0 *Unacceptable*
Characteristics						
Brood viability						
Temperament						
Spring buildup						
Overwintering						
Pollen hoarding						
Cleaning/hygiene						
Honey production						
Disease	–1	–2	–3	–4	–5	Eliminated
Type						
Mites	–1	–2	–3	–4	–5	Eliminated
Type						
Color	Golden					Black
Queen[a]						
Workers						
TOTAL SCORE						

Source: Adapted from Susan Cobey, Apiarist, University of California, Davis.
[a] *Note:* Lighter queens are easier to see, but that may not be important to you.

check the Internet or your local bee club for more information.

GRAFTING METHODS

Equipment for Queen Rearing

You will need to purchase or make grafting spoons (to pick up young larvae), queen cups (from which the bees make queen cells), and grafting bar frames (a frame that holds queen cups).

Grafting means to physically remove a 1- to 2-day old larvae (age is extremely important) with a grafting tool, and place it into a prepared cup with or without royal jelly. This procedure is tricky to do and takes much practice. Nimble fingers and keen eyes are needed when grafting these small larvae. This is where taking a class on queen rearing would be invaluable.

Doolittle Method

To graft worker larvae into artificial queen cups, you might employ the Doolittle method, named after G.M. Doolittle, a beekeeper who wrote extensively on the subject around the turn of the nineteenth century; his book has been reprinted recently (2008). He found that priming the queen cups with royal jelly and feeding the starter colonies were imperative to rearing good, healthy queens. Here are the major steps:

Step 1. Fit an empty frame with two or three bars to hold the queen cups. Some frames have a 3-inch (7.6 cm) strip of foundation above the bar (see the illustration of the Doolittle frame on p. 143).

Step 2. With melted beeswax, attach the wooden bases of queen cups (or plastic cups) to the underside of the wooden bar. If you are using plastic

Doolittle Frame

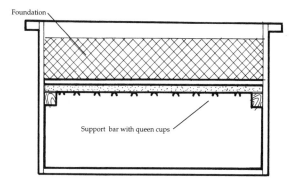

Foundation

Support bar with queen cups

Grafting Tool

equipment, drill holes in the bars in which the cups will fit. You can insert this frame with cups into a colony for the bees to clean and polish.

Step 3. Two days before you transfer larvae into these cups, dequeen a strong, two-story hive and feed it with syrup.

Step 4. The next day, shake the bees off every brood frame in the dequeened hive and remove all queen cells; these cells will provide you with royal jelly. Collect the jelly with a small spoon, eye dropper, or siphon; you can store the royal jelly in a sterilized jar in the freezer.

Step 5. Once you have enough jelly, remove a frame of larvae less than three days old (larvae 24 hours or younger are ideal) from one of your breeder colonies; you can cage the queen in a push-in cage overnight to ensure the correct-aged larvae can be collected. The correct-aged larvae are about the size of an egg; if you can easily see the larvae, they are too old.

Step 6. Prime the queen cups with the stored royal jelly; place it in the queen cups, covering the bottom. Dilute the jelly, if it is too thick, with about 25 percent boiled, distilled water. Do not be tempted to lick the spoon or grafting tool.

Step 7. With a grafting tool (see illustration of this tool on this page) or toothpick (carve one end flat and curve it slightly upward), transfer the larvae from worker cells into the primed queen cups. Place the larvae *on top* of the royal jelly, being careful not to drown them in it, and in the same

position they were found before the transfer. Because these larvae are about the size of an egg, it may help you to use a lighted magnifying glass to see them. Grafting should be done in a hot, humid room to keep the larvae from drying out or chilling. Cover the frames with a damp towel.

Step 8. Insert the finished frame, with the now-filled queen cells, into the dequeened starter colonies.

Starter Colonies

A day before you add the queen cups, make up a hive full of queenless young bees, capped brood, and lots of food.

Rearrange the frames in the dequeened colony so that the lower chamber has mostly sealed brood; the upper chamber should have (in order) a frame of honey, two frames of older larvae, a frame of young larvae, space for a frame with queen cells, a frame of pollen, one of older larvae, and one of honey.

Feed pollen patties and syrup to make up for any shortfalls. Give 40 to 50 queen cups to the starter, keeping them there for 24 to 36 hours.

Because this colony is queenless, it will not be strong enough to cap all of these cells. The purpose of the starter colony is to provide the queen larvae with copious amounts of royal jelly. If you are planning on reusing the starter colony, join it to a queen-right colony until you need it, or give it three or four queen cells to make their own queen.

Once all the queen cells you introduced are full of jelly, remove the frame of cells from the starter colony and divide them into cell-builder/finisher colonies.

Cell-Builder/Finisher Colonies

Each cell-builder/finisher colony should be a queen-right colony of two or three deep bodies, full of healthy bees. Prepare this colony a few days ahead of time by confining the queen to the lowest super below a queen excluder. Give each cell builder a bar with about 20 queen cells, right above the broodnest over the excluder. Feed the colony heavily with both pollen patties and syrup and place the frame holding the queen cups between frames of capped, emerging bees and young nurse bees. Pollen is critical for the nurse bees to produce the royal jelly. Check in a few days to see if the cells are getting enough jelly. Sacrifice a few capped queen cells to check the jelly levels—there should be lots of jelly in the cell, and the queen larvae should look plump and glistening white. Be sure that no miticides are used in this colony, as any chemicals can deform the queen larvae.

After five days, the cells should be capped. Queen pupae are **extremely delicate** at this time. **Do not** bump, cool, or overheat them, or your queens will be deformed. Know the age of the queen cells (keep a diary) so that one queen will not emerge too soon and kill all the others. Do not handle the cells until they are 9 or 10 days old (after grafting). Place ripe cells in a mating nuc (or in an incubator if you don't have enough nucs); each cell can be placed in a small vial to keep the queens isolated.

Mating Nucs

Because the first queen can emerge 11 days after grafting, make up the mating nucs a few days before they are needed. Bees added to (or making up) the nucs should be treated beforehand so they are disease and mite free; finish any treatment about 30 days before adding a new queen. Do **not** treat nucs with miticide strips as the chemicals can deform the queens. (Handle queens like you would honey: medicate or treat 30 days before supering for honey.) If mites or small hive beetles are a problem, use alternative control measure in the mating nuc. Move the nucs to the mating yard at least 1 to 2 miles away (1.6–3.2 km) from your home apiary, if they have the following:

- No queen.
- Two or more frames of mite-free brood/bees.
- Two or more frames of disease-free honey/pollen.
- An entrance that can be closed or restricted.
- Sugar syrup and pollen patty, if food is not incoming.

Transfer one or two queen cells into each mating nuc by cutting the cells from the frame bars. With a heated knife, cut the cell bases free from the wooden bars. Place cells into an insulated box (like a Styrofoam cooler) with a warm water bottle inside to transport them safely to the mating yard. After you select a nuc, wedge the cells between the top bars of two frames, making sure they get covered by bees immediately. Check food stores of each nuc and feed if necessary. Feed your nucs with fumagillin to control nosema if it is a problem or if the weather turns wet and cold. While you should not apply miticide strips to your nucs, you can use oil patties if tracheal mites are a problem (do not use in areas of small hive beetles). Close the entrance with grass or an entrance reducer to prevent robbing and make sure the boxes will not get chilled or overheated.

Mating Yard and Drone Mother Colonies

The nucs with the queen cells should be in or near your drone mother colonies. Again, these colonies must be free of varroa and tracheal mites before the queen cells are placed in the nucs.

With the reduction of feral bees, a result of the parasitic mites, it is important to have enough drones available to mate with your virgin queens. Because half the genetic material of your workers comes from their fathers, select drone mother colonies using the same criteria you use for your queen breeders. Choose only those colonies that show good characteristics (high-scoring colonies) but with a genetic stock different from that of your breeder queens, to avoid inbreeding.

Your goal is to produce large healthy drones, free of mites and diseases, that were adequately fed when they were larvae. Timing is everything so plan on having mature drones in your mating yard when the virgin queens are ready to mate; drones mature 10 to 15 days after they emerge.

Encourage drone production by placing one to three frames of drone foundation or comb into your selected drone mother colonies so the queen has ample room to lay drone eggs. You can insert frames half filled with regular worker foundation, or shallow frames in deep hive bodies, and let the bees draw out the rest of the frame. Most likely, they will draw out drone cells. Ensure there is enough food coming in, and if not, supply extra pollen and syrup. Large amounts of pollen will stimulate colonies to rear drones.

NONGRAFTING METHODS FOR REARING QUEENS
Miller Method

The Miller method of queen rearing, named after C.C. Miller, may be the easiest for the beginner and requires no special equipment. Prepare an empty brood frame by fitting it with a sheet of non-wired foundation. Once this frame is installed in the frame's top bar, cut triangles on the lower half of the foundation (see the illustration of the modified Miller frame on this page). An alternative method is to cut strips of foundation, roughly 4 inches wide and 6 inches (10 × 15 cm) long, and attach them to an empty brood frame. Cut the unattached lower half of each strip of foundation to form a triangle with its apex pointing downward (see the illustration of the alternative Miller frame on p. 146). Now follow this procedure:

Step 1. Move two frames of sealed brood and queen from a breeder colony into a nuc or specially prepared breeder hive.

Step 2. Insert the prepared frame between the two frames of the sealed brood.

Step 3. Make sure the queen is on one of the frames and that there are a lot of young workers.

Step 4. On either side of the brood frames, fill the hive with frames of honey and pollen (there should be no empty cells in these frames; otherwise, the queen may lay in them).

Step 5. The queen will be forced to lay eggs in the prepared frame as soon as cells are drawn (feed the colony if necessary).

Step 6. About one week later, remove the prepared frame; trim away edges of the newly drawn pieces of foundation until you encounter cells with

Modified Miller Frame

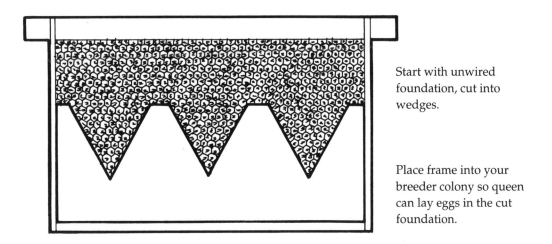

Start with unwired foundation, cut into wedges.

Place frame into your breeder colony so queen can lay eggs in the cut foundation.

After one week in a queenless colony, the frame should look like this:

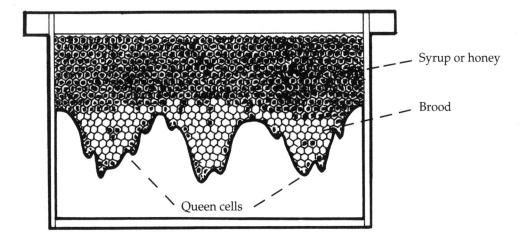

Syrup or honey

Brood

Queen cells

Alternate Miller Method

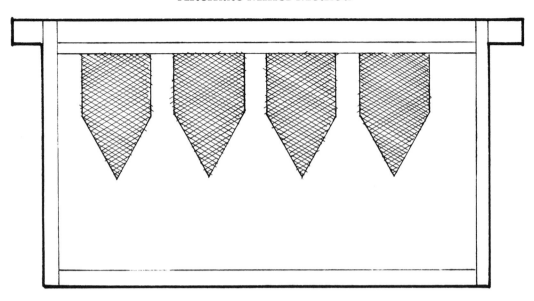

small larvae (preferably less than one day old, but never more than two days old).

Step 7. Give this frame to a cell-builder colony. You may have to trim the bottom of the triangles to reach the eggs and young larvae. These larvae should be on the lower edges of the foundation. You can insert another Miller frame into the breeder colony, or return the queen to her original hive. Or use this special hive to start a new colony, requeening the parent hive.

Step 8. Nine days after you insert the Miller frame into the cell builder, remove the sealed queen cells by cutting them from the Miller frame and attaching them to combs in queenless hives or nuc boxes, and place them in the mating yard.

Queens in these nuc hives will emerge and mate. These queens can then be left in the hives from which they mated or used for requeening after they have begun to lay eggs.

Note: Because much of the pure beeswax foundation today is contaminated with pesticides, you may need to use wax that your bees have drawn out themselves, with a commercial foundation base. If you have a lot of queen failures, consider having your wax foundation tested.

Queen-Rearing Kits

Bee supply companies now sell queen-rearing kits that do not require grafting and cutting comb. These come with good instructions, or you can search the Internet for instructions; read through the instructions first to determine if the kit will fit into your beekeeping operation. With it you can rear many queen cells, which you must treat like grafted queen cells, giving them to cell builders, finishers, and nucs. Some of these kits are the Nicot and the EZI-Queen systems; check bee suppliers and websites.

In general, the kits are composed of a plastic container the size of a comb honey section box. On one face of the box is a removable queen excluder, on the opposite side are evenly spaced holes into which a series of removable plastic queen cups are placed. The box, once fitted with the queen cups, is inserted into a frame, and then placed into a colony for a few days so the bees can clean and polish the plastic cells.

Make sure the colony is fed heavily for a week prior to and during this procedure. Then the breeder queen is introduced into the box, where she will deposit eggs into the queen cups. After the eggs have been deposited, usually within a day or two, the queen is removed from the box and released back into her colony. The next step is to take out all the cups containing eggs and attach them to a cell bar in the same way you would for hand-grafted cups; then place the bar into a cell builder colony.

The advantage of this system is that you do not have to hunt for frames containing the right-age larvae. All you need to do is check the plastic cups to see if the confined queen has laid eggs in them. Be aware that you have the potential to rear over 100 queens at once, which could tax your facilities if you are not prepared. The down side of this system is its cost and the many little pieces of equipment you need, which are easily lost or misplaced. Ask more experienced beekeepers about their successes or failures with the system before you invest in one.

Check for availability or for new products from bee suppliers or websites and follow the instructions that come with each kit.

 **Notes**

CHAPTER 11

Special Management Problems

This chapter covers some of the challenges in management that you are likely to encounter when working your colonies. Again, there are different approaches to tackling problems that confront you while inspecting your colonies. The information that follows will help guide you on how to recognize and resolve these challenges.

WEAK COLONIES

The goal of most beekeepers is to have colonies bursting with bees and storing surplus honey. However, it is not realistic to expect this to prevail in the beeyard. Weak colonies (low bee populations) will be encountered in every apiary. The percentage of weak colonies versus strong ones may reflect the beekeeper's management practices, or may be a result of other factors, such as a poorly mated queen (limited genetic diversity and/or insufficient sperm count), a failing queen, parasites, pathogens, pests (skunks), or adverse environmental conditions (weather prohibits cleansing flights and limits nectar gathering). Another factor is that hives may be located in poor foraging areas.

Prior to the advent of mites and other pathogens, a routine practice was to unite two weak colonies or a weak one with a strong one. Beekeepers are compassionate individuals and will take every step to save their bees from failing. Nonetheless, given that parasites and diseases are now common, before any consideration is given to uniting weak colonies or requeening them, you must determine the cause(s) for the decline. If it is due to diseases and/or parasites, further work with the colony will be wasting your time and money and will be inadvertently promoting

the spread of these problems to healthy hives. In situations in which the weakened colonies are diseased, check with your local apiary inspector for guidance (and see Chapter 13 on Diseases and Chapter 14 on Pests).

Uniting Weak Colonies

Weak colonies that are disease-free, including established hives, nucs, splits, newly installed packages, or swarms, can be united. These manipulations should be done by early summer at the latest in order to give united colonies the time to grow and put away reserves (honey and pollen) for the winter.

Weak colonies can be strengthened by adding a swarm, by joining two weak colonies, by requeening without joining, or, after joining, by adding a queenless package of bees or adding a frame or more of capped brood. However, there must be sufficient numbers of bees in any weak colony to be able to care for the added capped brood. Be judicious as to how you allocate capped brood to a weak colony.

When you encounter a weak colony or colonies late in the summer and during the fall, attempts to rescue them will in all likelihood fail. The best opportunity for rescuing these bees is to unite them to strong colonies. Once the bees become a single colony, empty hive bodies should be removed and stored.

A question that arises is what to do with any spare queen resulting from uniting colonies. If you have tackled the problem of uniting weak colonies by early summer, you may consider making some nuc colonies using the "spare" queens. This approach is a good way to determine the quality of the queen and if she was a factor in the colony's decline. If you unite the colonies and let the queens battle it out, you may lose the better queen. It would seem reasonable, all factors considered, that the queen from the weak colony may have been responsible for its weakened condition.

If you need to unite colonies later in the year—early fall, for example—be sure the joined colonies have ample stores of honey because they will not be able to collect nectar. Either feed them or provide frames of food from other colonies. Usually, beekeepers will knock out all the bees from a failing colony in front of a stronger one so they don't have to fuss with trying to baby a weak colony through the winter.

Whenever two colonies are being united, remember that each colony is capable of distinguishing between its members and those from another colony. Hive odors will be different, and unless some precautions are taken, the bees will fight. During a honeyflow, bees' ability to distinguish foreign bees is diminished, but in the absence of a flow, methods to separate the odors are necessary. If after these manipulations the existing queen fails to improve, the colony should be requeened. Also, continue to check for the presence of mites or other pathogens.

Newspaper Method

The most successful and least time-consuming method of uniting colonies is the newspaper method. This method works because the paper separates the hive odors, and as bees chew away the paper, the hive odors gradually blend. By the time the paper is gone, the bees act as one unit.

Uniting colonies using newspaper involves putting a single sheet of newspaper over the top bars of the stronger colony (see the illustration on p. 150). Follow these steps:

Step 1. Make sure the colonies are disease and mite free.

Step 2. If the weather is warm, make a few small slits in the newspaper with your hive tool to improve ventilation.

Step 3. Set the weak hive on top of the paper and cover; field bees from the weak colony's original hive site will drift to other hives. If these bees are defensive, they should be destroyed by collecting them in an empty super at the original site, then killing them with soapy water. Bees will slowly

eat through the newspaper, with most of it being chewed up within a week. Shredded paper will appear at or near the hive entrance.

Step 4. If the weather is extremely hot (day temperatures over 90°F [32°C]), wait for a cooler day or unite the hives during the late afternoon.

Step 5. If there is a dearth of nectar and pollen and the bees are unusually defensive, decrease the possibility of fighting by feeding syrup to the stronger hive for a few days before uniting.

Step 6. You can also sandwich a weaker colony between hive bodies; place one sheet of newspaper below and one sheet above the weaker hive and slit both sheets.

If after these manipulations the colony fails to improve, check again for the presence of mites or other pathogens. If such are present, either destroy it or treat it accordingly (see Chapter 14).

MOVING AN ESTABLISHED COLONY

Preparing the Colony

Sometimes it is necessary to move a colony of bees. Follow the procedures described below to move an established colony safely. It is a general practice to move a hive at least 3 miles (4.8 km) from its old site; if it is moved a distance of less than 3 miles, many field bees will return to the old location or drift to other colonies. The best time to move hives is in the spring, when populations are the lowest and the hives are light in weight.

You must first check and comply with all legal requirements pertaining to moving and selling bees. Have the state inspector certify that the bees are free of diseases and mites, especially if you are moving bees out of state, or selling or buying them. Then make your new location ready to receive the new hives—that means the apiary is mowed and/or fenced (see "The Apiary" in Chapter 4).

If the hive consists of more than two hive bodies, remove the extra honey supers (providing they are bee free) to make it easier to lift. Other methods for moving colonies larger than two deeps are to remove bees by using a blower and to shake or brush the bees from the frames onto the grass in front of the hive. Of course, these manipulations must be done a few days before the hive is moved. You can also get rid of bees from supers by using bee escapes in the inner

Uniting Colonies Using Newspaper Method

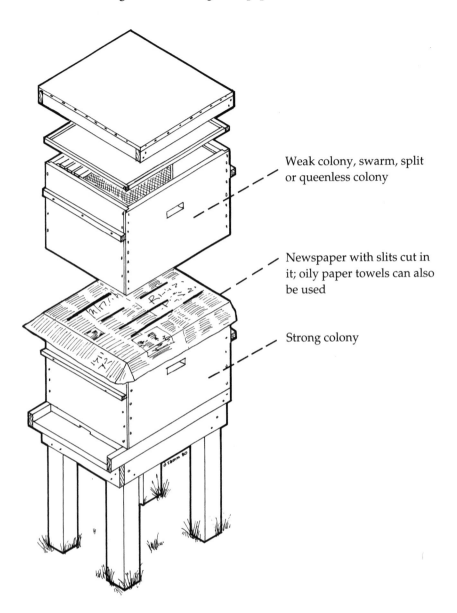

Weak colony, swarm, split or queenless colony

Newspaper with slits cut in it; oily paper towels can also be used

Strong colony

cover, a fume board, or escape board (see "Removing Bees from Honey Supers" in Chapter 9). If a colony is very congested, the bees may suffocate or overheat if confined without proper ventilation.

Move a strong colony with the help of a mechanical lift or extra strong (and willing) volunteers, or in manageable pieces, leaving extra hive bodies at the original site to pick up stragglers. Alternatively, leave a weak colony to collect the stray bees and move it later.

On the day of the move, early in the morning, smoke bees inside and close all entrances. You can also move the colonies at night.

Step 1. A few days before the move, staple bodies with hive staples or tie the bodies together with metal or nylon ratchet straps or bands. Use plenty of smoke. The very best way to hold a hive together, especially if you have a lot of hives to move, is to invest in these straps; this way you can strap the hive furniture together and even use the straps to lift it. This method is far superior to fastening hive bodies with nails or staples, which do not adequately secure the hive equipment (but can do in a pinch) and leave holes in your equipment.

Step 2. Tape or screen all holes and cracks in the

Moving an Established Colony

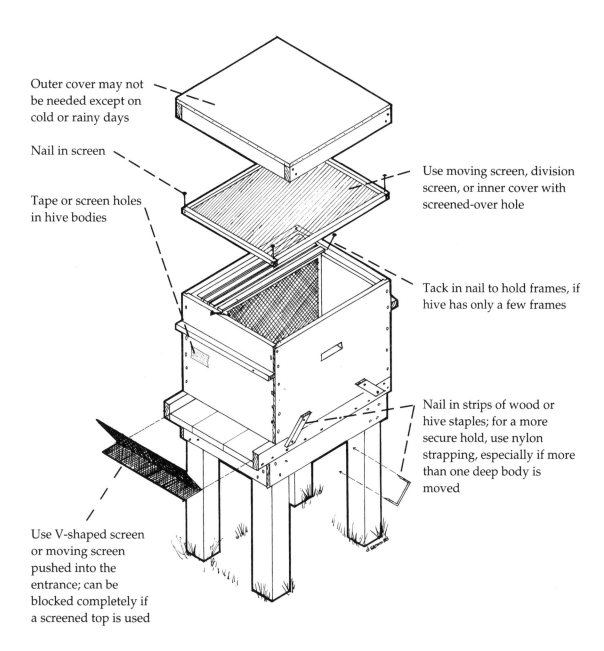

Outer cover may not be needed except on cold or rainy days

Nail in screen

Tape or screen holes in hive bodies

Use moving screen, division screen, or inner cover with screened-over hole

Tack in nail to hold frames, if hive has only a few frames

Nail in strips of wood or hive staples; for a more secure hold, use nylon strapping, especially if more than one deep body is moved

Use V-shaped screen or moving screen pushed into the entrance; can be blocked completely if a screened top is used

hive bodies (see the illustration on moving an established colony on p. 151).

Step 3. If the hive is very populous despite using the methods above, or if the weather is very hot, add a shallow super with empty frames above the top hive body to collect the overflow of bees; otherwise bees might be hanging outside the hive when you return to move them.

Step 4. On the day of the move, close the entrance with a roll of newspaper, stuffed grass, or an entrance reducer kept in place (and covering the

hole) with duct tape. You can also use a V-shaped piece of screen, but it may need to be taped or stapled in place.

Step 5. If the weather is warm, place a screened board (like a division screen) or screened inner cover on top of the hive under the outer cover.

Step 6. If you are moving colonies during hot weather, take off the outer cover and secure the screened top with the bands. Replace the outer cover until you are ready for the move, and then remove it when you move the hive. You can also use a screened bottom board to increase the ventilation (check with bee suppliers for this).

Step 7. In the evening, near dusk, smoke the entrance to drive the bees inside and use a piece of screen the length of the entrance and about 5 inches (13 cm) wide to close off the entrance. Slide a V-shaped piece (or a roll) of screen into the entrance so it will spring against the bottom board and hive body, and secure it with duct tape, nails, or staples. Entrances can be closed completely if a screened top replaces inner and outer covers. You can move the hive that evening or early the next morning.

Loading and Unloading Colonies

Once the hive bodies are strapped together, load the bees onto a truck. **Do not** move a hive in the trunk of your car or in the back of a closed van—loose bees inside a vehicle are dangerous to the driver.

If the weather is hot, remove the outer cover while in transit so the screened top is exposed. The hives should be packed close together on the truck, with the frames parallel to the road; this will prevent the frames from sliding together if the truck stops suddenly. While you are loading and unloading the hives, keep the engine running because the vibration of the vehicle will help keep the bees in their hives.

Once all the hives are loaded, secure them to the truck with ropes or other tie-downs. The object is to keep the hives from shifting while you drive them to their new homesite. Hives that shift, especially off their bottom boards, will make unloading difficult. If you are moving the colonies great distances, purchase a special net that will cover the hives and confine all the loose bees. Make sure the lids are secure as well, as they can blow off easily if they are not tied down; rocks or other weights can shift and fall off, so strap them or tape them down with duct tape.

Once you are at the new location, make sure the site is ready to accept the hives and you can drive close to it. Smoke the entrances just before you unload the hives and then just before you remove the entrance screens. Unload them all, and **only then** remove the screens, as you are leaving. If you moved during the day, fill the hive entrance loosely with grass to slow the bees' exit and to keep them from drifting. If they exit slowly, a few bees should be able to come out and scent at the entrance; this will help any loose bees to relocate at the hive. You don't really need to do this if you move at night; just open the entrance after the colonies are all in the new location.

If this is the final site for these hives, replace the top screened board with the outer cover. Even if this is a temporary spot, replace the outer cover, to keep rain out. Inspect the hive after a few days to see if all is well.

Problems associated with moving hives are as follows:

- Hives could shift off their bottom boards, or hive bodies break, permitting bees to escape. If moving old, leaky equipment, cover all the hives with a traveling screen or other netting to contain loose bees.
- Bees can suffocate if weather is too hot.
- Queen could be killed, injured, or balled.
- Hive bodies and combs could break.
- If moving in winter or very early spring (when temperatures are below 50°F [10°C]), the winter cluster could break apart. Bees could then recluster on empty combs and starve, or existing brood could become chilled before the bees have a chance to cover the brood.

Moving Short Distances—I

Follow these procedures to move an established hive *less* than 3 miles:

Step 1. During the day, move the original hive off its stand to the new location.

Step 2. In its old location, place a nuc box or one deep hive body with a bottom board and top cover.

Step 3. Fill the hive or nuc box at the old location with dry comb frames and one frame of brood from the original hive, with or without a caged queen or queen cells. Field bees from the original hive will return to the new box at the old location; this becomes a new, smaller colony that can be moved in about a week.

Step 4. When ready, move this small hive from its original location to a new site at least 3 miles away.

Step 5. After about two weeks, this hive may be moved to the desired location; you have split the old colony and created another.

Moving Short Distances—II

Move all the hives, leaving no stragglers behind, over 3 miles away for three weeks and then move them again to the desired location.

It is often recommended that when you are moving established hives very short distances, each hive be moved 1 to 4 feet (0.3–1.2 m) every few days until they are at the desired location (few if any bees will return to the original location). But this process is slow and not recommended unless the distance is less than 30 feet (10 m).

ROBBING

You will be amazed how quickly bees can start snooping around to see if they can get some free food. Hence, occasionally, bees will collect nectar and honey from other colonies. This type of bee behavior, referred to as *robbing*, usually occurs when bees are unable to obtain food from flowers. Whenever weather is suitable for flight, foragers will set out in search of food. If plants are dormant or blooming but not yielding nectar (because it's too hot or too windy), bees continue to search and are attracted to the odor of ripening nectar and honey stored in other colonies. You can recognize robbing bees by their weaving flight activity at the hive entrance (see "Robbing Flight" in Chapter 2) and by the presence of fighting bees on the landing board.

Nucleus hives, queen-mating boxes, and other colonies low in population, as well as hives with openings resulting from disrepair, are the most at risk of being robbed. If you observe increased activity at the entrances of these hives, you may have a robbing problem.

The robbing bees return home and communicate (by dancing) the location of the target hive from which the food was taken, and soon additional foragers, recruited to the unlucky colony, remove the remaining stores. The weakened colony, fighting off the thieves, may be severely affected, even die. Some robbing likely takes place in apiaries whenever there is a dearth. Be alert to any weakened colony when you visit the apiary. If you find one, then check that it and the queen are still intact, then reduce the entrance of the hive and move it to a location with fewer strong hives.

The best way to prevent robbing activity is to maintain colonies of equal strength, keep the entrances of all weak or small colonies reduced, avoid open feeding of sugar syrups or honey (as this stimulates robbing), and minimize or avoid hive inspections during a dearth. If examination of colonies is imperative, cover any exposed super with a *robbing cloth* (an old sack or sheet) or place removed frames in a covered, empty hive body. Moving frames also allows honey from burr or brace comb to drip, attracting more bees. Quickly finish your examination to keep robbing bees to a minimum.

If you must feed during a dearth, do so with in-hive feeders (see Chapter 7) and use sugar syrups instead of honey. In the fall after you take off honey, bees are also prone to robbing, as broken combs can drip honey, attracting robbers. If newly extracted "wet" frames are to be cleaned (see "After Extracting Honey" in Chapter 12), place them on strong colonies only in the late afternoon. Or put wet supers on all the colonies, to keep them busy cleaning, and not robbing each other. Some beekeepers put all the wet supers out in the open for all bees to clean, but this may not only spread foulbrood but also entice other robbing insects, such as yellow jackets and hornets (and bears), which might not differentiate between empty supers and colonies of bees. In the fall, supers of freshly extracted combs placed above the broodnest should be removed after they are cleaned, since the bees may move up into them over winter and remain there until they starve.

If a robbing frenzy has started on a weak or smaller hive, close up the weak hive and dust robbers with flour. By following the dusted bees, you will find the robbing colony (or colonies). Severely reduce the entrance of the robbing colony and of the colony being robbed, and change the direction of the entrance of the offending colony. Again, remove the weaker hive(s) to another area if robbing persists, keeping entrance reduced until robbing stops. Many beekeepers have their nucs or other weak hives in a separate yard.

Some beekeepers use a *robbing screen*, a device that forces bees exiting their colony to fly upward instead of straight out. Robbing bees tend to land on the flight deck of the bottom board and won't figure out how to get into the colony. Another way to help control robbing is to maintain bee strains with low robbing tendencies, such as Carniolans and Caucasians; Italian races and Africanized bees tend to rob more aggressively.

SWARMING

By reproducing, organisms perpetuate and protect their kind from extinction. Social insects like honey bees can reproduce new individuals (workers) within the colony unit, but compared to other animals, they produce relatively few female reproductives (queens). The reason this works for honey bees is that these insects take very good care of their offspring and progeny; animals that produce few children take better care of them, so they can grow up to have their own children. To pass the genes on to the next generation, daughter queens must be produced.

Honey bee colonies (and tropical stingless bees and some ants) do this by *swarming*. This activity of dividing the nest with new reproductives is very expensive. A colony divides, and part of it leaves for a new homesite, usually with the old queen, while the remaining members continue at the original site with a newly emerged—and later mated—queen. In this manner, a single unit becomes two.

An abundance of food, a high worker population, and the formation of many queen cells, often called *swarm cells*, indicate that swarming preparations are under way. Shortly after the swarm cells are sealed, the colony will cast a swarm. Bees will exit as a swarm on any warm, windless day; usually between 9 a.m. and 3 p.m. (earlier or later if the weather is favorable). Occasionally, bees will swarm when the weather is less than favorable.

After the swarm issues, some of the bees will alight on a nearby object and begin fanning with their scent glands exposed to attract the remainder of the swarm and the queen. Soon a *cluster* of bees forms. It is this cluster—readily visible to the casual observer—that is correctly called a swarm (see the illustration of a swarm on this page). Scout bees will go out to locate a new home and then return to dance on the cluster; this communicates the location of new homesites. When one site has been agreed upon (within a few hours to a few days), the swarm flies to the new homesite, guided by scenting bees. Research is continuing into how bees decide to swarm, how they reach a decision on site locations, and what makes a good home.

Bees in a swarm are usually quite gentle. Before leaving their old hive, they engorge on honey, which seems to contribute to their gentleness. Another reason for their gentleness might be that because the homeless cluster is only a temporary situation, the division of labor—including guarding—that prevails in a normal hive is either nonexistent or not as prevalent.

Swarming versus Productive Hives

Swarming was once considered a sign of "good and productive" beekeeping, for the beekeepers could increase their holdings from the numerous swarms available. Straw skeps, logs, and other types of cramped hives have been used to house bees since the 1600s, but these containers quickly became overcrowded and thus promoted the swarming of bees.

Today, swarming is viewed as a sign of a beekeeper's negligence because it means a loss of both bees (unless the swarm is captured) and the production of honey. Although most beekeepers make efforts to prevent or control swarming, it is not an easy task. The picture is further complicated by the fact that most methods used for controlling or preventing swarming result in manipulations that reduce the colony size (which is what happens when the colony swarms).

Thus, although swarming can be controlled or prevented, in doing so the goal of maintaining populous colonies for the honeyflow is somewhat sacrificed. Nevertheless, this is far better than having the colony cast a swarm that may leave the apiary site before you can recapture it.

Since the 1990s, swarms in some areas have become less numerous as a result of predation by the parasitic bee mites. Colonies weakened by mites are

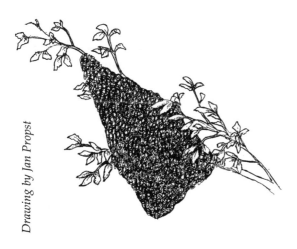

Drawing by Jan Propst

Swarm

not strong enough to swarm, and feral colonies, which were once common, have been killed by mite infestations. Sometimes, the bees filling this niche are Africanized bees.

Reasons for Swarming

Honey bee colonies swarm for any one or more of these reasons:

- Congestion.
- Unbalanced numbers of different-aged workers.
- Overheating (perhaps due to lack of noontime shade).
- Defective or old combs (those with too many drone cells or cells that are irregular, thick, damaged, or otherwise not suitable for the queen to lay in, reducing broodnest capacity and increasing congestion).
- Queen's egg laying becoming restricted as empty cells are filled with honey.
- Inclement weather, which keeps bees confined to the hive and causes congestion (bees hanging out of colony).
- Failing queen—instead of superseding the queen, the colony may swarm.
- Decline of queen pheromone production—the level of pheromone being distributed throughout a highly populous colony is insufficient to control swarm preparations.
- Genetics or race of bees.
- Idle nurse bees.

Other Reasons Why Bees Leave

Under certain conditions, the entire original colony may depart its home. This is called *absconding* and could be caused by:

- Disease or mites.
- Starvation.
- Wax moth (or other pest) infestation.
- Fumes from newly painted or otherwise treated hive equipment.
- Poor ventilation.
- Excessive disturbance of the colony by the beekeeper or vandals.
- Excessive disturbance by animal pests such as skunks and bears.

Signs of Swarm Preparation

Signs that a colony is in some stage of swarm preparation are clearly visible during routine hive inspections. The list below presents a rough chronology of the various signs you might see in a colony that may ultimately swarm:

1. Rapid increase in worker population occurs (especially in spring, after a minor honeyflow and before a major honeyflow).
2. Drone rearing begins as worker numbers increase.
3. Broodnest (area where eggs, larvae, and pupae are located) cannot be expanded due to combs already occupied with brood and/or honey.
4. Queen cup construction along the bottom edges of the frame becomes evident.
5. Queen deposits eggs in these queen cups; larvae are present.
6. Queen's egg laying tapers off and amount of young brood decreases.
7. Queen is restless (thinner and has lost weight so she can fly).
8. Many queen cells are present and contain larvae that vary somewhat in age.
9. Field bees are less active and beginning to congregate at hive entrance; this can also happen if weather is hot or colony congested.
10. Swarm cells are capped or sealed.
11. Swarm is cast.

Signs of Imminent Swarm

A colony that has been making swarm preparations can be expected to issue a swarm:

- After queen cells (swarm cells) are sealed over.
- When wax has been removed from the tips of queen cells, exposing the cocoon (referred to as a "bald spot").
- When few bees are foraging (little flight activity of bees at hive entrance) compared to other hives of same strength.
- When bees are clustered near the entrance, not due to hive congestion or warm temperatures.
- Usually on the first warm, sunny, calm day following a short period of cold, wet, cloudy days when congestion in the hive is aggravated.

Clipped-Queen Swarms

A clipped queen (or old queen whose wings are frayed or damaged) will attempt to leave the hive with a swarm, but being unable to fly, she will not accompany the other bees in flight and will be left behind, usually on the ground near the hive from which she attempted to swarm. The swarming bees, without a flying queen, may return to the hive while they are still airborne, cluster on the ground with the queen, or cluster on a branch nearby. After a brief time they will return to the parent colony. Eventually, the bees will swarm, accompanied by a virgin queen that can fly.

Sometimes swarms will issue, but the queen remains inside the colony. These bees may remain airborne or cluster temporarily before returning to the colony; bees without a queen will return.

If you witness any of these events, take the following steps to discourage their recurrence:

Step 1. Find and cage the queen either before or after the swarm returns.
Step 2. Move the parent hive from its stand and replace it with a new hive of foundation or dry drawn comb.
Step 3. When the swarm returns, let the queen walk in with them. If the swarm has already returned to the parent hive at the old location, shake half of the bees in front of the new hive; the bees will enter this hive. Release the queen so she can walk in the hive entrance with the bees.
Step 4. Check after 10 days.
Step 5. Requeen the colony with new stock; the queen might have swarming instincts and is probably old. Any virgin queens emerging from the original colony will likely have this swarming instinct.

Another method would be to let the swarm return to the original colony after you remove the queen cells. Check after 10 days to remove any additional queen cells, or *demaree* the hives; see "Demaree Method" below. Requeen the colony later with non-swarming stock.

Prevention and Control of Swarming

Successful swarm prevention means that you are able to keep bees from initiating queen cup construction that may lead to swarming. You practice swarm control when you find and remove queen cups and cells and other signs of swarm preparations already evident. Although the times for initiating swarm prevention and control are different, the manipulations are the same and include:

- Relieving congestion by adding more room in which the queen can lay eggs.
- Providing storage space for the growing bee population.
- Separating the queen from most of the brood.
- Interchanging weak colonies with strong ones.

Reversing

Reversing the brood chambers at regular intervals, or as needed beginning in the spring, is one method used to relieve congestion in the hive. Through the winter, the colony and its queen move upward through the hive bodies (see "The Winter Cluster" in Chapter 8). By spring, the cluster is usually in the topmost hive bodies, and because the queen is more reluctant to move down, the brood will be confined there. Unless the queen, broodnest, and bees are put on the bottom with the empty hive bodies on top, the colony is likely to become congested and will eventually swarm, even though there is available expansion space below. Even if the broodnest is not congested, still reverse the hive bodies so plenty of empty combs are available for the queen.

Here is a quick outline for reversing hive bodies (two deep and one shallow; see the illustration on reversing on p. 157):

Step 1. Take an extra bottom board to the beeyard. Move the congested hive off its stand or from its location and place the extra bottom board in its place. You can also put the bottom board down next to the old location. Take the topmost super (S1) and place it on the new bottom board. Now take the hive body containing the queen, most of the bees, and brood (2) and put it on top of the first body (S1).
Step 2. Place at least one deep hive body (1) above the broodnest (2).
Step 3. Clean the original bottom board and go to the next hive.
Step 4. Repeat the procedure until all the hives are reversed.

Reversing Hive Bodies

Reversing with Two Deeps and a Shallow

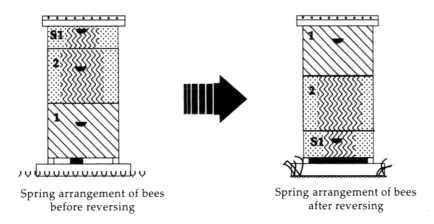

Spring arrangement of bees
before reversing

Spring arrangement of bees
after reversing

Reversing with Three Deeps

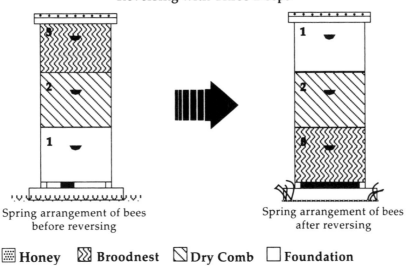

Spring arrangement of bees
before reversing

Spring arrangement of bees
after reversing

▦ **Honey**   ▨ **Broodnest**   ◺ **Dry Comb**   ☐ **Foundation**

Step 5. If the queen is reluctant to move up after a week, exchange a frame of brood from the broodnest with an empty frame and move the brood frame up into the second hive body.

Step 6. If three deep hive bodies were present, the order after reversing is (1) top, (2) middle, and (3) bottom as shown in the illustration.

Other Ways to Relieve Congestion

Hives that are very congested due to poor combs or inadequate space for brood are more likely to swarm. Listed below are some techniques for relieving such conditions:

1. Add extra frames or supers full of foundation.

2. Use a screen bottom board or prop up inner cover to allow for more ventilation.

3. Separate brood and queen:
   - Place the queen, with unsealed brood, eggs, and bees in the lowest body.
   - Above this, place a super with foundation.
   - Above this, place a body filled with capped brood and the rest of the bees.

Decrease the number of bees or brood in the hive by splitting the hives to make additional ones, called *increases* or *splits* (see the illustration on feeding a weak colony on p. 158). This means, of course, that you are making additional colonies and expanding the size of your apiary. If you do not want to increase

Feeding a Weak Colony (Split, Nuc, or Swarm)

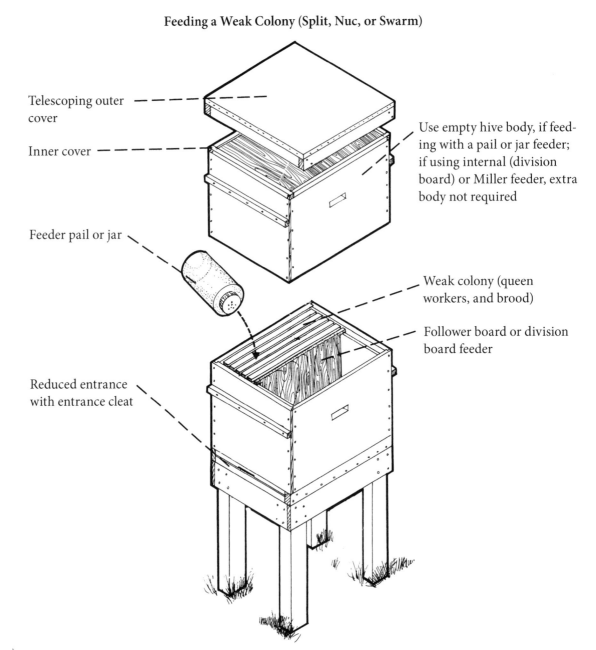

Telescoping outer cover

Inner cover

Use empty hive body, if feeding with a pail or jar feeder; if using internal (division board) or Miller feeder, extra body not required

Feeder pail or jar

Weak colony (queen workers, and brood)

Follower board or division board feeder

Reduced entrance with entrance cleat

the number of your colonies, you can always sell your extra hives or donate them to your local bee club to help get someone else started. Alternatively, splits made to control swarming can be reunited later in the season to consolidate colony numbers.

To make a split, follow these steps:

Step 1. Move frames of capped brood, honey, and bees from the congested hive into a new box (e.g., one deep body). You should try to leave the frame with the queen in the original hive, the one you are splitting. You can also put these frames into nuc boxes and requeen to have spare nucs available for requeening later (see Chapter 10), to augment the number of your colonies or to sell to others.

Step 2. If you are combining frames of capped brood and bees from different colonies, spray each frame of bees with syrup to reduce any fighting among the bees.

Step 3. Give the new hive a frame of open brood (either a frame of eggs or one of newly hatched larvae from your best hive), so they can make their own queen. Requeen the split with a new queen, or provide some queen cells (usually swarm cells).

Modified Demaree Method of Swarm Control

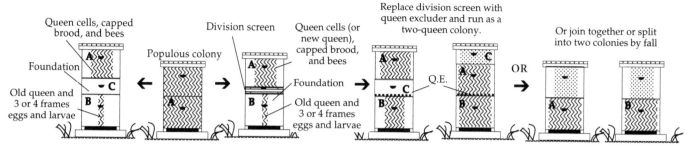

WITHOUT INCREASE — Queen cells, capped brood, and bees; Foundation; Old queen and 3 or 4 frames eggs and larvae; Populous colony; Division screen; Queen cells (or new queen), capped brood, and bees; Foundation; Old queen and 3 or 4 frames eggs and larvae

WITH INCREASE — Replace division screen with queen excluder and run as a two-queen colony. Q.E. Or join together or split into two colonies by fall OR

⊠ Broodnest   ⊘ Dry Comb   ⊞ Honey   ☐ Foundation   - - Queen excluder (Q.E.)

Whatever is provided, place it in the middle of frames of emerging brood.

Step 4. If you made an increase in a deep hive body, it should have the frame of open brood; frames of capped, emerging brood; frames of foundation; and frames of honey and pollen, or empty drawn comb filled with syrup. Use your judgment on how to combine the categories of frames; if no honey-flow is anticipated, put in an extra frame of honey.

Step 5. Reduce the entrance to discourage robbing; check after one week.

Step 6. If you are using nuc boxes, a good rule of thumb is to have each contain two frames of bees and brood, two of food, and one of foundation.

Demaree Method

The Demaree method, described first by George Demaree in 1884, makes it possible to retain the complete population of a colony while practicing swarm prevention and control. Basically, it separates the brood from the queen and decreases the congestion. Here is one way to demaree (see the illustration of the modified Demaree method on this page):

Step 1. Select a populous colony (A). If you are not making an increase, follow the left side of the diagram (without increase). If you are increasing, follow the right side of the diagram (and go to *Step 9*).

Step 2. Have ready a hive body (B) filled with frames of dry drawn comb (from which brood has already emerged) and a shallow (C) with foundation. If no honeyflow is on, use less foundation, because the bees will chew it; if you have only

foundation, feed bees with syrup so foundation will be drawn out.

Step 3. Place the empty hive bodies beside the hive to be demareed (A). Find the queen and place her on a frame containing very young larvae and eggs. Make sure there are no queen cups or cells on the frame with the queen; if present, remove them or replace the frame.

Step 4. Remove two frames or foundation from the middle of B and place the frame with the queen and clinging bees there.

Step 5. Remove A from the bottom board for the moment and set B in its place.

Step 6. Add two or three frames of honey and pollen from A to B and fill the remaining spaces with dry comb or foundation.

Step 7. Place a queen excluder above B, and then place super C (full of foundation or dry comb) above the excluder.

Step 8. If you do not want to make another colony (an increase), add three or four frames of eggs and larvae plus two or three frames of honey in A and place on top of C. Fill in the rest of the space with frames of foundation.

Step 9. If you are making an increase or split, separate the old queen in B below (fill up the space as described in 8) with a division screen. Above the division screen, place A, filled with queen cells (or a new queen), capped brood, and bees. If you are requeening A with a purchased queen, remove all queen cells from the brood frames. Any extra frames can be given to other colonies.

Step 10. Once you have two established queens in A and B, separate them with a honey super (C) placed above a queen excluder, and run the colony as a two-queen unit. (Remove division screen.) You have created a two-queen colony. Add additional supers as needed.

At some point in the season you may wish to return the two-queen colony into a single queen colony or into two separate colonies. The latter can be achieved by placing the upper hive body containing the queen on its own bottom board and adding appropriate equipment to it. By taking this step, you will have made an increase from one colony to two; or the queen excluder can be removed and the two queens will decide which of them will prevail. In this case, you have united the two-queen colony and returned it to a single-queen colony.

Variations of this method are used to rear queens (in warm weather), run a two-queen colony, or make increases; a division screen can be used in place of a queen excluder.

Advantage

● Population kept at right size to take advantage of any honeyflow.

Disadvantages

● Must find the queen.
● Many manipulations necessary.
● Time-consuming.
● Many trips to the apiary needed.

Interchanging Hives

In an apiary where the hives are in long rows (never a good idea), the bees tend to drift toward the row ends. As a result, the colonies in the middle may be weaker than the colonies on the ends; see "Hive Orientation" in Chapter 4. The stronger colonies may become more congested and begin swarm preparations.

If a hive is very populous and seems likely to swarm at some point but has not yet made preparations to do so, interchange it with a weaker hive. More incoming food-bearing foragers will return to the stronger hive's location but will enter the weaker hive, augmenting its population. Conversely, the strong hive will have a sudden decrease of incoming field bees, and any idle bees that might have normally initiated swarm preparations will begin foraging.

Foreign bees entering the switched hives should not fight if there is a honeyflow in progress. To de-crease chances of fighting, wait for a good honeyflow before interchanging the hives. Very weak colonies or nucs should not be strengthened in this manner, unless the queen is caged and candy plug exposed; otherwise, the incoming strange foragers can overwhelm and kill her.

Other Factors

The following factors may also be of importance in helping to decrease swarming in some hives:

1. Provide young (less than a year old), vigorous queens from non-swarm stock or hybrid queens with non-swarming tendencies.
2. Ventilation to increase airflow within a hive:
   ● Hive bodies can be staggered.
   ● Inner or outer cover can be propped up.
   ● Screened bottom boards can be placed on top of the bottom board; this gives bees more ventilation and can be kept on the hive year-round.

The first two ventilation techniques might encourage robbing when the honeyflow is over; thus, only strong colonies should be manipulated in the ways described. To avoid all this extra work, we suggest requeening any colony that has an old queen.

CATCHING SWARMS

Bait Hives for Capturing Swarms

It is generally not possible to check your home apiary on an hourly basis, and to give such attention to your out yards throughout the swarm season is nearly impossible. Despite good management procedures for swarm prevention or control, a given number of colonies will cast swarms. Swarms will remain clustered during good weather for brief periods, spending a few hours to several days clustered within the area of an apiary before departing for a new home site.

The loss of any swarms from your bee yard is costly, given the fact that colonies that cast swarms take time to rebuild their population and may produce less surplus honey. In addition, often colonies that cast swarms are likely to be healthy and prosperous ones whose genetics are worthy of retention.

As a consequence of these circumstances, beekeepers have turned to using bait hives to recover swarms before they take an abode elsewhere. Bees captured in bait hives may come from swarms emanating from your apiary, from someone else's yard, or from bees residing in any number of cavities not under the management of a beekeeper.

Bait hives can be defined as containers whose cavity size is sufficiently attractive to scout bees for them to "convince" the swarm to move to it. In addition to providing the proper cavity size, the addition of combs (when appropriate), lures, and the physical position of the bait hive (in reference to both its height and the compass direction of the entrance) will play a role in enhancing its attractiveness to scout bees.

The cavity size of bait hives is approximately the dimensions of a deep super (Langstroth brood chamber) or 1.4 cubic feet (40 L). Beekeepers may therefore assemble a bait hive from standard hive parts, using an old deep super (nail boards for a top and bottom and add some kind of projecting wood hanger to attach the box to a tree). Use only a few nails, so you can open it easily when it is filled with bees. See the extension publication: Seeley, T.D., R.A. Morse, and R. Nowogrodzki, Bait Hives, Cornell University Bulletin #187, 1989; website: http://www.ecommons.cornell.edu/bitstream/1813/2653/2/Bait%20Hives%20for%20Honey%20Bees.pdf. It provides dimensions for the construction of another type of bait hive. In addition, a modified peat or fiber flower pot is available from beekeeping supply catalogues for use as a bait hive.

Research on bait hives gives insight on the selection process by the scout bees, which you can use to place your bait boxes where you will get the most success. Bait hives should be located up to 15 feet (5 m) above the ground, although ones located lower have succeeded in attracting scout bees. Bait hives should not be located in full sun and can be placed a few hundred yards (meters) from any existing apiary.

When using a standard Langstroth brood chamber, add two or three drawn empty brood combs (void of honey, pollen, and brood) in the hive with the remainder of the cavity filled with frames of foundation. The entrance to the brood chamber should be reduced to ¼ the size of the standard entrance because small entrances are preferred by scout bees over larger entrances. For example, the entrance to the Cornell bait hive is 1¼ inches (3.2 cm) in diameter! Entrances should be located on the lower front and face south-southwest.

Note: wax, propolis and other odors will also attract mice and wax moths; therefore such bait hives should be removed and stored properly at the end of the swarm season.

Once the bait hives containing movable frames are occupied by bees, the frames can easily be transferred to another hive body. In addition, another asset to using fully drawn wax combs is that the odors emanating from them, and the propolis deposits on them, serve to lure the scout bees to these bait hives. Today various lures (pheromones) make the bait box more attractive, and they are readily available from most beekeeping catalogues. These lures can be applied inside and outside of these hives.

The wood fiber flower pot and the Cornell bait hive do not lend themselves to the inclusion of drawn comb and therefore create a problem when you have to transfer the bees and their comb into movable frame boxes. Cutting the comb and wiring them to empty frames is cumbersome and messy, but effective.

Note: The use of bait hives in areas inhabited by Africanized bees is not recommended.

Types of swarm boxes include:

1. Decoy or bait hives—with drawn comb or foundation—can be placed at various distances and directions from the apiary. Wax, propolis, and other odors may attract the scout bees and, ultimately, the swarm, but they might also attract mice and wax moths, so empty bait hives should be removed and stored properly at the end of the swarming season. Empty hive bodies or nuc boxes can also be used. Place these bait hives up to 15 feet (5 m) in a tree, and face the entrances to the south-southwest.
2. Lures, using queen or Nasonov pheromones, can be purchased and placed inside empty boxes or special wood fiber or peat pots. While designed originally for trapping Africanized bees, these fiber pots can also be used for northern swarms.
3. Low, dark objects close to the ground—such as a burlap or cloth bag wrapped around a low branch in a rough sphere—may attract a swarm to cluster there.

Advantages to bait boxes

- Easy to construct or obtain
- If placed in your apiary(ies), may reward you with swarms that may otherwise be lost, possibly into another beekeeper's bait box.

Disadvantages

- Bees settled in a box with no frames need to be transferred into standard equipment, and if they have already made comb with brood and honey, this could be difficult.
- Bait boxes need to be securely fastened to trees, poles, or buildings; they could fall down.
- Boxes will also attract hornets and wasps (even birds) or Africanized bees; therefore, caution is advised when checking or collecting the boxes.
- Need to be monitored regularly, especially during the prime swarming season in your area.

Swarm-Collecting Containers

To be prepared, you should always have extra hive bodies or nucs full of foundation for hiving swarms. If the swarms have to be collected some distance from the apiary, bring along a single deep hive body or nuc box, with the bottom board nailed on. Large swarms may not fit inside a nuc box, although such large swarms are not common today. Shake the swarm in front of the hive, and if you see the queen, cage her and put her inside the collecting box. After most bees have entered, close the entrance with a piece of screen. The hive can then be either carried off or left there unscreened until the evening so that any stray bees can rejoin the swarm; its entrance should be screened when the hive is retrieved in the evening. Other containers that can be used for collecting swarms are:

- A cloth bag (not a plastic bag); shake the swarm into the bag and transport it to the apiary, or if the swarm is on a tree branch, envelop the swarm with the bag, tie it closed, and cut the branch.
- An old basket with a secure cover.
- A cardboard box that can be closed.
- A well-ventilated 5-gallon plastic pail; burn a lot of holes in the sides with a hot nail and fit the lid with a screen to close.
- A screened box, like an old bee package or a cardboard box; shake the swarm into the box and carry the box to the apiary.
- Remember to take some empty queen cages to contain swarm queens if you find them. That way, the rest of the bees will cluster wherever the queen (or caged queen) is placed.

Swarms, especially large ones, need plenty of ventilation and must be kept out of direct sunlight. Often bees are "cooked" or smothered when collected in an inappropriate container (one that is too small or too airtight). Bees in a cooked swarm will look wet and will be crawling on the bottom (similar to a "cooked" package of bees); these bees will not recover.

A swarm in a temporary container should be stored, like a package of bees, in a cool, dark place until it can be placed in proper hive furniture. But hive it as soon as possible, because the bees will quickly begin comb construction, making removing them messy.

Collecting and Hiving a Swarm

Beekeepers are often called by homeowners, other individuals, humane societies, and police and fire departments to retrieve swarms. If you wish to collect swarms to enlarge your apiary or strengthen weak hives, notify these agencies by letter, email, or phone each spring and ask to be put on the "swarm list." Most beekeepers are thankful to get swarms and consequently pick them up without charge. In today's age of lawsuits, however, when you enter another's property, you may be liable for damages, real or imagined, that you might cause in collecting swarms. Thus, it would be wise to restrict your swarm collecting to your own beeyard, or become a professional bee remover; this may be a new career in areas where Africanized bees are a problem.

If, however, you choose to collect swarms, forewarned is forearmed. You will need to ask precise questions about the swarm before you go out and explain what you can and cannot do. Ask the following questions first. They will save you time and grief later on:

1. How high off the ground is the swarm? If the swarm is higher than 10 feet (3 m), forget it, unless you are prepared and equipped for tree climbing or are skilled at using a ladder.
2. Can I drive to the swarm's location? What is the swarm clustered on (e.g., tree, post, drain, or building), and if it is in a tree, can the branch on which it is hanging be cut off? Bring pruning or lopping shears.
3. How long has the swarm been there? If the swarm has been there more than a day or two, either it

will be very hungry and therefore volatile, or it will have departed by the time you get there.

4. How large a swarm is it? The size of a basketball? A softball? Smaller swarms may contain a virgin queen or be queenless, or Africanized, and may not remain at the cluster site by the time you arrive.

5. Are you sure these are honey bees? If there is a gray, paper-like nest, then the insects are hornets and a professional exterminator should be contacted.

6. Are these bees yours to give away? Or did they come from the neighbor's hives?

Once these questions have been answered to your satisfaction, get instructions on how and where to find the swarm and with swarm-capturing equipment, off you go. It may be helpful to get a map off the Internet or use a GPS unit.

Bees in swarms are usually well engorged with honey and therefore gentle. But sometimes a swarm is bad tempered, especially if it has been clustered for several days and the bees are hungry. In any case, it is prudent to wear a veil when collecting swarms.

Some beekeepers carry spray bottles of syrup, often medicated syrup. Bees sprayed lightly with the syrup will engorge the food and become gentle and easier to handle (see illustration on collecting a swarm on p. 164). These are the basic steps for collecting and hiving a swarm:

Step 1. Spray the cluster gently with sugar syrup from your spray bottle. If the swarm is clustered on a tree limb, with the owner's permission, cut away excess branches, leaves, or flowers. Avoid shaking or jarring the cluster.

Step 2. If the swarm is jarred and the bees begin to break the cluster, spray the bees and wait for the cluster to re-form.

Step 3. While you steady the limb or branch with one hand, saw the limb or clip it free from the tree.

Step 4. Shake the swarm into a hive or collecting container prepared for the bees or, if possible, put the entire cut limb into the collecting container.

Step 5. If the swarm is on a post or flat surface, brush or smoke the bees into a hive or container, directing them gently with puffs of smoke. Look for and try to capture the queen and put her in a queen cage; this will make collecting the

swarm much easier. After you put the caged queen in your container, the bees will soon enter it.

Step 6. A piece of cardboard can be used like a dustpan to scrape bees gently into the container or in front of the hive entrance.

Step 7. If feasible, leave the collection container and come back in the evening to make sure all the bees are in the box. Close up with a screen and take home. Sometimes, bees may be left behind or a small cluster of bees has re-formed higher up. If possible, spray these bees with soapy water to kill them, as many homeowners are upset if all the bees are not collected. In some instances, you may need to close up the collecting container and take the bees immediately.

Step 8. Back at the apiary, shake the bees from the container into a hive filled with foundation or unite the swarm with a weak colony. If you used a collecting box with frames, wait a few days with the entrance open and install it as you would a nuc or unite it to a weak colony. Bees in a swarm are gorged with honey and nectar, which stimulates their wax glands. Put frames of foundation on a hived swarm, and you will be amazed at how fast they draw out the comb.

Once you have hived the swarm, you should attempt to evaluate the queen's performance. After a month or so, if she lives up to your standards (e.g., the number of frames of brood, the amount of honey or pollen collected) keep her (if she is good) or requeen the colony (see "Breeder Queens" in Chapter 10). Requeen the colony by the next year with a queen from non-swarming stock, because swarm queens are older and may have genetic components that favor swarming.

If you caged the swarm queen, you can unite the bees to a weak colony that has a young queen, and make a nuc with the swarm queen. Swarms united to colonies should be placed in a super with foundation and then placed over the colony with which they are to be united. (See "Uniting Colonies," p. 148.)

If you know which colony swarmed, check it for additional queen cells. Some colonies can cast several swarms. The first, or *prime swarm*, is accompanied by the original, old queen. Subsequent swarms, called afterswarms, will contain virgin queens and are generally smaller than the prime swarm.

Collecting a Swarm

Clear excess branches from the swarm cluster.

Spray with sugar syrup, and then cut carefully.

Lay swarm in front of new hive body or swarm-collecting box. Look for queen as bees walk in, and cage her.

Precautions

Always treat swarms as if they are diseased or are carrying mites. Because you will not know this beforehand, install all swarms on foundation and sample the bees to see if varroa mites are present. If mites are found, sprinkle the bees with powdered sugar, and while they are scraping off the sugar, they will also scrape off some of the mites. You can do this when the swarm is still clustered in a temporary container,

especially a screen-bottom nuc, which will allow the mites to fall completely outside the hive and off the bees.

If a swarm is put on drawn comb, the bees may regurgitate drops of nectar or honey containing disease spores from their honey sacs into the drawn comb. By being installed on foundation, the bees will digest the honey along with the spores rather than depositing the honey/nectar and spores into the cells.

Destroying a Swarm

If the swarm must be destroyed because it is diseased, it is in a location dangerous to people, it is full of mites, or it is Africanized, a thorough spraying of soapy water (½–1 cup liquid detergent or soap per gallon of water) will kill the bees quickly. It knocks them to the ground immediately, even if the bees are airborne.

QUEEN SUPERSEDURE

Supersedure is how the colony replaces an old or inferior queen with a young queen. The workers in the colony build a few queen cells, and when a new queen emerges, she destroys the other queen cells and may destroy the old queen (sometimes mother and daughter queens coexist for a short time). Swarming does not usually take place when a queen is superseded.

Some of the reasons for the supersedure are listed below:

- Queen is deficient in egg laying.
- Queen produces inadequate amounts of queen substance (pheromone) due to age, injury, lack of nourishment when a larva, or other physiological problems.
- Queen is injured as a result of clipping, fighting among virgins, or temporarily balled by workers when released.
- Queen was injured when removed from or placed into queen cage.
- Queen is defective, not raised under ideal conditions, or poorly mated.
- Queen did not receive enough nourishment as a larva (which may contribute to some physiological defects).
- Colony and queen have nosema disease or tracheal mites.
- Weather has been inclement for extended periods (other than winter).
- After installation of a package, when numbers of adult bees decline and no new ones emerge for 21 days, the remaining older workers may undertake supersedure activities.
- Queen's pheromone levels are low.

Supersedure / Emergency Cells versus Swarm Cells

Young queen larvae can begin their development in queen cells or in worker cells. Queen cells developed from worker cells are either supersedure or *emergency queen cells*. The sudden loss of a queen usually forces the bees to modify worker cells into emergency queen cells, and as the larval queens develop, the cells' edges are slowly enlarged by the added wax. This eventually will form the peanut-shaped vertical, capped cell that is characteristic of all queen cells (see the illustration of swarm and emergency queen cells on this page).

Queen cells that begin as queen cups usually located along the bottom edges of frames and are referred to as *swarm cells* (cells containing queens that will replace the resident queen after she departs with a swarm). These queen cells are numerous, some containing larvae, some pupae all of different ages. By

Swarm cells

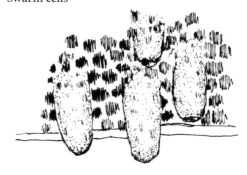

Supersedure or emergency queen cells

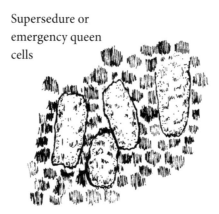

Swarm Cells vs. Supersedure (or Emergency) Queen Cells

contrast, *supersedure/emergency queen cells* are few in number, contain larvae or pupae of the same age, and are similar in size. In addition, they can be found near the central region of the brood comb. A queen issuing from an emergency queen cell replaces one that is no longer present! In the case of the queen emerging from a supersedure cell, she replaces a queen who may still be present!

The state of the colony (e.g., low bee population or bubbling over with bees), the time of the year, and the location of the queen cells should provide enough information for you to know exactly what the colony's intentions are.

In supersedure

- Few (between one and five) cells are made.
- The colony is usually not very populous.
- The brood pattern is scattered or almost nonexistent because the queen is injured, diseased, or failing.
- Queen cells converted from worker cells are located near the center of the comb.
- The age of queen larvae is the same, so all the queen cells mature nearly at the same time.
- Drone brood often appears in worker cells.
- Queen cells are present after the normal swarming period.
- Wax to make cells darker in color.

In swarming

- Numerous (between 10 and 40) queen cells are present on the lower edge of combs.

- The colony is populous.
- Numerous frames of capped brood are present, and there is a diminished number of cells with uncapped brood.
- The ages of queen larvae are varied, which may result in multiple swarms as different queens emerge over time.
- Swarming season, that is, early spring (March to June, depending on the latitude and elevation), is on. Sometimes swarming preparations may begin in the later summer.
- Light color to wax in swarm cells.

In emergency replacement

- Few (between one and three) cells and will be roughly the same age.
- Queen cells are usually constructed on the face of the comb amid capped worker brood.
- Attempts to make queens from older worker larvae may result in intermediate forms of queens on maturation.

Note: emergency or supersedure queens rarely look as robust as queens developed during the swarm season.

LAYING WORKERS

When a colony loses its queen and is unable to rear a replacement as a result of a lack of eggs or young larvae (less than three days old), some workers may start to lay eggs. The ovaries of these females will

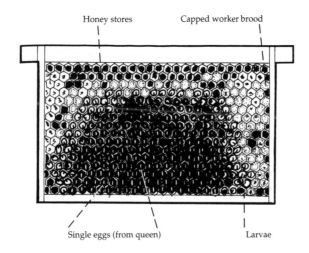

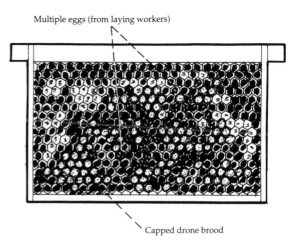

Brood Frame from Queenright and Laying Worker Colonies

mature, and after these bees are fed royal jelly, eggs will mature. Since workers are incapable of mating, the eggs they lay will be unfertilized or drone eggs. The worker population within a colony that has laying workers, therefore, will slowly decline because the rearing of new workers stops with the loss of the queen.

Laying worker colonies can be recognized by the number of eggs seen in each cell. If more than one egg is present and they are on the sides of the cell walls, instead of on the bottom, laying workers should be suspected (see the illustration of the brood frame of a laying worker on p. 166). To correct this situation, several methods have been tried, some of which involve the introduction of a queen, queen cells, or a frame containing larvae less than three days old. Often bees within a laying worker colony are first

shaken from their frames at a distance of 100 yards (100 m) or more from the parent colony just before the introduction of a queen, queen cells, or young larvae. This action supposedly leaves behind the laying workers, which are too heavy to fly, causing the rest of the bees to accept a new queen readily.

Unfortunately, attempts at rescuing a colony of laying workers from inevitable doom have never worked to anyone's satisfaction. Frequently the colony will reject the introduced queen or queen cells, or it will rear workers rather than queens from the introduced larvae. You also lose valuable honey-making time. So the best solution is to unite a colony of laying workers with a queenright hive. Experimenting with the other methods or inventing your own, however, may be worth the experience.

 **Notes**

CHAPTER 12

Products of the Hive

HONEY

The worldwide definition of honey reads: "Honey is the natural sweet substance produced by honey bees from the nectar of blossoms or from secretions of living parts of plants or excretions of plant sucking insects on the living parts of plants, which honey bees collect, transform and combine with specific substances of their own, store and leave in the honey comb to ripen and mature" (CODEX Standard 12-1981 Rev. 1 [1987] http://www.fao.org/docrep/w0076e/w0076e30.htm CODEX). In most cases, floral nectar is the sweet substance, primarily composed of sugars and water secreted by nectary glands of plants. These glands are either inside a flower or on the outside of the flower (extraflora nectary). The sugary secretions are one of nature's designs to attract animals to flowers for "sexual favors" (i.e., to pollinate the flower). Potential pollinators include, but are not limited to, birds, bats, beetles, flies, butterflies, moths, wasps, and bees. The association of flowers and pollinators is found in the fossil record, which shows changes in some flower structure (such as nectaries) depending on the particular pollinators.

Nectar

Physical Changes

The water content of nectar is high (up to 60 percent), and reducing it is one of the steps necessary in converting nectar into honey (which is 18 percent water). As foragers collect nectar, it passes into their honey stomachs where it mixes with small amounts of beneficial bacteria living there and the enzyme in-

vertase (secreted by the bees' salivary glands). Upon arriving at the hive, this mixture is transferred by the foragers to the house bees and they continue the process of converting nectar into honey. More invertase is added to the mixture, breaking down the sucrose molecules found in nectar into two simple sugars: glucose and fructose. At the same time, small amounts of the nectar are regurgitated onto the base of their proboscis (tongue) where the solution comes in contact with the warm air within the colony, evaporating water from the nectar. After repeated exposure of the nectar to the warm airflow, small amounts are deposited into cells, where further evaporation and chemical alteration of the nectar continues. As warm air circulates in the hive, fanned by other bees, the evaporation rate of the water is increased.

The elimination of water from the nectar serves two purposes; it reduces the volume of water in the nectar (by half), enabling the bees to store a more concentrated food (sugars) in less space, and it increases the volume of sugars, creating an environment that is inhospitable to organisms capable of altering (or spoiling) the final product—honey.

During a strong honeyflow, you can stagger honey supers by about ½ inch (1.3 cm) or so to increase ventilation and hasten the ripening process in your colonies, especially if your colonies are in a humid location. Be careful to monitor this extra space, because when the honeyflow is over, the additional openings can encourage robbing.

During dearth periods, foragers continue to search for food. At these times it is not uncommon to observe bees robbing weaker colonies, removing liquids from fruit with bruised skin, or sipping liquids from open containers of sugar solutions (such as soda cans). A mixture of these solutions can be converted to "honey" with off flavors and colors. With an Internet map search, you can obtain a general idea of what kind of resources may be available for your bees within flying distance from the apiary; for example, heavily forested, extensive meadows, hayfields (alfalfa), fields of crops that yield an abundance of nectar, orchards, and so on. You may also discover a transfer station, a local dump, or a sugar refinery within flight distance of your bees. These locations can result in bees collecting contaminated sweets. Internet searches will also give you the opportunity to find apiary sites away from your property.

A definition of honey from the National Honey Board is "the substance made when the nectar and sweet deposits from plants are gathered, modified and stored in the honeycomb by honey bees . . . without additions of any other substance." Definitions are in place to protect against the sale of substances that look like honey. Often these imitations are packaged in containers normally used for packaging honey. Such imitations must be clearly labeled.

If you have a unique floral source in your area (such as citrus, sunflower, fireweed, or sourwood), be sure to separate the supers that contain these unique honeys; this will enable you to label them as such when selling your crop. If you want to verify that the honey is from a monofloral source, you can send samples of your honey to independent labs, and (for a price) they can extract the pollen in the honey and determine the floral source. A list of labs can be found on the National Honey Board website (*www.honey.com*) and see the References). (Note: The Honey Board is going to be changed to the "Honey Packers and Importers Research, Promotion, Consumer Education and Industry Information" or "PIB" for "Packer-Importer Board.")

Chemical Changes

The enzymatic activity in nectar represents its chemical alteration into honey. Nectar from flowers generally consists of up to 60 percent water, up to 50 percent sucrose or table sugar, and other components in minor amounts. The enzyme invertase breaks sucrose into the two simple sugars (carbohydrates) called *glucose* (dextrose) and *fructose* (levulose or fruit sugar); see "Sugars" in Chapter 7. These two sugars are the principal components of honey, with fruc-

---

Chemical Properties of Honey

Principal components	Percentage
Water	17.2
Fructose or fruit sugar	38.2
Glucose or grape sugar	31.3
Sucrose or cane (beet) sugar	1.3
Maltose and other reducing disaccharides	7.3
Trisaccharides and other carbohydrates	4.2
Total sugars	99.5
Vitamins, minerals, amino acids	0.5

Enzymes

1. Invertase (inverts sucrose into glucose and fructose, from bees).
2. Glucose oxidase (oxidizes glucose to gluconic acid and hydrogen peroxide plus water; from bees).
3. Amylase (diastase; breaks down starch to dextrins and/or sugars; may aid in bee's digestion of pollen).
4. Catalase (converts peroxide to water and oxygen) and acid phosphatase (removes inorganic phosphate from organic phosphates; from plants).

Aroma constituents

Alcohol, ketones, aldehydes, and esters.

Sources: J.W. White, Jr., M.L. Riethof, M.H. Subers, and J. Kushnir. 1962. Composition of American honeys. Washington, DC: USDA Bulletin 126. E. Crane. 1990. Bees and beekeeping: Science, practice, and world resources. Ithaca, NY: Comstock. National Honey Board, http://www.honey1.com and Bee Product Science, http://www.bee-hexagon.net.

---

tose usually predominating over glucose. Other sugars that remain after invertase activity include small amounts of sucrose and other complex sugars, depending on the type of nectar (see the sidebar on chemical properties of honey).

In addition to invertase, the enzyme *glucose oxidase* is found in honey. Glucose oxidase converts glucose to *gluconic acid* and *hydrogen peroxide*. These substances endow ripening honey with antibacterial properties and high acidity (low pH), but they are very delicate and heat sensitive. Honey's high sugar content (over 80 percent) and low pH make it a hostile environment for bacteria.

When honey is reduced to ash, trace amounts of minerals are found: calcium, chlorine, copper, iron, magnesium, manganese, phosphorus, potassium, silica, sodium, and sulfur. Other components of honey are amino acids, and vitamins—all in trace amounts (see the sidebar on the properties of honey).

Forms of Honey

Honey is packaged and sold in several forms—as all liquid, a combination of liquid and comb, as all comb, or in a granulated or crystallized form. Note that most honey will eventually granulate, or crystallize (becoming semisolid) (see "Granulation" below). Correct storage of honey will delay this process. *Extracted honey*, the liquid form, is removed from the comb and packaged in bottles or jars. Honey is classified into colors ranging from water white (clear) through amber (gold) to dark (black). Light-colored honey tends to be mild in flavor, whereas darker honey has a more pronounced, stronger flavor. At times you may notice different colors of honey in your frames. If you want to take the time, you can segregate these frames and extract the honey from them separately, to produce honeys of different colors. Most honey is blended, however, and if your operation is small, it may not be worth this extra effort. For a complete description of the standards for extracted honey, refer to the Pure Food and Drug Laws on the USDA website and "Honey and Honey Products" in the References.

Comb honey remains in the wax honeycomb, and it too can granulate. Granulated comb honey cannot be extracted except with heat, which will melt the wax and honey until it is warm enough to separate. Some honey (e.g., canola or rape seed and mesquite) granulates very quickly. To anticipate if your comb honey will granulate, record the blooming dates of your major honey plants so you can identify which flower nectars granulate quickly.

The basic types of comb honey are:

- *Section comb*, consisting of individual wooden, boxed sections; plastic sections called Bee-O-

Pak; Half-Comb boxes (designed by J. Hogg); or circular plastic sections called Ross Rounds (but designed by the late W.S. Zbikowski).
- *Bulk comb*, in which the entire comb on a frame is wrapped, or sections are cut out of a frame (cut comb) and packaged.
- *Chunk comb*, in which sections of cut comb are placed in a bottle, which is then filled with liquid honey.

Comb honey is the purest form of honey; the delicate fragrances and flavors of the floral source are sealed in the comb and not dissipated by heating the honey. Unfortunately, many consumers today do not know how to eat honey in the comb. A technique many beekeepers use to improve sales is to include recipes or have taste tests (see "Selling Your Hive Products" below, and "Honey and Honey Products" in the References).

Honeydew

Honeydew refers to the sugary excretions produced by plant-sucking insects, such as aphids (Hemiptera: Aphididae) or other insects. These insects, feeding on plant sap, are usually found on the tender shoots of evergreen conifers such as pines (*Pinus* spp.), fir (*Abies* spp.), spruce (*Picea* spp.), and larch (*Larix* spp., a deciduous conifer), and the leaves of many deciduous trees such as oaks (*Quercus* spp.), maples (*Acer* spp.), willows (*Salix* spp.), ash (*Fraxinus* spp.), basswoods (*Tilia* spp.) beech (*Fagus* spp.), mesquite (*Prosopis* spp.), and black locust (*Robina* spp.).

The aphids suck the sap from these plants to extract the amino acids and other nutrients, and excrete the excess water and sugars onto the leaves and twigs. If the secretions are profuse, honey bees eagerly collect them because they are rich in sugars normally found in honey and contain some other complex sugars. "Pine Honey" or "Forest Honey" (*waldhönig*) is popular in some parts of the world, especially Europe. This honey comes from honeydew found on some evergreens in European forests.

Honeydew honey comes in various flavors and colors, depending on the plants on which the aphids have fed. Dark honeydew honey can be bitter in part because of its higher mineral and ash content. Proponents of this honey claim it has higher antibacterial activity.

One characteristic of some honeydew honey is its

rapid crystallization. It is possible that many honeys contain some honeydew, especially if bees are collecting nectar in areas next to plants with heavy aphid populations. Microscopic components in honey can be used to identify its source; these include pollen for nectar honey and algae and sooty mold particles for honeydew honey.

Another type of plant liquid that bees collect is a dew-like droplet from leaves of corn (*Zea mays*) and cotton (*Gossypium* spp.), to name just a couple. Corn seedlings, for example, will sometimes secrete drops of water, especially in the early morning when the soil moisture is high; this is called *guttation*. Recent research in Europe has shown that when the corn (or other seeds) are coated with a systemic pesticide, it is transported through the plant to the leaves and into the water droplets. Bees collecting these droplets have been poisoned.

THE HONEY HOUSE

A sanitary honey house for extracting, bottling, or otherwise handling honey is very important, and in some states it must meet certain health and food safety requirements. It is imperative that the honey house be kept clean—running water and a washable floor are necessities—and be as insect-proof as possible. Bees and other pests attracted to the smell of honey or wax can get into everything, including newly strained or bottled honey. Fecal material from insects and animals is also a problem in any shed that is not tight.

If you have only one or two hives, you can extract the honey in the kitchen or a clean basement. (Clear this with the family first to avoid conflicts!) First, prepare the space. Put down some newspaper or painting cloth to keep the floor underneath from getting sticky. Bring in a few frames or a complete super and cut the cappings off a frame or two of honey. Let the honey drip into a pan; if the room is warm, the honey will drip out faster. If working in the basement, be careful that it is not too humid; if humidity is high, your honey may start to ferment.

For those who have more than a few hives (10–20), you may have to commit one place for all the extracting and associated tasks. If you decide to move in this direction, use these guidelines in planning a good honey house. For starters, include the following necessities:

- Electricity.
- Hot and cold running water.
- Restroom facilities.
- Washable floor (concrete or ceramic tile) with center drain.
- Washable ceiling and walls, coated with "food approved" paint.
- Hose and water source to wash floors.
- Easy loading/unloading area for moving in heavy supers.
- An uncapping/extracting area where sticky, wet frames are handled and honey is strained.
- A hot room, to warm unextracted honey supers and dry down those not fully cured.
- Dehumidifier, if area is excessively damp.
- Storage space for empty, dry supers in an unheated portion of building.
- Dust-free space for bottling, labeling, and storing honey.

If you are using metal equipment for honey processing and storage, use only stainless steel, because the acid nature of honey will react with all other metals. Food grade plastic is safe to use. Though not essential, the following equipment may also be stored in a honey house, if the structure is large enough:

- Uncapping knives.
- Extractor(s).
- Cappings tank or tray.
- Honey pumps.
- Nylon mesh straining cloths.
- Capping baskets.
- Screen to drain cut comb honey.
- Bottles and labels.
- Holding tanks for bottling and barrels for bulk storage.

If your area is large enough, you can use part of it as work space for constructing and repairing frames, supers, and such, during the time when you are not extracting. Also, you can store your smokers, hive tools, gloves, suits, and veils here as well. You could also think about putting in an extra washing machine and a refrigerator or freezer devoted to your beekeeping operation.

Some excellent floor plans have been published in various books. Check out the sections "Books on Bees and Beekeeping" and "Honey and Honey Products" in the References; the 1992 edition of *The Hive and the Honey Bee* has some basic floor plans, as does *The ABC and XYZ of Bee Culture* (2007). Any book by E. Crane is also a good source, as she lists other references. Talk with other beekeepers, go online and look at various sites, and visit honey houses in your area, large and small, to see how other beekeepers handle their hive products. Some bee suppliers also sell floor plans for extracting rooms.

Beekeepers with fewer than 20 hives can easily manage by using any sanitary space (garage, basement) and an extractor, uncapping knife, and some of the equipment listed above. Or consider getting together with other beekeepers in your area, or members of a bee club, to buy or share equipment and space. An arrangement whereby beekeepers help one another pull and extract honey supers, and divide the finished product proportionally, should be made ahead of time, to avoid misunderstandings.

If you have 100 or more hives, congratulations, you are no longer a hobby beekeeper. However, more recently beekeepers with a small number (1–20) of colonies are now considered as **small or part-time** beekeepers rather than hobby beekeepers (compared to full-time or commercial beekeepers). **Note:** Dr. N. Ostiguy from Penn State University reports that this change in wording from "hobby" to "part-time" is a result of comments by congressional members when approached to support research on honey bee diseases and the current decline in honey bee colonies. Congress doesn't provide money for research on hobbies, but it does provide money for research irrespective of the size of operations.

Consider going into this as a business and purchasing more professional, commercial equipment. If you are heading in this direction, work with some bigger or commercial beekeepers to see how they run their operations before investing a lot of your hard-earned money.

Precaution: Now that small hive beetles are here, it is recommended not to store honey supers, because this pest can invade and destroy honey still in the comb (see "Small Hive Beetle" in Chapter 14). You can alleviate this problem by circulating the air via fans on your supers, or putting the supers into a cooler; better still, extract your honey immediately and place the wet supers back on the hives.

Extracting Honey

Once a frame of honey has been at least three-quarters sealed with wax cappings, it can be removed from the hive and processed. Supers with frames of honey ready to be extracted should be placed in a bee-tight room or honey house. If the room temperature is between 80° and 90°F (26.6° and 32.2°C), the honey can be extracted with ease. If you are removing honey supers in cold weather, let the supers stand in a warm room until the honey is room temperature. So as not to promote the granulation of honey, avoid storing supers at temperatures below 57°F (13.9°C); or you can **freeze** honey supers if you have the room or a commercial freezer.

The wax cappings that seal the honey in the cells are commonly cut away with a heated or an electric uncapping knife or rotating chains or blades. Hobbyist beekeepers generally use an uncapping knife,

Radial Honey Extractor

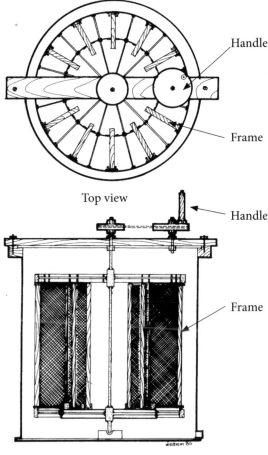

Handle

Frame

Top view

Handle

Frame

Side view

Basket Extractor

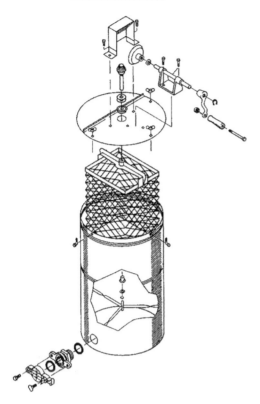

Used with permission of Dadant & Sons, Inc.

whereas the commercial outfits have more mechanized equipment. Cut the cappings off both sides of the frame, letting them drop into a screened basket or into some other container that will permit the honey to drain off. These cappings are later melted down to separate them from any remaining honey (see "Wax Cappings" below).

Place the frames into an *extractor* (radial or basket-type) such that frames of equal weight are opposite each other (see illustrations of honey extractors on p. 172 and 173). If the weight is distributed unequally, the extractor will wobble and vibrate wildly, due to the unbalanced load.

If you are using a *basket-type* extractor (frames are parallel to the sides), start with a slow spin and gradually increase the speed. Spin the frames on one side for three minutes, then reverse them and spin on the other side for three minutes. Do this until all the honey is removed.

If using a *radial* extractor (frames are perpendicular to the sides), there is no need to reverse frames—both sides of the frame are extracted simultaneously because of their perpendicular placement. Start with a slow spin of about 150 revolutions per minute (rpm)

and gradually increase the speed to 300 rpm. Spin at the maximum rate for about 15 minutes. Honey can be draining from the extractor, into a strainer on a bucket while it is spinning, but keep your eye on the bucket, as it will quickly fill up and overflow. Follow the instructions of your particular extractor.

If there is brood in a frame of honey, do not extract honey from that frame. Store such frames in a separate super, and at the end of the day place this super onto a weak colony. It is also very important that you have not fed antibiotics or applied any mite controls to colonies from which you are removing honey. **No medications should be applied to bee colonies that have honey supers.** There is **no tolerance** for these materials in honey for public consumption.

After Extracting Honey

Extracted honey should be strained immediately, to remove wax, bees, and other debris, and may require a second or a third straining through finer mesh. The strainer can be made of nylon or metal screen—any material that is easy to wash and will not become easily clogged. Stay away from cloth, such as cheesecloth, for straining honey, because it can leave lint and other foreign matter in your honey. Remember, warm honey strains much faster than cold honey.

After extracting:

Step 1. Place the strained honey into a storage container or into a holding tank until it can be put into the final jars or bottles. Honey in a storage tank should sit overnight, covered, in a warm room before being bottled. During this time, fine air bubbles and wax particles rise to the top. Skimming off this layer before bottling results in a better, cleaner finished product.

Step 2. Remove the empty, wet frames from the extractor and place them in empty supers; return these to hives at dusk to allow bees to clean the wet frames. To avoid robbing problems from yellowjackets and other bees, place duct tape over any holes or cracks to block such entrances. If no other honeyflows are anticipated, remove the cleaned supers after a week or so and store them. Wax moths will destroy combs that are not properly stored, that is, if they are not stored at freezing temperatures or otherwise protected (see "Wax Moth" in Chapter 14).

Step 3. Extracted combs should not be stored wet because any remaining honey will crystallize in the cells, providing the "seeds" that will hasten the granulation of next year's crop. Wet combs can also ferment and mold, in which case you should melt them down and install fresh foundation. Now is a good time to cull combs that are distorted, broken, or otherwise not good.

Step 4. Remove as much honey as possible from the cut wax cappings and then melt the wax cappings down (see "Wax Cappings" below).

MORE INFORMATION ABOUT HONEY YOU SHOULD KNOW

Changes That Can Occur to Honey

Honey is a supersaturated solution of sugar and water, and as such has various properties, such as hygroscopicity, fermentation, granulation, thixotropy, and the formation of hydroxymethylfurfural (HMF).

Hygroscopicity. Hygroscopicity is the ability of a substance, especially fructose, to absorb moisture from the air. Honey, which contains different amounts of fructose, can absorb moisture in a moderately humid area. In low humidity honey will give up moisture to the air. Conversely, in areas of high humidity, the surface of the honey will pick up the moisture, which will gradually diffuse into the rest of the honey; during long-term storage, honey will absorb enough water as to raise the overall content of water to above 17 percent. If honey incorporates too much water, sugar-tolerant yeasts will spoil the honey by causing it to ferment (honey that contains over 18 percent water will certainly ferment). This is why you should never consider the long-term storage of honey in a damp area, such as a basement. Also, make sure the containers are tightly closed.

The good side of this property is that it benefits the baking industry because it helps keep baked goods that contain honey moist and soft. Honey is also used in cosmetics for the same reason.

Fermentation. Sugar-tolerant or *osmophilic* yeast spores (genus *Zygosaccharomyces*), under high moisture conditions, are able to germinate in honey and metabolize its sugars. These are "wild" yeasts, not the ones found in bread or beer. As the sugars of honey are metabolized, the yeasts produce the by-products alcohol and carbon dioxide, which will spoil honey. Originally, these yeasts may have been used to produce early alcoholic drinks; today "cultured" yeasts give more predictable results and are used to make mead (honey wine) and honey beer. Probably all honeys contain osmophilic yeasts in the form of dormant spores; given the right conditions, these spores will germinate and multiply.

Conditions that will influence fermentation are:

- Water content of the honey.
- Temperature at which the honey is stored.
- Number of yeast spores in the honey.
- Granulation of the honey, which results in an increase in the water content of the remaining liquid portion.

Fermentation of honey can be prevented if its moisture content is less than 17 percent, if it is stored at temperatures below 50°F (10°C), or if it is heated to 145°F (63°C) for 30 minutes, which kills the yeast. Heat the honey rapidly (it must not reach a temperature of more than 180°F [82°C]), to prevent it from burning and darkening; stir carefully while heating. Cool honey quickly to retain the delicate flavors. Once honey has fermented, it cannot be saved and should not be fed to bees because the alcohol content may poison them.

Granulation. Most honey will granulate after it is removed from the comb; some kinds of honey granulate in just a few days after being extracted (pure canola, aster, mesquite, or goldenrod honey), whereas other types remain liquid for weeks, months, and even years (alfalfa, sourwood, and others). Different flower nectars have different proportions of sugars. Honey consisting of a greater proportion of glucose to fructose will granulate faster, because only the glucose crystallizes. Glucose forms two kinds of crystals; one is a simple sugar crystal and the other is a glucose hydrate crystal. The latter contains some water, and because of this, when honey crystallizes, the chances increase that it will also ferment. Honey that granulates naturally can also produce undesirably large crystals (e.g., canola honey) or very fine crystals (e.g., mesquite honey).

In order to keep stored extracted honey in a liquid state, it has to be heated to 145°F (63°C) for about 30 minutes. To keep out any "seed" or particulate matter, air bubbles, wax particles, or pollen, commercial packers force "flash-heated" honey through special filters to strain out any foreign particles. The cleaned honey is then "flash-cooled" and stored or bottled.

Partially granulated but unfermented honey, often considered "spoiled" by the uneducated consumer, is, in fact, perfectly good to eat. Larger honey producers make a form of controlled granulated honey, or "cremed" honey, by a method called the Dyce process (described in this chapter) and sell it as a spread. (Note the spelling of "cremed." It is missing the "a" as in "cream" so as not to suggest this product contains any dairy ingredients.) This honey has very fine crystals and does not ferment. The ideal temperature for honey to granulate is 57°F (14°C). Unless the object is to produce this kind of honey, store honey above this temperature. To prevent granulation during long-term storage, honey may be kept in a freezer.

To liquefy small quantities of granulated honey, place the container in a pan of warm water until the crystals are melted. Do not let the honey overheat, as many of the flavors and aromas of honey are volatile and are destroyed by heat. Five-gallon buckets or barrels of honey are generally liquefied in a "hot room." Another method is to put an uncapped container of honey into the microwave and heat it for one minute and then for additional 30-second increments until it is liquefied. Never leave honey containers unattended when heating.

Thixotrophy. A few rare honeys have thixotropic characteristics. In the comb, the honey appears to be solid and cannot be extracted due to its thick, viscous nature. However, if the honey is subjected to vibration with a special type of extractor, or as honey is being spread on bread, it will liquefy. As soon as the vibration stops, the honey reverts back to a thick, gel-like solid. Thixotropic honey contains more proteins, which impart the honey with this unique property. The most famous thixotropic honey is from ling (*Calluna vulgaris*), commonly found growing on the moors of Europe. Another is pure grapefruit honey (*Citrus paradisi*) and Manuka honey (*Eucalyptus* spp.) from New Zealand.

Formation of Hydroxymethylfurfural (HMF). HMF is produced when fructose breaks down in the presence of high temperatures and some metals. HMF is measured in honey sold on the world market, and if the levels are too high, the honey may be rejected. Factors that favor HMF formation are:

- Fructose (starting material); HMF will not form in sucrose syrup.
- Heat.
- Time.

- Low pH (acidic conditions).
- Metal ions (stabilize HMF), for example, if the honey (or high-fructose corn syrup) is stored in an untreated metal container.

If HMF levels are very high, such as when honey or high-fructose corn syrup has been stored in the sun, do not feed it to bees. If you rehydrate the syrup by adding water, the HMF breaks down and forms levulinic and formic acids. Both of these acids are toxic to bees and make the honey or syrup unpalatable. It is best to throw this honey or syrup away.

Medicinal Properties of Honey

Honey is known for its many healing properties; some have been scientifically studied, and some have not. Various groups maintain that honey is helpful in:

- Retaining calcium in the body.
- Counteracting the effects of alcohol in the blood.
- Providing quick energy.
- Deterring bacterial growth, especially on burns.

Medicinal applications of honey (such as for burns and wound healing) have been reported, showing that honey:

- Contains an anti-inflammatory agent.
- Stimulates new tissue growth.
- Promotes healing, preventing scarring.
- Provides a protective thick layer.
- Has antimicrobial properties (due to the low pH and low water content and the presence of hydrogen peroxide) and beneficial microbes (if the honey is fresh or uncured.)
- Contains antioxidants (Science News September 12, 1998; www.sciencenews.org); darker honey contains more antioxidants than lighter honey.

Check the website for the National Honey Board (http://www.honey.com); it should have the latest information on the medicinal properties currently studied in honey. Also, very good information is in *The ABC and XYZ of Bee Culture* (Shimanuki et al., 2007).

Various Types of Honey Containers

1. Plastic squeeze skep
2. Glass hexagonal jar
3. Plastic squeeze honey bear
4. Plastic queenline jar
5. Glass queenline jar

1.   2.   3.   4.   5.

Bottling Extracted Honey

After all your honey is extracted, you should begin bottling it within a month. Otherwise, it may crystallize while it sits in a large holding tank, and then you would need to heat it to make it free-flowing again. Before you start to bottle, have everything ready in a dust-free and clean room. If you have not done so already, heat the honey to 140° to 160°F (60°–71°C) to kill yeast spores, and then begin to fill the bottles. Cool the honey quickly. Bottling dispensers or gates at the bottom of tanks or buckets make filling small bottles easy; for larger operations, a nondrip and accurate bottler is worth the investment; check with bee suppliers or visit a honey packing plant to see how it operates and what equipment it uses.

Jars used for honey are usually glass, called "queenline" jars, and have a distinctive shape (see the illustration of the various types of honey containers on this page). They are the best because they can be recycled and heated easily. Some plastic jars are also good, but they may collapse when heated. Honey bears now come in many styles and are a popular and recognizable container (there are also "honey angels"). Containers can be ordered from suppliers advertised in bee journals or online.

Fancy jars are also available and can be used to promote unique honey or for a specialized clientele. Jars, even newly purchased jars straight out of the box, must be washed before being filled, to eliminate dust and other contaminants. (Also check the boxes for other insect pests.) Wash them in a dishwasher and dry upside down so all the water will drain out. Make sure your caps are also clean and dust free, and use some kind of gasket or sealer to ensure honey-tight lids. Different styles of tamper-proof lids and seals are also available, and if you are selling honey to commercial stores, you need to investigate these kinds of seals. Labels can be printed from your computer or purchased preprinted if you want a professional look.

A typical mid-sized to large bottling operation should include:

● Jars (stored in cardboard boxes).
● Dispenser or filling machine with automatic shutoff.
● Automatic capper.
● Labeling area and labels.
● Packing area, where filled jars are placed into the boxes.
● Shipping and storage area for boxes of jars.

The shelf-life of bottled honey is around six months, depending on the kind of honey (some types crystallize faster than others) and storage temperatures. If your bottles are stored longer and the honey begins to crystallize, you can put the jars in hot water baths to reliquefy the honey; then label the jars and package them. For more information on selling your product, go to the end of this chapter.

Cooking with Honey

Honey is a natural sweetener, often used to replace sugar in cooking. Because honey is a combination of sugars that have been broken down into the simple sugars (fructose and glucose), it is very digestible. Honey is also able to keep baked goods moist because of its hygroscopic property (discussed earlier).

When you are substituting honey for sugar in any basic recipe, you should observe the following rules:

- Measure honey with a greased utensil.
- Honey has 1½ times the sweetening power of sugar.
- Reduce the liquid in the recipe by ¼ cup for each cup of honey used to replace sugar.
- Use a mild-flavored honey, unless the flavor of the honey is a necessary part of the product.
- Some people add ¼ teaspoon of baking soda per cup of honey to counter honey's acidity.
- Reduce the cooking temperature of the final product by 25°F.

If honey is stored in a warm place (not the refrigerator) or if it is frozen, it will granulate (crystallize) much more slowly and will not ferment. Honey stored this way could last for years. For more information, see "Honey Cookbooks" in the References or search the Internet for honey recipes.

Cautionary note: It is not advisable to feed honey to infants less than one year old. Spores of the bacterium *Clostridium botulinum* exist everywhere and are able to survive in honey. In the digestive tracts of infants younger than one year, the spores may progress to the vegetative stage of their life cycle and produce toxins that could prove fatal or injurious to infants. This disease is known as infant botulism. Again, check the National Honey Board website for more information.

Preparing Comb Honey for Market

If you are cutting comb honey from a frame (cut comb honey), use a warm, sharp, thin-bladed knife and cut the comb on some kind of screen to allow the honey from the sliced cells to drip off. (Wire fencing material over a large flat pan works well.) Use a template when you cut the honeycomb to fit into your containers; that way all the sections will be uniform. Allow the honey to drip out overnight and carefully package each section in a clear plastic cut-comb box or other clear container. Make sure no boxes leak when they are displayed or sold. Other leakproof containers are available from a variety of sources; check the Internet, bee journals, and beekeeping suppliers. Special labels are also available for these boxes. These cut comb sections can also go into liquid honey jars to make chunk honey.

Section comb honey, whether in plastic, wooden boxes, or plastic cassettes (or other comb honey section boxes), is relatively simple to prepare for the consumer. It takes no special equipment except labels to fit onto the boxes. One problem you will have is what to do with stained, damaged, or half-filled sections. Half-filled ones can be used the next season to bait new section supers. Stained or damaged sections can be sold as seconds, cut out and put in a jar and filled with liquid honey (to make chunk honey), or crushed to extract the honey.

All the sections need to be as clean as possible, free from debris, propolis, and other foreign matter. Use a sharp knife to scrape off the propolis that accumulates on the rim of the sections. This job is both fast and easy, but make sure you do not nick the comb or allow debris to fall inside the finished product. Nicking the comb or otherwise damaging it, or having it detach from the section box or holder, will cause the honey to leak into the final package. Leaky section boxes not only are unattractive but also do not sell well and discourage future customers.

Store your comb honey sections in the deep freezer (either boxed or in the section super) to kill any ants or other hitchhikers and to keep the comb from granulating. Be careful not to bump or drop these delicate honeycomb sections; the slightest bump will cause a crack in the cappings, which will surely result in leakage when the honey warms up.

If you are using the old-fashioned wooden sections, as an extra precaution place each section in a

Other Measurements of Honey

- Specific gravity of 15% moisture @ 20°C: 1.4225.
- Acidity: average pH for honey = 3.9; contributes to honey's antibacterial nature (pH for vinegar, Coca Cola, beer = 3.0; for tomatoes = 4.0).
- Osmotic pressure: more than 2000 milliosmols/kilogram (makes honey a hyperosmotic solution); also contributes to antibacterial activity.
- Refractive index: 1.55 if water content is 13%, 1.49 if water content is 18%.
- 1 gallon (3.785 liters) weighs 11 pounds 13.2 ounces (5.357 kg).
- 1 pound (0.453 kg) has a volume of 10.78 fluid ounces (3.189 milliliters).
- Caloric value: 1380 calories/pound.
- 1 tablespoon = 60 calories.
- 100 gram = 303 calories.

Thermal Characteristics

- Specific heat: 0.54–0.60 cal/g/°C at 20°C (68°F).
- Conductivity of honey at 21% moisture at 21°C: $12.7 \times 10-4$ cal/cm2/sec/°C.
- Conductivity of honey at 17% moisture at 21°C: $12.8 \times 10-4$ cal/cm2/sec/°C.
- Freezing point at 15% moisture: from –1.43 to 1.53°C (32.4°F).

Sweetening Power

- 1 volume of honey = 1.67 volumes of granulated sugar.
- 1 pound = 430 grams (0.95 lb.) of sugar.
- 1 gallon (3.785 liters) = 9.375 pounds (4.25 kg) of total sugars.

Color

Color	Pfund scale (mm)	Optical density
Water white	< 8	0.0945
Extra white	9–17	0.189
White	18–34	0.378
Extra light amber	35–50	0.595
Light amber	51–85	1.389
Amber	86–114	3.008
Dark amber (molasses color)	> 114	—

Sources: Honey Technical Glossary, http://www.arkadiko-meli.gr/default.asp?contentID=568; Airborne Honey Company, http://www.airborne.co.nz/manufacturing.shtml; E. Crane, ed. 1975. Honey, a comprehensive survey. London: Heinemann Ltd. H. Shimanuki, K. Flottum, and A. Harmon (eds.). 1992. The ABC and XYZ of Bee Culture. Medina, OH: A.I. Root. Also check References under "Honey and Honey Products."

plastic sandwich bag and then store the sections in a box placed in the freezer. Once at room temperature, slip each section inside the cardboard display box (that you have purchased). Do not freeze the wooden sections in the cardboard boxes because the boxes may get wet and fall apart when defrosted.

If you have frozen your other comb honey sections, the same rule applies: attach the labels after the boxes have come to room temperature, or the labels will wrinkle on the wet containers. Print the price and weight on each section and stamp it with your name, address, and zip code. Whichever section comb system you buy, make sure you purchase the correct labels and other accoutrements. Be sure to use indelible ink to keep it from washing off. In general, comb honey has a shelf life of two to three months.

Honey Products

Most extracted honey is sold for table use or for use in baked goods. However, there are many other products that contain honey that can be sold as unique or value-added items. You can also experiment with your own unique brand of manufactured goods. Some suggestions provided here have been on the market for a long time and sales continue to grow. Check online for other ideas for honey products.

You can make honey beer, honey wine (mead), and honey vinegar. Other products include honey butter and candy made with or filled with honey, honey ice cream, honey drinks, and dried fruit and nuts stored in or mixed with honey. Some of these products are very perishable and will need to be refrigerated. If

Dyce Process

E.J. Dyce developed and patented this process for making cremed honey in 1935. If it is made properly, the final product is a granulated honey with very fine and smooth crystals that spreads like peanut butter. To make cremed honey, follow this procedure.

Have on hand a "seed" or starter honey. The starter should be from store-bought crystallized (cremed) honey that has a very fine-grained consistency. You can also reuse your own granulated honey, as long as the crystals are very fine (by grinding or naturally occurring). Heat honey that has a water content between 17.5 and 18.0 percent until it reaches 150°F (66°C), stirring constantly. This pasteurizes the honey to kill any yeast spores that would otherwise cause the product to ferment. Strain through two or three layers of very fine nylon mesh to collect all pollen grains, sugar crystals, wax particles, and other impurities. This straining is very important; if you do not do it, the honey may "seed" on these particles instead of on the seed honey you add. Cool honey rapidly to 70° to 80°F (21–27°C), stirring slowly. (Put container in ice water.) Do not add any air bubbles into the honey; you can prevent this from happening by stirring slowly without breaking the surface of the honey.

When the temperature is between 70° and 80°F, add 10 percent by volume the starter seed (i.e., if you have 10 lb. of honey, add 1 lb. of starter). Incorporate this into the honey without the addition of air bubbles, as bubbles will create a frothy top on the finished product.

Let this settle for a few hours, skimming froth off the top (air bubbles will produce a frothy appearance) if needed, and then pour the mixture into the final, marketable containers. Now store this honey at temperatures between 45° and 57°F (7–14°C) for a week until it is completely crystallized. Label and sell. Do not allow these containers to overheat, because that will cause the honey to become liquid again. Store crystallized honey in a cool, dry place until ready to sell.

Source: H. Shimanuki, K. Flottum, and A. Harman (eds.). 2007. The ABC and XYZ of bee culture: An encyclopedia of beekeeping (41st edn.). Medina, OH: A.I. Root.

you add other food to honey, it then becomes a "food product"; check and comply with all federal and state food laws before selling these products.

BEESWAX

The domestic wax industry can obtain only two-thirds of the beeswax it needs from U.S. beekeepers; the rest must be imported. Beeswax is used for cosmetics, candles, and foundation. Beekeepers probably are the most common user of beeswax, selling their wax back to bee supply companies to make into wax foundation for their frames. Minor uses include pharmaceuticals and dentistry concerns, followed by foundries and companies that make floor polishes, automobile wax, and furniture polish. Beeswax is a minor ingredient in some adhesives, crayons, chewing gum, inks, specialized waxes (grafting, ski, ironing, and archer's wax), and woodworking supplies and for arts and craft projects (such as batik).

Because of the ability of beeswax to become contaminated with wax-soluble pesticides, some "organic" beekeepers are letting the bees make their own comb, instead of using commercial foundation. They are worried that commercial foundation may contain a low level of chemical compounds, as well as what bees come in contact with while foraging. Beeswax can hold many such impurities for decades (see Pesticides in the References).

Wax foundation is expensive and beeswax is a valuable hive product; you should make every effort to save all cappings, old combs, and bits and pieces of extra wax scraped from frames and other hive parts. Melt these pieces in the solar wax melter and trade or sell wax blocks to dealers. Separate out old dark combs from lighter cappings or new wax because the lighter shades can be worth more when selling wax.

Cappings, old combs, and wax scrapings should be kept in airtight containers or frozen until you are ready to process them. Wax moths can quickly infest

Characteristics of Beeswax

- Melting point: 142–151°F (61–66°C); wax from cappings: 146.7°F (63.7°C). Paraffin melts at 90–150°F (32–66°C).
- Solidification point (liquid wax becomes solid): 140–146°F (60–63.5°C).
- Flash point (wax vapor ignites): 490–525°F (254–274°C), depending on purity of the wax.
- Temperature of plasticity: 89.6°F (32°C).
- Relative density at 68°F (20°C): 0.963 (water is 1.0, so wax floats).
- Saponification value: 95.35.
- Acid value: 16.8–24.0.
- Ester value: 66–80; ratio of ester to acid is 3:4.3.
- Refractive index (light bending property) at 176°F (80°C): 1.4402.
- Electrical resistivity: 5–20 × 1012 ohm m.
- Shrinkage: If melted to 200°F (93.3°C), shrinks 10 percent (by volume) when cooled to room temperature.

Solubility: Insoluble in water; slightly soluble in cold alcohol; soluble in benzene, ether, chloroform, and fixed or volatile oils. Beeswax is stable for thousands of years.

Beeswax Components

Formula: $C_{15}H_{51}COOC_{30}H_{61}$

Component	Percentage of total
Monoesters	30–35
Diesters	10–14
Hydrocarbons	10.5–14
Free acids	8–12
Hydroxy polyesters	8
Hydroxy monoesters	4–5
Triesters	3
Acid polyesters	2
Acid esters	< 1
Free acids	12
Free alcohols	< 1
Unidentified	6

such wax if it is not protected, or their eggs may be already present, waiting to hatch. Melt cappings separately from the old combs, since the latter contain non-wax substances that would impregnate and reduce the value of the almost pure-wax cappings. Use extreme caution when melting wax—it ignites easily, and wax fires are difficult to put out (see the information on beeswax in the sidebars).

If you have used chemical miticides in your colonies, use your wax to make products such as ornaments and furniture wax or polishes, or sell bars for craft or sewing projects. To be sure your wax is clean, pay to have it analyzed (see Pesticides in the References), and then you can use that information as an advertising point.

Melting Beeswax

Use extreme care when processing beeswax as it is highly flammable and can cause serious burns.

Follow these simple precautions when melting beeswax:

- Avoid extended exposure to high heat over 185°F (85°C), as it will darken the wax.
- **Never** use an open flame; use a double boiler over electric heat.
- Use only aluminum, nickel, tin, or stainless steel containers to melt wax, as other metals will discolor it.
- Never use direct steam to melt wax, as it permanently changes the composition of wax.
- Separate dark, old wax from light, new wax and never add propolis to wax.
- Don't melt plastic frames in the wax melter; they can warp.
- Never store pesticides near beeswax or combs because such chemicals are lipophilic and will bind with wax. This includes miticide strips.

Separate white, new wax and cappings from the older, darker wax from old combs and scrapings. Lighter wax is more valuable; both colors of wax can be melted with one of these devices:

- Cloth bag set in boiling water.
- Electric wax melter.
- Solar wax melter.

- Double boiler made of aluminum or stainless steel; iron or copper will darken the wax.
- Steam chests or hot boxes.
- Wax press that is steam heated.

Older, dry comb that may have wax moth silk or lots of pollen or fermented or crystallized honey can be soaked in soft rainwater. Place the old comb and scraps in a burlap or cloth bag and submerge it in a tub or barrel of rainwater or acidic water (**never** alkali water, as the wax will become spongy and not be fit for other uses). Place stones or bricks in or on the bag to help keep it submerged. Heat the water between 150° and 180°F (66° and 82°C) for several hours, occasionally poking the bag with a stick to allow the wax to move through the fabric to the surface of the water. After the wax has melted, remove the tub from the heat and allow the water to cool. The wax will solidify on the surface of the water and most of the debris will fall to the bottom of the tub. Render and filter the wax cake a second time.

None of the methods will be sufficient to render all the wax found in old combs; the remaining mixture (*slumgum*) of wax and debris should not be discarded, but saved and taken to a dealer who has the special equipment needed to render it.

Solar Wax Melter

The solar wax melter is essentially a box painted black outside and white inside and covered with a piece of glass, Plexiglas, or other plastic sheet and made airtight. It is put in a sunny location and tilted at a right angle to the sun's rays. The sun heats the interior of the box, something like it would a greenhouse, melting the wax inside, which collects into a pan. For greater heating efficiency, use two pieces of glass or Plexiglas, separated by a ¼-inch (6.4 mm) gap. The inside of the box contains a metal tray, fashioned from sheet metal, onto which the wax comb and scraps are placed (see the illustration of the solar wax melter on p. 182).

The melter will render cappings, new burr comb, and old comb, but it will not melt frames of old comb completely. After old, dark comb has been in the melter for a few days, collect the black, gummy remains (*slumgum*) and either compost it (it is rather acidic) and use as a soil amenity or discard it. If you have a lot of it, and it is feasible, take the slumgum to a bee supply dealer who has equipment to extract the remaining wax chemically. The amount of wax you can derive from broken pieces of foundation, cappings, and old or burr comb is significant and well worth the effort.

Wax Cappings

When extracting honey, you will generate buckets of wet wax cappings and honey. If you have only a few buckets, you can place them into a hot room (where it heats up to about 143°–151°F [62–66°C]). Some beekeepers make an insulated box heated with a single light bulb; this will melt all the wax in 24 hours. The remaining honey, if not overheated, can be eaten or bottled or fed back to bees (if it contains no foulbrood spores). If the remaining honey is overheated and dark, it will be unacceptable to both people and bees, so discard it, or use it to feed wax moths, if you are rearing them (see Appendix F).

For larger amounts of cappings, you may want to invest in a cappings spinner (an extractor to separate out most of the honey), and some sort of water-jacketed stainless steel melter. The best way to determine what you need is to visit several beekeeping operations and see how they handle their wax.

Cappings wax is worth a lot of money, and recov-

Solar Wax Melter

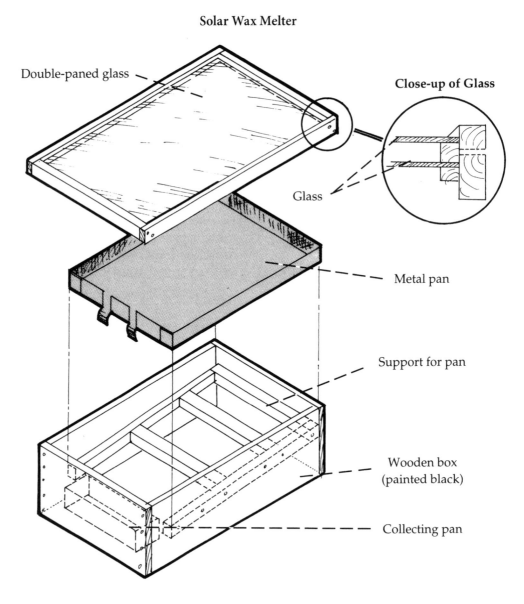

Double-paned glass

Close-up of Glass

Glass

Metal pan

Support for pan

Wooden box
(painted black)

Collecting pan

ering it is well worth the effort. After you separate out the honey, melt the wax and pour it into plastic tubs or loaf pans (first coated lightly with talcum powder or cornstarch). Weigh the finished block so you can estimate how many pounds of wax you have. Again, if you have not used any chemical compounds for mite controls, you can sell the wax in the blocks, or render it into saleable products, such as lotion bars; otherwise, make candles or ornaments. Again, having your wax tested for contaminants is a good idea.

Rendering Wax: Final Filtering

Once you have collected the chunks or blocks of wax from your wax melter, you must further refine it to remove the fine particles of dirt, honey, and other impurities. Melt first in water using a large double boiler or other wax melter (check bee supply catalogs), and then refine the wax by filtering it through several layers of cotton sacking, an old sweatshirt, or paper towels. If you are using this wax for value-added products (such as candles), you will need to filter it again using coffee filters or paper towels.

When you are melting small amounts of wax to pour into molds, or the like, use an electric frying pan, an old electric coffee pot, or double-boiler arrangement (in which you place the wax in a small can or pan within a larger pan of hot water). Buy used metal pans at a thrift store and use them only for wax. For the final filtering, pour hot wax into a filter-covered can and, to keep the wax liquid, put the

Honey Bee Soufflé

½ cup butter
½ cup sifted flour
1 ½ teaspoons salt
½ teaspoon paprika
Dash of hot pepper sauce
2 cups milk
½ pound sharp cheese, grated
8 eggs, separated
½ cup marinated bees (larvae and young pupae
    in 1 cup soy sauce, ¼ cup sake or sherry,
    1 garlic clove, 1 dried hot pepper, 2 table-
    spoons ginger root, grated)

Melt butter in double boiler over boiling water and add next four ingredients; mix well and gradually add milk, stirring until sauce thickens. Add cheese until it melts; remove from heat. Separate egg whites from the yolks. Beat egg yolks until lemony and stir slowly into cheese sauce. Beat egg whites until stiff and fold sauce into whites. Layer bottom of 1-quart greased soufflé dish with bees and cover with sauce. Bake at 475°F for 10 minutes; reduce oven temperature to 400°F and bake 25 minutes longer.

Source: R.L. Taylor and B.J. Carter. 1992. Entertaining with insects; or, The original guide to insect cookery. Yorba Linda, CA: Salutek.

can into a warm oven until all the wax has drained. This fine-filtered wax can then be used for candles, in batik, or for whatever you decide (see "Beeswax" in the References).

BEE BROOD

Generally an unexploited product of the hive, bee brood is rich in proteins, fats, and other substances required in our daily diet. Bee brood consists of over 15 percent protein, 4 percent fat, and about 77 percent water, compared with beef, which is 23 percent protein, 3 percent fat, and 74 percent water. It has over 100 times the vitamin A present in milk and over 60,000 times as much vitamin D.

The value of this hive product does not yet compensate for the cost of removing brood from comb and the reduction of the adult colony population that ensues if too much brood is removed. Honey bee brood is currently used, on a small scale, as food for birds, reptiles, and fish. Drone larvae are often used for fish bait; if you cut out drone brood for varroa control, you may be able to sell it in these markets. Bee larvae also make a good quiche or soufflé (see sidebar).

BEE VENOM

Bees require pollen in their diets in order to synthesize some of the components of venom. The synthesized venom is stored in the poison sac of worker

Honey Bee Brood Components

Component	Larvae	Pupae	Beef	Milk	Egg yolk
Water (%)	77	70.2	74.1	87.0	49.4
Ash (%)	3	2.2	1.1	3.5	
Protein (%)	15.4	18.2	22.6	3.9	16.3
Fat (%)	3.7	2.4	2.8	< 5	31.9
Glycogen (%)	0.41	0.75	0.1– 0.7	—	—
Vitamin A (IU) —	89–119	49.3	1.6	32.1	
Vitamin D (IU) —	6130–7430	5260	0.41	2.6	

Source: E. Crane. 1990. Bees and beekeeping: Science, practice, and world resources. Ithaca, NY: Comstock.
Notes: % = Percentage of fresh weight. IU = International units/gram fresh weight.

Analysis of Bee Venom

Class of molecule	Component	Percentage of dry weight (%)	Effects
Enzymes	Phospholipase A_2	10–12	Bursts cells by degrading cell walls. Decreases blood pressure and inhibits blood coagulation. Causes pain; activates arachidonic acid to form prostaglandins (which regulate inflammatory response). Wasp toxin contains phospholipase A_1.
	Hyaluronidase	1–3	Dilates the capillaries, causing the spread of inflammation. Hydrolyzes connective tissue. Helps spread other components.
	Acid phosphomonoesterase	1.0	
	Lysophospholipase	1.0	
	Alpha-glucosidase	0.6	
Other proteins and peptides	Melittin	40–50	Releases histamine and serotonin from mast cells. Bursts blood and mast cells. Depresses blood pressure and respiration. Induces production of cortisol in the body.
	Apamine	1–3	Increases cortisol production in the adrenal gland. Is a mild neurotoxin.
	Mast cell degranulating peptide (MCD)	1–2	
	Secapin	0.5–2.0	
	Procamine	1–2	
	Adolapin	1.0	Is anti-inflammatory and analgesic because it blocks cyclooxygenase.
	Protease inhibitor	0.8	
	Tertiapin	0.1	
	Small peptides (with less than 5 amino acids)	13–15	
Physiologically active	Histamine	0.5–2.0	Causes allergic response: itching, pain.
Amines	Dopamine	0.2–1.0	Increases pulse rate.
	Noradrenaline	0.1–0.5	Increases pulse rate.
Amino acids	Gamma (γ)-aminobutyric acid	0.5	
	Alpha-amino acids	1.0	
Sugars	Glucose and fructose	2	
Phospholipids		5	
Volatile compounds		4–8	

Sources: E.M. Dotimas, K.R. Hamid, R.C. Hider, and U. Ragnarsson. 1987. Isolation and structure analysis of bee venom mast cell degranulating peptide. Biochimica et Biophysica Acta, 911:285–293. R.A. Shipolini. 1984. Biochemistry of bee venom. In: A.T. Tu (ed.), Handbook of natural toxin, vol. 2, pp. 49–85. New York: Marcel Dekker. E. Crane. 1990. Bees and Beekeeping. Ithaca, NY: Comstock. See "Venom" in the References.

and queen bees (see Appendix A). Venom contains a complex array of chemical substances, such as water, amines, amino acids, enzymes, proteins, and peptides, as well as some sugars and phospholipids. Venom also contains histamine, which reacts adversely with the body chemistry of some individuals. (See "Analysis of Bee Venom," p. 184.)

To collect substantial amounts of venom, either for medical or for other uses, an electrical grid is placed near the entrance of a hive. This special grid produces a mild shock, and bees that land on it react by stinging a sheet of nylon taffeta below this grid. The venom is deposited on and collected from a glass plate located below the nylon portion of the device.

Research is still in progress concerning the benefits obtained from honey bee venom for persons with rheumatoid arthritis and other diseases. Bee venom therapy, including *apitherapy*, is becoming more popular today. In addition, more recent research indicates that some of the components of venom are much more effective than other serums in desensitizing persons who are allergic to bee venom (see Appendix C and "Products of the Hive (Other Than Honey)" in the References, or search the Internet for more information).

ROYAL JELLY

Royal jelly consists, roughly, of 66 percent water, 4.5 to 17 percent ash, 11 to 13 percent carbohydrates, 12 percent protein, 5 percent fat, and 3 percent vitamins, ether extracts, enzymes, and coenzymes. Royal jelly, the sole food of queen larvae, is manufactured by young nurse bees from two glandular secretions: a white jelly from the hypopharyngeal and mandibular glands, and a clear jelly, which is a hypopharyngeal secretion mixed with honey.

Royal jelly has long been collected and used in Asian countries for medical purposes. Modern uses include cosmetics, lotions, and dietary supplements. None of the curative properties of royal jelly have been extensively studied in the United States, nor should any be claimed on product labels. In addition, some people can be allergic to this product, and an appropriate warning should be included on any label.

Queen breeders often collect and freeze royal jelly to use during queen-rearing operations. If you

are considering buying royal jelly from outside the United States, **be warned** that it can contain viruses that may be detrimental to bees; this discovery was reported by researchers studying the causes of colony collapse disorder.

PROPOLIS

Propolis is a resinous mixture used by bees to seal cracks in the hive. The word is derived from the Greek words *pro* (before) and *polis* (city), and refers to its use by bees to reduce the hive entrance ("in front of" or "before" the "city") for protection against winter elements and defense against ants.

Forager bees gather gums and resins that are in a semiliquid state from the bark, buds, and wounds of trees and shrubs—such as the alders, poplars, mesquite, and some conifers—or from abandoned equipment (such as mite- or disease-killed colonies). Foragers transport the material back to the hive on their pollen-collecting structures. Inside the hive, house bees pull it off, since it is too sticky for the foragers themselves to dislodge, and then mix it (40–60 percent) with beeswax and a third, unknown substance. It is this mixture that is called propolis.

Propolis is used by bees to:

- Coat the inside of the hive walls.
- Seal the inner cover, bottom board, and outer cover to the hive body.
- Strengthen comb.
- Glue frames together.
- Seal hive bodies together and caulk any small crack or crevices.
- Embalm foreign objects too difficult to remove (e.g., dead mice, snakes, and large insects).
- Fill small, tight spaces and cracks that are smaller than the bee space.

To beekeepers, propolis is a sticky, gummy mess in the hive and makes separating the hive bodies difficult, even when using a hive tool.

Studies have documented antimicrobial activity of propolis, which may help bees keep the hive clean and kill foreign microbes that would otherwise live in small cracks. Dead animals coated with propolis are virtually mummified and will not decompose.

Some of the components identified in propolis include flavonoids, benzoic acid, cinnamyl alcohols and acids, alcohols, ketone, phenols, terpenes, sesquiterpene alcohols, minerals, sterols, volatile oils, beeswax, sugars, and amino acids. The compounds in propolis that have some pharmacological (antiviral, antibacterial, antioxidant, antifungal, anti-inflammatory, and antihistamine) activities are quercetin, pinocembrin, caffeic acid, phenethyl ester, acacetin, and pinostrobin.

Beekeepers who collect propolis lay a special plastic screen (check bee supply stores) on the top bars of the uppermost honey super, and allow the bees to fill it with propolis. Once filled, the screen can be frozen and the propolis knocked off, cleaned, and bottled. Propolis is even appearing in capsule form in health food stores, as a health supplement. None of its curative properties have been extensively studied in the United States, nor should any be claimed on the label. As with other bee products, some people can be allergic to propolis, thus an appropriate warning should be given when selling this product.

Because propolis is not water soluble, use rubbing alcohol to remove it from your hands and clothing. Other people find that a mechanic's hand cleanser works well. You can scrape it or burn it off hive tools.

POLLEN

Pollen, the protein-rich powder produced by the male parts of flowers, is collected and sold by beekeepers to health food stores, to pollination businesses, to bee dealers (for bee food), and to allergy patients (as a desensitizing agent). Pollen traps are put on hives to remove and collect pollen pellets from foraging bees. Collected pellets should be stored properly; see "Pollen" in Chapter 7 and "Products of the Hive (Other Than Honey)" in the References. Be careful that the pollen collected comes from flowers that have not been sprayed with pesticides, because they can contaminate it.

Pollen contains a number of nutrients, such as carbohydrates, proteins, and minerals. The composition varies according to the plant species, and bees are not able to distinguish from the less nutritious ones. Pollen constitutes 10 to 35 percent of the protein in the diet of bees. Remember, pollen that bees convert into bee bread supplies the essential amino acids needed to manufacture their own proteins and enzymes and

Pollen Components

Component	Percentage of total (%)
Water	7.0–16.2
Crude protein	7.0–29.9
Ether extracts	0.9–14.4
Carbohydrates	
Reducing sugars	18.8–41.2
Nonreducing sugars	0–9.0
Starch	0–10.6
Ash	0.9–5.5
Unknown	21.7–35.9

Other components include lipids, organic acids, free amino acids, nucleic acids (plant), enzymes, sulfur, nitrogen, flavonoids, and growth regulators (plant).

contains such vitamins as A, C, D, E, B_1, B_2, B_6, and B_{12} as well as minerals, such as sulfur, nitrogen, phosphorus, and many minor elements; see the description of pollen components in the sidebar. Some substances found in pollen have yet to be identified.

Pollen pellets are sold in health food stores for human consumption. Some claim that pollen is a superior food with curative powers and provides extra energy. Authoritative reports as to the value of pollen as human food are contradictory. To survive the rigors of sun and rain and preserve its precious cargo, pollen grains are protected by a wall of intine and exine. Some pollen grains over 10,000 years old have been excavated intact (good enough to identify plant species) from acid bogs. Bees have special stomachs, which humans lack, to grind off some of the exine layer and digest the protein content of the pollen grain. Recent studies indicate that the protective layers of some pollen can be ruptured and the proteins released into human digestive systems. Whatever the claims, there is a market for pollen, and it may be lucrative for beekeepers to collect and sell it, since the importation of pollen pellets from overseas is not permitted (because they could contain chalkbrood, chemical toxins, and other pathogens).

Before you sell or consume pollen, be certain that it was not contaminated by pesticides. Beekeepers trap pollen while some crops are being sprayed with insecticides, to reduce the amount of contaminated pollen in the colony. Such pollen should never be sold for human (or bee) consumption. Know the source of pollen before buying or selling it. There are labs, books, and websites that will identify the plant species of your pollen, which is a fascinating and informative activity. Other labs will test pollen (and other hive products) for contaminants, which may be worth the cost if your business depends on selling a pure product. The National Honey Board has a list of labs on its website.

Some companies will buy bee-collected pollen, so check the bee magazines for advertisements. But remember, none of its curative properties have been extensively studied in the United States, **and some people are hypersensitive to this product and could suffer a severe allergic reaction. Add a warning when selling pollen.**

Any bee products can cause an allergic reaction in some people, and you can face potential lawsuits. Look into getting product liability insurance if you plan to sell bee products. **Do not rely on your homeowners insurance.**

WAX MOTH LARVAE

Rearing wax moths for the fish-bait industry or as pet food can be a lucrative and successful side business to beekeepers. Rear the moths in an area away from your bee operations so you don't add more pests to infest your bee equipment. For information on rearing waxies, see Appendix F.

SELLING YOUR HIVE PRODUCTS

There are many avenues open to you to sell your hive products: the office, local retail outlets, roadside stands or orchards, flea or farmer's markets, fairs, your local church, or organizations to which you or your friends belong. Keep it simple at first; large grocery chains are not a good place to start because (1) they require a year-round supply; (2) it's not easy getting your foot in the door; and (3) you have no control over the price.

Here are some rules, researched by the National Honey Board, which you should follow to be successful in keeping your customers happy. In addition,

check with the local agriculture extension offices, other beekeepers, and the Internet for other ideas.

- Have a reliable supply of quality product.
- Local honey or hive products are best. Use slogans such as "Produced by Bees in Your County" for regional appeal, especially in souvenir shops or highway markets.
- Keep your jars clean; people do not like sticky jars.
- Do not heat your honey higher than 120°F (49°C), or you will ruin the delicate, floral flavors.
- Make sure your bee products are clean and free from debris.
- Use eye-catching, attractive labels; in general the public does **not** like to see bees or bee hives on the label.
- Use consumer-friendly words, such as "**pure,**" "**natural,**" or "**raw**" on natural hive products. Honey can be labeled "Raw" if it has been minimally filtered or heated.
- Conduct taste tests, give out free recipes, dress honey bears in holiday outfits, make gift baskets, or use other "value-added" techniques to help sell your products.
- Use "**organic**" only if you comply with the USDA organic food standards.
- Try making "cremed" honey using the Dyce process (described in this chapter).
- Use different kinds of honey jars (see the illustration of various honey containers on p. 176).
- If you wish to mix honey with dried fruits, nuts, or flavorings, you must label these carefully as these are considered a "food product" and must meet federal and state standards. Instead, you can give small jars of this away as gifts.

Advertising is also an important part of selling your products. Take advantage of promotions available from the National Honey Board or other organizations, many of which are free. Many beekeepers simply sell honey from their home, but before expanding your operation to a point at which you need an outside sign, check with your homeowners insurance policy and any local zoning codes: forewarned is forearmed. A small classified ad in local papers may be all you need to have a steady supply of customers. You can also explore Internet sites.

Honey Labels

Jars of honey have to conform to state and federal pure food laws. The guidelines below will meet most requirements, but check with your state agriculture department for more information.

The following information must appear on each jar or section comb of honey you sell:

- Name, address, city, state, zip code of manufacturer, packer, or distributor.
- Name of product (e.g., Orange Blossom Honey).
- Net weight in pounds, ounces, and grams in the lower 30 percent of the display label (e.g., Net Wt. 8 oz. [227 g]).
- Additional labels or enhancers: Local Honey, Award Winning, Made by US Bees, or nectar source labels (Pure Fireweed, Oregon Wild Flower Honey, or Sourwood Honey).

Labels can be purchased from most bee supply dealers, or label companies that advertise in the journals. You can have a printer make up your own label preprinted with your name, address, and so on, but the initial cost is rather high. Instead, you can buy blanks and have a stamp made with your name and other information that you can later imprint with indelible ink. Labels for round- or square-section comb honey are also sold by bee supply companies or label companies that advertise in the bee journals.

You can also print your own labels on your home computer. Make sure you use special label paper and indelible ink. In this way you can change the label to fit the season, and personalize it, with pictures of flowers.

It is becoming common for people to have beehives on their property under the care of a beekeeper. When the honey is bottled, personalized labels are printed, and they provide their close friends with their honey. You can also personalize your labels even more for special customers with their pictures or names—be creative!

Nutritional labels are required, especially if you claim a nutrient content or health-related statement. Again, check with the National Honey Board or your state food and labeling regulatory agency for particulars. Some beekeepers put on additional labels giving instructions on how to reliquefy honey that has crystallized. Unfortunately, any instructions you supply could make you liable for any injury to a person following those instructions. Carefully label your products. When in doubt about wording on the labels, start by buying several jars of honey (or other hive products) from three different national distributors; it is likely that their labels and product information will cover most if not all the basic requirements of the law. You can model your labels on these. You may also want to check with a lawyer about the wording.

 **Notes**

CHAPTER 13

Pathogens and Parasites of Honey Bees

Honey bees, like all animals, have diseases and pests that can destroy individuals as well as the colony if not given a little help. Any domesticated animal that is placed together in large numbers is subjected to diseases that would otherwise not be a problem. Commercially raised bees are especially vulnerable to disease because many commercial beekeepers keep thousands of colonies in close proximity and often move them many miles to pollinate fruits and vegetable fields. If you are a hobbyist or side-line beekeeper, some of these conditions may not apply; you will have to decide if you want to treat your colonies with all the available drugs or if you want to treat only the most serious diseases. Many countries ban the use of drugs, electing instead to either destroy infected colonies or use other options. If you choose not to treat, check with your state apiary inspection office; you do not want to be the source of an infection in your area, or break local laws.

COLONY COLLAPSE DISORDER

Bee losses have been recorded as early as 1896 and have been called by various names such as May disease, spring dwindling, disappearing disease, and autumn collapse. Since the 1970s, there has been a further decline in the number of "domesticated" as well as feral colonies. Factors attributed to this decline include urbanization (loss of forage), pesticide use, introduced parasitic mites (tracheal and varroa), and an aging beekeeper population going out of business.

In the spring of 2006, large numbers of overwintering bee colonies in North America were found with virtually no adult bees. While such disappearances have occurred throughout the history of apiculture, the term colony collapse disorder (CCD) was first applied to these bee disappearances. Other countries have not been immune; that same year, European beekeepers began to observe similar symptoms, especially in Spain. The causes are still not fully worked out because only survivors are tested (since most of the bees are not in the colony), but diseases, including the new nosema disease and viral infections, and varroa mites have been found in dying colonies. Some other possibilities include environmental stresses, poor nutrition, and exposure to pesticides, as well as the rigors bees go through in migratory beekeeping operations. Other unproved possibilities (and some wild speculations) implicate cell phone radiation and genetically modified (GM) crops. But weather is also a factor; too much or too little rain will influence the amount of nectar flowers secrete and ultimately the amount of honey stored and the ability of bees to overwinter successfully.

A survey by the U.S. Department of Agriculture's Agricultural Research Services (USDA-ARS) and Apiary Inspectors of America (winter 2008) revealed that in the United States, 864,000 (36%) of 2.4 million hives were lost to CCD. The survey was completed by 300 (20%) of the 1500 commercial beekeepers, and showed an increase of 11 percent over the 2007 losses and 40 percent over the losses of 2006. Reports of large losses in other countries have also been published, but keep current with the bee research in the journals, as this phenomenon needs historical information to put it in perspective. Check websites, especially those from the USDA-ARS and Penn State

University, to name a few (see Colony Collapse Disorder in the References). Most of the colonies affected by CCD were in large-scale commercial operations, not in hobbyist apiaries.

In bee colonies that succumb to CCD, generally all of the following conditions occur simultaneously:

- Adult bees are absent, with little or no accumulation of dead bees in or around the hive.
- In these abandoned colonies, there are food stores, both honey and pollen, but:
  - The stores are not immediately robbed by other bees.
  - If hive pests are present (such as wax moth and small hive beetle), damage is slow to develop.
- Some capped brood is found in colonies (bees will not normally abandon a colony until the capped brood have all hatched).
- Queen is evident, but all the workers appear to be the same age (young) and they are slow to feed on supplemented food (syrup and pollen patties).

As of March 2007 the Mid-Atlantic Apiculture Research and Extension Consortium, or MAAREC, recommends the following when beekeepers notice these symptoms mentioned above:

- Do **not** combine collapsing colonies with strong colonies.
- When a collapsed colony is found, store the equipment where bees will not have access to it. **Equipment from a collapsed colony should not be used again; burn it, combs and all, unless you have access to a gamma irradiation chamber.**
- If you feed your bees sugar syrup, use fumagillin.
- If you are experiencing colony collapse and see a secondary infection, such as European foulbrood, treat the colonies with Terramycin, not Tylan (see foulbrood disease, below).

Check the Internet and especially the Colony Collapse Disorder Working Group, Penn State University, for updated information.

DISEASES OF ADULT BEES

Nosema Disease

Nosema, the most common adult bee disease, is caused by a microscopic protozoan or microsporidia (now considered to be fungi). Nosema disease is common in insects and has been used as a biological control for locusts and grasshoppers (*N. locustae*) and wax moths (*N. galleriae*). Other species of *Nosema* also affect silkworm moths, bumble bees, and leafcutter bees. Nosema species are specific to the insect they attack; therefore, grasshopper *Nosema* will not infect honey bees.

Currently two Nosema species are found in honey bees. One is *Nosema apis* Zander, which has been found in all races of European honey bees. A second, perhaps more virulent species, *N. ceranae*, was first described in China in 1994; in 2004 it was found in European bees in Spain where it was responsible for large colony losses. Originally described from Asian honey bees, *N. ceranae* is thought to be one of the contributing factors in CCD. More information on this new disease is being released, so keep current in the bee journals.

Both species are spore-forming microorganisms that invade the digestive cell layer of the midgut of bees. Up to 30 million spores can be found in a single bee one and a half weeks after initial infection. The spores of both species are very difficult to tell apart, requiring molecular techniques. If you want a positive identification, send samples of bees to the USDA-ARS Bee Research Lab in Beltsville, Maryland (log on to its website for information on how to send samples).

Nosema is now distributed worldwide, wherever bees have been introduced; it is a serious problem in temperate climates. Recent reports state that *N. apis* has been overrun by the new *N. ceranae*. However, we will list both species here. Normally, *N. apis* is most prevalent in the spring, especially after winter weather has confined bees to their hive. If it does not kill the colony over winter, nosema greatly reduces the life span of all castes of adult bees, reduces honey yields, and is a factor in the supersedure of package bee queens, further delaying the growth of a hive in which packages are installed.

Life Cycle of *Nosema apis*

Both Nosema species have similar life cycles. The only way for bees to become infected is from picking up the Nosema spores from other bees or from contaminated food or water. It takes only a few spores to infect bees. When the spore enters the ventriculus (midgut), a filament forms and then attaches and protrudes into the cells. Once inside, the filament in-

Nosema apis Spores

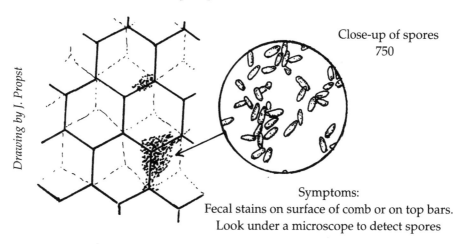

Close-up of spores
750

Drawing by J. Propst

Symptoms:
Fecal stains on surface of comb or on top bars.
Look under a microscope to detect spores

Nosema apis Symptoms

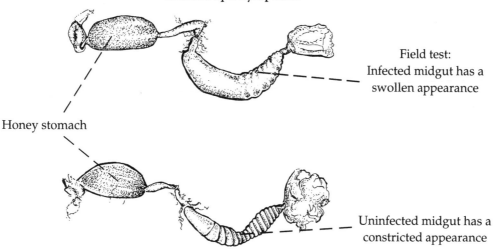

Field test:
Infected midgut has a
swollen appearance

Honey stomach

Uninfected midgut has a
constricted appearance

creases in size and multiplies. This vegetative state is not infective, but it produces more spores within 5 to 10 days, which can burst the cell walls, releasing the spores into the bee's midgut. These new spores either reinvade other cells within the bee, or pass out of the host bee with the feces. If the spores reinfect the same bee, digestive function of the bee can be repressed within two to three weeks.

The spores are viable for up to one year in the fecal material and can be found outside the colony, especially at common drinking areas, within the colony on the comb, and in pollen and honey. The disease is spread by drifting and robbing bees coming in contact with fecal material on frames and combs. Package queens or purchased, caged queens commonly have Nosema spores and should be treated.

Nosema can be a serious disease if not checked, especially now that *N. ceranae* is more prevalent; it can kill bees within a week, weakening foragers to the point where they don't return to the colony, leaving a skeleton population behind. Because nosema is often confused with symptoms of other diseases (such as pesticide poisoning, amoeba or mite predation), diagnosis is important to properly treat this disease. Some effects of severe Nosema infection are:

- Reduced longevity of workers (by 50%).
- Reduced honey yield (by 40%).
- Queen supersedure, as egg laying in queen is adversely affected.
- Reduced function of hypopharyngeal (food) glands, resulting in poor brood-rearing ability, because the nurse bees are unable to produce enough brood food.

A Bee's Encounter with Pathogens and Parasites

Schematic by D. Cox-Foster

- Disruption in hormonal development, causing bees to age faster and forage earlier in life than normal.
- Disruption in the secretion of digestive enzymes, causing bees to starve to death.

Although bees with this disease display no specific symptoms, listed below are some signs to watch for. Remember, these symptoms can **also** be associated with pesticide poisoning, mite damage, or other pathogens. If most of these symptoms are observed in the spring of the year after winter confinement, nosema should be suspected. While these symptoms could be apparent at any time of the year, they are most noticeable in spring and fall.

- Bees unable to fly or able to fly only short distances.
- Bees seen trembling and quivering; colony restless.
- Feces on combs, top bars, bottom boards, and outside walls of hive; also correlated with dysentery or diarrhea (see "Dysentery" below).
- Bees seen crawling aimlessly on bottom board,

near entrance, or on ground; some drag along as if their legs were paralyzed.
- Wings positioned at various angles from body (also called *K-winged*); that is, wings are not folded in normal position over abdomen but with the hindwing held in front of the forewing.
- Abdomen distended (swollen).
- Bees not eating when fed syrup.
- Bees abandoning the colony, leaving the queen and a few emerging workers.

The only way to diagnose positively for nosema disease is to dissect the bee. A field test, which is not very reliable but good in a pinch, is to pull apart a bee until the viscera are visible. If the midgut (ventriculus) is swollen and a dull grayish white, and the circular constrictions of the gut (similar to constrictions on an earthworm's body) are no longer evident, then nosema is the culprit (see the illustration on nosema spores on p. 191). The normal gut is brownish red or yellowish, with many circular constrictions.

A better way is to use a hemacytometer, which is a special slide with a grid etched onto it. Here's how:

- Collect 30 to 60 bees from a suspected colony.
- Make sure you collect older, forager bees (at the entrance).
- Store the bees in alcohol or place in a plastic bag in the freezer.
- Cut off the abdomen of the bees and grind in a mortar and pestle.
- Add 25 ml water per 50 bees.
- Grind again and remove a sample to look through the microscope.

For complete instructions, go to the University of Minnesota Extension website (http://www.extension. umn.edu/Honeybees/components/pdfs/poster_167_ nosema_spores_24x33.pdf), "Testing for Nosema Spores using Hemacytometer" by Gary C. Reuter, Katie Lee, and Marla Spivak.

If you place the sample on a plain microscope slide, Nosema spores, if present, will be seen clearly at a magnification of about 40X. They are small, smooth, ovoid bodies, much smaller than pollen grains, which are also present in the gut. It is best to look at the spores under higher magnification, over 100X, they measure 4 to 6 micrometers long and 2 to 4 micrometers wide. If you have any questions, and for a positive identification, send samples to the USDA-ARS Bee Research Lab in Beltsville.

As we have already mentioned, infection with *N. ceranae* is the more serious disease, and treatment is highly recommended, especially if your apiary is near a commercial operation, or if you have purchased queens or package bees. We need to repeat that the symptoms are not always clear and often confused with those of other diseases, pesticide poisoning, and mite predation; therefore, diagnosis is important in order to treat nosema properly (see "Diseases and Pests" in the References).

Differences between *N. apis* and *N. ceranae*

In general, the biggest difference between the two Nosema species is that the Asian variety can cause a bee colony to die within eight days after exposure, much faster than with *N. apis*. It appears to affect foraging bees the most, killing them while they are outside and leaving the home colony weak. In addition, this new parasite presents risk year-round in all climates, unlike *N. apis*, which usually is most visible in the early spring. When both species are present, researchers find that *N. ceranae* outcompetes *N. apis*.

Nosema Spores

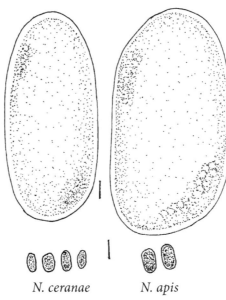

N. ceranae   *N. apis*

Bar = 5 micrometers

The best way to determine the difference between the two species is to examine the spores under a microscope. In general, *N. apis* spores are slightly larger than *N. ceranae* spores (see the illustration of the two spores on this page). Again, for a positive identification, you will have to send samples to the USDA-ARS Bee Research Lab in Beltsville (see References).

Treatment

Good management practices, including apiary sites that protect bees from cold winds and shady conditions in winter, and feeding the antibiotic fumagillin, can help control the disease and ensure healthy colonies. To control nosema and prevent it from spreading:

- Provide fresh, clean water; individually feed water via Boardman feeders to each colony, if you have a small apiary.
- Provide a young queen.
- Locate hives at sunny sites, sheltered from piercing cold winds but with good ventilation.
- Maintain adequate stores of pollen, honey, or cured sugar syrup; if stores are short, bees should be fed a heavy (2:1 sugar : water) medicated syrup in early fall and protein supplements.
- Keep only clean combs; sterilize or dispose of those that are soiled with fecal material.

- Provide upper hive entrance during winter to improve hive ventilation.
- Reduce stress to bees, such as moving bees often for pollination. Again, make sure the bees have adequate honey and pollen stores during inclement weather.
- If bees are not feeding on the syrup, spray them (if the weather is above 60°F [15.6°C]) so they can clean it off each other and thus ingest it.

Combs with Nosema spores must be sterilized, especially if you are going to reuse frames from dead colonies. If you do not disinfect the frames, you are just spreading the disease. Many methods to fumigate combs have been described; however, before you start, check with your state apiary inspector or department of agriculture, and comply with state and federal regulations, especially if you have hundreds of supers. In addition, some states have fumigation or gamma chambers for disinfecting diseased combs. Irradiation of hive furniture contaminated with disease spores, including Nosema, is an effective way of reducing other diseases in older combs. Be sure to check this out.

You can kill spores by heating the supers with combs in a hot room or autoclave to 120°F (49°C) for 24 hours; combs should be free of honey and pollen (temperature should not get above 120°F or the wax will melt). Some compounds that have been found to kill Nosema spores are 10 percent bleach and 60 percent acetic acid.

If it is not possible to fumigate combs, you can replace old, fecal-stained and dark combs with new foundation. A good rule of thumb is to replace frames when you can no longer see through them when you hold them up to the sun. Set up a cycle to replace them every two to three years by marking the top bars every time you install new foundation, so you can easily determine the age of each frame. In other words, as you go through your hives in the spring, pull out the older combs and replace them with new foundation.

Chemotherapy

The drug used to control both Nosema species is bicyclohexylammonium fumagillin (isolated from the fungus *Aspergillus fumigatus*), or fumagillin, and is sold as Fumagilin-B (and other names). It is used as a preventive measure and can be purchased from suppliers of bee equipment. It is not effective against Nosema spores, and therefore must be fed over several weeks in medicated syrup. This drug is viable for two years, which can be extended if unopened bottles are stored in the freezer.

To make the medicated syrup, use warm water, 95° to 122°F (53°–50°C), and then add the medication and the sugar (2 parts sugar to 1 part water), in that order. It is fed in the formulation of 25 mg fumagillin per liter (0.26 gallon of sugar syrup). **Read the instructions carefully** and follow them. Also, bees infected with *N. ceranae* may not be cured with this medication, or the medication dosage may have changed; be sure to stay current on the most up-to-date procedures.

Package bees are especially vulnerable, and it is recommended to feed one gallon of medicated syrup after installation. Research is still investigating whether doubling the amount of fumagillin is effective against *N. ceranae*, so check the current literature. Fall feeding may be more effective than spring feeding.

Amoeba Disease

Another protozoan is the amoeba in the kingdom Protista (which includes mostly single-celled eukaryotes such as the slime molds, algae, and the phylum Sarcodina or amoebae). The amoeba that affects honey bees is called *Malpighamoeba mellificae* Prell, which forms resistant spores called *cysts*.

A single bee can have a half-million cysts within three weeks of the initial infection. All adult bee castes and drones can be infected. Bees ingest the cysts from infested food, water, fecal material, or elsewhere in the hive. The cysts germinate in the intestines (or rectum) and make their way to the Malpighian tubules (the kidneys) from the hindgut within 24 hours; what the amoebae are feeding on is not known. However, some researchers report that once the infection is underway, the abdomen becomes distended and the tubules cease functioning, taking on a glossy appearance when filled with the cysts. These spherical cysts measure 6 to 8 micrometers, big compared to the smaller Nosema spores, which are elliptical (and not found in the tubules). Cysts also migrate down the tubules into the hindgut, where they are voided with feces.

Amoebae are found mostly in bees infected with nosema disease, and the cysts are observed in fecal material along with Nosema spores. Although they are mostly present in workers, queens can be infected as well.

There are no clear symptoms other than dwindling populations, as bees die away from the colony. The effect of heavy infestation results in reduced honey yield and impaired functioning of the tubules. Development time of the cysts is slowed in cool weather (68°F [20°C]) but increases as the temperature in the broodnest reaches 86°F (30°C). Therefore, spring is the time when amoebae infections are most severe, peaking in May in the Northern Hemisphere.

The only control is maintaining hygienic conditions in the apiary and at the water source, and decontaminating frames (as in Nosema) or replacing equipment; fumagillin has no effect. Some reports indicate requeening has been successful in saving an infested colony (as long as the queen is not infected).

Minor Diseases in Adult Bees

Septicemia

Septicemia is caused by several different bacteria found in the hemolymph (blood) of bees, the most common bacteria being *Pseudomonas aeruginosa* (=*P. aspiseptica*). Although septicemia rarely if ever debilitates bee colonies, it destroys the connective tissues of the thorax (legs, wings) and antennae. The disease can be recognized by these symptoms:

● Dying bees are sluggish and the hemolymph turns white (instead of clear).
● Dead bees decay rapidly.
● Dead bees become dismembered when touched as the muscle tissue degenerates.
● Dead bees have putrid odor.

Bees come into contact with the bacteria in soil, water, and infected bees by way of their breathing tubes (tracheae). It is still not clearly understood how the disease is transmitted or how to treat it, but some success has been found by requeening colonies and placing hives in locations that are sunny and dry and have good air drainage.

Spiroplasma

Spiroplasma is a coiled, motile prokaryote without a cell wall. It was first described in the nectar of some honey plants, but infection with it is rarely diagnosed.

There are other minor diseases not mentioned here, which many beekeepers will never see. For a complete look at all the diseases of bees, see "Diseases and Pests" in the References.

OTHER PROBLEMS: DYSENTERY

Dysentery is not caused by a microorganism and is not a disease. It is primarily the result of poor food and long periods of confinement. This condition is generally seen in northern states in late winter or in overwintering cellars when bees are placed in controlled environments. Even though workers can retain 30 to 40 percent of their weight in fecal material, if bees are fed improperly or because of other conditions, they will defecate inside the colony, resulting in spotty combs and streaked entrances. In general, dysentery is caused by:

● Fermented stores.
● Diluted sucrose syrup fed in fall.
● Syrup with impurities such as those found in "raw" or brown sugar.
● Dampness near hive or in apiary due to poor drainage.
● Long periods of confinement.
● Too much moisture in the hive.
● Honeydew in stores.

Symptoms

The symptoms of dysentery are similar to those of other adult bee diseases:

● Sluggish bees.
● Swollen abdomens.
● Staining on the hive furniture, with yellow to brownish fecal material.

Because this is not a disease but a condition, the only way to treat dysentery is to:

● Provide a winter exit, so bees can take cleansing flights on warm winter days instead of defecating inside the hive.
● Provide good winter stores of low water concentration (properly cured honey and sugar syrup).
● Feed thick syrup (2:1 sugar : water) in fall if bees need more stores going into winter.

- Medicate (as for nosema) as a preventive measure to help control the diarrhea.
- Clean dirty combs and hive bodies by power washing and rotating combs with new foundation every three to five years.

THE PESTICIDE PROBLEM

Farmers, growers, and orchardists apply appropriate chemicals to protect their crop investment; most of these—fungicides and herbicides—are labeled safe for bees and presumably cause few problems. But bees are insects and can be killed by a variety of pesticides. Honey bee pesticide problems arise when insect pests or fungal (or other) diseases threaten crops that bees are working, and the growers use chemicals to protect those crops. Recent investigations into the cause of CCD have led researchers to analyze pesticide residues in beeswax, pollen, and bee bread, and unfortunately, low levels of pesticides (mostly fungicides and acaricides used for mite control) are becoming more prevalent contaminants, even in new foundation wax. If you look at the breakdown list of pesticides (see the sidebar on types of pesticides), you

Types of Pesticides

Acaricides
Algicides
Antifeedants
Avicides
Bactericides
Bird repellents
Chemosterilants
Chitin inhibitors
Fungicides
Herbicides
Insect attractants
Insect repellents
Insecticides
Mammal repellents
Mating disrupters
Molluscicides
Nematicides
Plant activators
Plant growth regulators
Rodenticides
Synergists
Virucides

can see that we are flush with chemicals in our environment, used for lawns, in Christmas tree farms, and for rat bait as well as to kill bugs. Some products break down fast when exposed to sun and water, but others (e.g., DDT) can remain in the environment for years. In addition, some of the breakdown products or a combination of pesticides may be **more** toxic to bees; much more research needs to be done here. (See References: Pesticides and Colony Collapse Disorder.)

Bees come in contact with many chemicals while foraging. The best way to minimize the interaction of bees and pesticides is to know what problems bees could encounter within the 2-mile (3.2 km) foraging distance from your apiary. Get a map of your location and scout problem areas (see the illustration on forage areas for honey bees on p. 32, in Chapter 2). Google Maps and other Internet sites make this task much easier nowadays; also drive around to get an idea where your bees are foraging.

If your apiary is located near orchards, tree farms, nurseries, or other areas such as parks that may be sprayed, your bees may be exposed to pesticides. Bees are also exposed to chemicals (such as at factory sites, dumps, and junk yards), and certain plants are sprayed during bloom (with fungicides) to protect the crops. Again, being vigilant on what is sprayed on which crops in your area will keep you better informed. Here are some situations when bees could come in contact with poisons:

- Inadvertent but direct application on flying bees. Bees may die in the field or after they return to the hive.
- Contact with recently applied compounds on target crop. Depending on the formulation, bees may die in the field or return and die in the hive.
- Consumption of contaminated water, nectar, or pollen (e.g., from roadside spraying). Field bees, hive bees, and larvae will die inside the hive.
- Misapplication of material, including direct application to non-target plants (drift from treated areas, the use of inappropriate chemicals or application methods). This will kill field bees, hive bees, and larva.
- Chemicals (e.g., fungicides) that are sprayed directly on blooming plants or are systemic in the plant tissues (nicotinoids and others). This may kill or disrupt foraging bees, plus they may be bringing back contaminated pollen and/or nectar.

It is important to know the characteristics of insecticide-treated bees as they can be easily confused with other disease symptoms. In general, you will notice:

- A sudden reduction in numbers (thousands of bees) in a previously strong colony in the middle of the summer season.
- Excessive numbers of dying and dead bees, within 24 hours, in front of the hive, on the bottom boards or on top bars.
- Dying larvae crawling out of cells.
- A break in the brood-rearing cycle, disorganization of hive routine.
- Inappropriate queen supersedure.
- Within four to eight weeks, brood becoming chilled because of the lack of workers, or dying due to disease or poisoned pollen.
- Frames of stores (pollen or honey) not being consumed.

Bees are also exposed to insecticides, and each kind of insecticide affects bees in a different way. Some of the kinds of insecticides in general use today include:

- Nicotinoids.
- Organophosphates.
- Chlorinated hydrocarbons.
- Carbamates.
- Dinitrophenyl.
- Botanicals.
- Pathogenic.

Pathogenic insecticides, unless specific for hymenopterous insects, are not toxic to bees. Some are used to control many lepidopteran insects (for example, the wax moth or gypsy moth), such as (1) *Bacillus thuringiensis* (Dipel, Biotrol, Thuricide), and (2) polyhedrosis virus.

A complete list of pesticides is online. See Alan Wood, Compendium of Pesticide Common Names; http://www.alanwood.net/pesticides/. The symptoms for bee poisoning can be found in recent editions of *The Hive and the Honey Bee* (1992) and *The ABC and XYZ of Bee Culture* (2007); check the References too. Also, investigate pesticides and bee poisoning on other Internet sites. New research on pesticides and bees is constantly being conducted—stay current with this topic.

What Beekeepers Can Do

First, register your bee hive locations with the state bee inspector's office; the office can alert you if pesticide application in the area of your apiary is imminent. If you aren't registered, they can't notify you.

To lessen the chances of hive exposure to pesticides, beekeepers can take the following steps:

Step 1. Careful selection of apiary location:
- Locate and meet farmers, landowners, or land renters within a 2-mile (3.2 km) radius of the apiary.
- Contact beekeepers in the area to learn about past problems.
- Check plat, county, or air photo maps or online maps, even soil-type maps, to assess apiary location in relation to areas that may be sprayed (parks, orchards, tree farms, residences, golf courses).
- Become familiar with the crops grown, production methods, rotation practices, and past pest or disease problems.
- Be aware of planting, blooming, and harvest dates of target crops.
- Become knowledgeable of the spraying schedules for that orchard, crop, or location.

Step 2. Assess your chosen apiary site; weigh these chemical danger potentials:
- Spray drift from nearby treated areas or irrigation ditches or ponds.
- Frequency of sprays during the season.
- Cyclic or unexpected outbreaks of insect pests (such as gypsy moths and mosquitoes).
- Need for sprays during crop blooming period (fungicides).
- Application methods used (air or ground, low volume, ultra low volume, standard, or electrostatic equipment).
- Herbicide or fungicide applications.

Step 3. Become familiar with:
- The identification of crop pests in your area.
- Know when other chemicals will be applied (e.g., after a rain).
- Pest population levels that require spray treatments (economic threshold).
- The types of pesticides used locally, their common names, and formulations.
- Registration procedures for apiary sites, so applicator can locate your hives.

Step 4. Know formulations of pesticides: Formulations with the designation WP (wettable powder), EC (emulsifiable concentrate), MC (microencapsulated), or D (dust) will kill on contact and may be picked up by bee feet or body hairs. Pesticides in these forms may also land on pollen or drift to water puddles near sprayed areas. The addition of stickers or spreaders may significantly reduce problems caused by these formulations, making the chemicals less accessible to bees by sticking to plants instead. However, recent studies suggest that some stickers themselves are toxic to bees, so keep current with the research.

Systemic pesticides, those that are absorbed into the plant tissues and kill insects that feed on the plants, are now common. Such pesticides are used as seed treatments or are placed in or on the soil; others are painted on tree trunks. The danger in systemic chemicals is that because they flow in the plant sap (to kill insects such as aphids and other sap-eaters), some of these compounds or their breakdown products are being found in both pollen and nectar as well as in beeswax. Another danger is the dust generated from mechanical planters; this dust can put the pesticides in the air, thus killing non-target insects, such as honey bees. The toxic dust can come from pesticide-treated seeds.

Step 5. Determine:
- Local weed/wildflower blooming periods; learn to identify local honey plants.
- Where bees are foraging at any particular time (by marking bees).
- Where bees will forage next (sequence of blooming plants).
- What time of day bees are on particu*lar target crops.*

Step 6. Anticipate:
- Changes in cropping practices.
- Scheduled and unscheduled sprays.
- Crop blooming periods and sequences.
- Potential pests (mosquito and bark beetle infestations).

Finally, when it happens, know whom to contact and what to do if bee kills are evident. Have handy the phone number and address of the local apiary inspector, if one exists in your state, or of your state or county cooperative extension or agriculture office.

Find out what legal recourse you have and how and where to take samples for analysis. In most cases, **if your bees are not registered with the state agriculture department**, you will have no recourse.

Remember, many times exposure to pesticides may lie **not** with the farmer or orchardist spraying but with a neighborhood homeowner killing those "pesky" insects on the backyard rose bush or "weedy" dandelions in the lawn.

Protecting Bees from Pesticides

If you have time and it is practical, the best protection method of all is to move your hives at least two miles from the target area. This is the most expensive but most successful protective method. If you can't move your hives, here are some general protective measures you can take before spraying occurs; use one or a combination of several methods:

- Make sure the applicators (local or contracted) know you and your apiary locations (supply maps).
- Check your county extension office or other authority for information on protection programs in your area.
- Make routine contacts with landowners, renters, applicators, and county agents for updates on pest problems.
- Post your name, address, and phone numbers (or a neighbor's number so you can be immediately contacted) conspicuously near your apiaries.
- Paint hive tops with a light color for easy aerial identification.

When spraying is imminent, here are some quick methods to protect your bees:

- Reduce hive entrance.
- Gorge hive with sugar syrup, by pouring it directly on top of the frames (bees will stop foraging to help clean it up); pour in about a quart twice a day for one day prior to a spray, and once a day for two or three days following a spray.
- If practical (a few colonies only), close the entrances with 8-mesh hardware cloth or screen, to confine bees, and place a screened cover on top, covered with a wet cloth or burlap. Keep this cover wet, especially if the weather is hot, with a

sprinkler or watering can, for **at least** 24 hours. This is a dangerous step to take, because even with wet burlap, hives could overheat and die very quickly.

- If weather is cool, you can safely screen entrances with 8-mesh hardware cloth without danger of bees overheating.
- Activate pollen traps to collect contaminated pollen (destroy this pollen afterward).
- Feed colonies with syrup, water, and clean pollen patties during **and** after spray period.

Once a kill has been experienced, you must immediately help the colonies that have been affected. After all, if it is early enough, they may still yield a surplus of honey, or at least store enough for the winter. Here are some things to do:

- Combine weakened colonies to increase populations.
- Requeen when necessary (if queen is dead), rather than waste time by letting bees rear new queens.
- Destroy contaminated stores, combs, and equipment; supply new equipment and clean combs or foundation.
- Feed syrup, pollen, or pollen substitutes to maintain colony and stimulate brood rearing.

All beekeepers should strive to cooperate with neighboring growers for the mutual benefit of each. But the ultimate responsibility for a colony's protection rests with you the beekeeper, not the farmer, landowner, or applicator. You need to make the selected sites as safe as possible and be alert to the expected problems while anticipating the unexpected. Utilize the numerous resources available to help you and your bees. Such sources of information, if still available, are:

- State and county extension offices (extension entomologists, agronomists, horticulturists) if they exist in your state, and their publications on crops, insect identification, insecticide lists, and formulations. Check websites as well.
- State apiary inspector, to register your apiary locations.
- Regional, state, and local beekeeping organizations.

- Libraries and city/state agencies, and their Internet sites, for maps and other references.
- Other beekeepers or your local bee association.

BROOD DISEASES

Brood diseases can be devastating to both novice and commercial beekeepers alike. Recognition of healthy and diseased brood is an important part of colony management, and awareness of how these diseases are carried and spread may prevent a serious outbreak. Become familiar with recognizing diseases by taking workshops and seminars.

Diseases are spread by:

- Beekeepers, moving between diseased and clean colonies and not cleaning hive tools, gloves, or other bee wear and equipment.
- Interchanging brood frames within or between apiaries.
- Buying old, diseased equipment or combs and interchanging or mixing them with clean, undiseased colonies.
- Honey, either fed directly (even store-bought honey) or robbed by bees.
- Pollen, honey, or royal jelly, sold commercially.
- Package bees or queens.
- Swarms.
- Foraging bees spreading virus and other pathogens on flowers.

Bee diseases are not mutually exclusive. A colony could have nosema and both foulbrood diseases at the same time (as well as mites). Some conditions, such as chilled brood, might resemble diseased comb. Careful attention to the symptoms and the condition, history, and mite levels of the colony is necessary to differentiate them.

Bacteria: American Foulbrood Disease

American foulbrood disease (AFB) is caused by the bacterium *Paenibacillus larvae* subsp. *larvae* (=*Bacillus larvae*), which exists in both a spore and a vegetative stage. The disease is transmitted by the spore, and the infected brood is killed by the vegetative stage, when the spore germinates in larval guts. This is the most destructive of the brood diseases and is the reason why apiary inspection laws were first passed.

American Foulbrood Disease

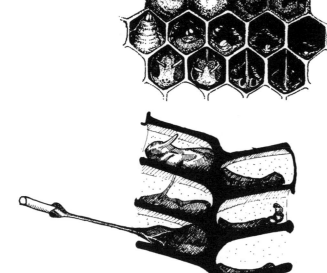

Symptoms:
Cappings are sunken, brood appears dark brown and "melted" down, and pupal tongue sticks up

Test:
See if the contents of the cells are sticky and will extend at least an inch when drawn out

Once the vegetative stages appear in a colony, the disease is spread rapidly and the colony weakens; in most cases, the hive will eventually die unless it is resistant to AFB. Spores can live in hive products (honey, wax, and propolis) for up to **80 years!!** It takes about 10 spores fed to a one-day-old larva for it to become infected. More spores are required to infect older larvae. Death of the developing bee occurs after the cell is capped over and not immediately visible. Once AFB has progressed, diseased larvae, partially uncapped by the bees, turn into black, sticky scales, often seen with their pupal tongues stuck out.

AFB is transmitted from hive to hive in these ways:

- Beekeepers, with diseased equipment, tools, or bee suits.
- Cells, in which larvae hatch, may contain spores.
- Spores, present in honey and/or pollen, are passed on to larvae by nurse bees feeding them.
- Cleaning bees, spreading spores throughout hive when attempting to remove dead brood or scales.
- Robbers from diseased colonies, entering an uninfected hive or bees robbing a diseased colony.
- Bees drifting from diseased to clean colonies.
- Swarms that have AFB.

Symptoms

The symptoms of AFB are varied and sometimes are confused with other diseases or even mite infestations. Here are some things to look for:

- Brood pattern is irregular rather than compact.
- Healthy larvae are glistening white; diseased ones lose this appearance and turn from light brown to dark brown. Larvae die upright, not twisted, in cells.
- Since the death of larvae and pupae often occurs after their cells are capped, the cappings become concave and some will be punctured by bees attempting to remove the dead brood (see the illustration on this page). These puncture marks are very prevalent.
- Surface of cappings will be moist or wet rather than dry.
- Larvae long-dead develop the consistency of glue and are difficult for bees to remove.
- Eventually dead larvae dry out; the dried remains or scales adhere to the bottom, back, and side walls of the cell and are difficult to remove as well.
- Some dead pupae, shrunken into scales, have their tongues protruding at a right angle to their scale or straight up. This may be the only recognizable characteristic, but it could also be missing.

Characteristics of Some Diseases and Pests

Common characteristics	Possible problem
1. Adult bees are absent with little or no accumulation of dead bees in or around the hive. 2. Any food stores are not immediately robbed by other bees. 3. Damage by other pests is slow to develop. 4. Some capped brood is present in colonies (bees will not normally abandon a colony until all the capped brood have hatched). 5. Queen is evident, but the workers appear to be all the same age (young). 6. Bees are slow to feed on supplemented food. 7. Large areas of brood are not covered by bees.	Colony collapse disorder (CCD)
1. Bees are unable to fly or able to fly only short distances. 2. Bees are seen trembling and quivering; colony is restless. 3. Feces is found on combs, top bars, bottom boards, and outside walls of hive. 4. Bees are crawling aimlessly on bottom board, near entrance, or on ground; some drag along as if their legs were paralyzed. 5. Wings are positioned at various angles from body (also called *K-winged*); that is, wings are not folded in normal position over abdomen but with the hind-wing held in front of the forewing. 6. Abdomen is distended (swollen). 7. Bees are not eating when fed syrup. 8. Bees are abandoning the colony, leaving the queen and a few emerging workers.	Nosema, virus, pesticides
1. There is a sudden reduction in numbers (thousands of bees) in a previously strong colony in the middle of the summer season. 2. Excessive numbers of bees are dying or dead, within 24 hours, in front of the hive, on the bottom boards, or on top bars. 3. Dying larvae are crawling out of cells. 4. There is a break in the brood-rearing cycle, and disorganization of hive routine. 5. Inappropriate queen supersedure occurs.	Pesticides
1. Brood pattern is irregular rather than compact. 2. Larvae turn from light brown to dark brown and die upright, not twisted, in cells. 3. Cappings are sunken and punctured. 4. Surface of cappings is moist or wet rather than dry. 5. Larvae long-dead develop the consistency of glue and are difficult for bees to remove. 6. The dried-out brood or scales adhere to the bottom, back, and side walls of the cell and are difficult to remove. 7. Dead pupae have their tongues protruding at a right angle to their scale or straight up. 8. There is an unpleasant putrid "foul brood" odor, which can permeate apiary if many colonies died over winter.	American foulbrood (AFB) disease
1. Larvae die coiled, twisted, or in irregular positions in their cells. 2. Discolored larvae are clearly seen. 3. Dry scales are easily removed from their cells, unlike with AFB disease. 4. Some larvae die in capped cells, scattered, discolored, concave, and punctured. 5. A sour odor may be present. 6. Dead larvae are rubbery and do not adhere to cell walls. 7. Drone and queen larvae are also affected.	European foulbrood disease
1. White, mummified larvae are found. 2. Infected larvae are usually removed from their cells. 3. Dried mummies turn dark gray to black. 4. Mummies are found in brood frames and on the bottom board.	Chalkbrood
1. Young workers (and drones) emerge twisted, deformed, and wrinkled. 2. Bees are smaller, underweight, and discolored. 3. Varroa mites are present.	Deformed wing virus

Characteristics of Some Diseases and Pests (continued)

Characteristics	Disease/Pest
1. Larvae are darkened from white to yellow; eventually they will turn dark brown. 2. Older larvae develop leathery skin and dark head regions. 3. Black-headed larvae are bent toward cell center. 4. Larvae fail to pupate and die with heads stretched out. 5. Diseased larvae are easily removed in liquid-filled sacs (the larval skin). 6. Scales are dry, brittle, and easily removed.	Sacbrood virus
1. Populations of bees are dwindling. 2. Weak bees are crawling on ground with K-wings. 3. Hives with plenty of honey stores are abandoned in the spring. 4. Hive bodies are spotted with fecal matter.	Tracheal mites, virus, nosema
1. Capped drone or worker brood is infested; cappings can be punctured, as in foulbrood. 2. Adult bees are disfigured, stunted, with deformed wings, legs, or both; there can also be crawling bees on ground. 3. Bees are discarding infested or deformed larvae and pupae. 4. Pale or dark reddish brown spots are found on otherwise white pupae. 5. Spotty brood pattern is seen and diseases are present. 6. Dead colonies are found in the early fall. 7. Queens are superseded more than normal. 8. Foulbroods and sacbrood symptoms are present. 9. AFB symptoms exist, but no ropiness, odor, or brittle scales.	Varroa mites, virus
1. Tunnels are observed in combs. 2. Silk trails are seen, crisscrossing one another. 3. Small dark specs are found on bottom board or in the silk trails in a hive. 4. Silk cocoons are attached to wooden parts. 5. Capsule-sized depressions are carved in wooden hive parts. 6. Piles of debris and larvae are on bottom board. 7. Moths may be present.	Wax moth
1. Whitish to tan larvae are present on cells or on bottom board. 2. No silk tunnels are present. 3. Honey is slimy and has particular orange odor. 4. Bees do not rob deserted colonies. 5. Dark, small beetles are seen on comb or hiding on bottom board.	Small hive beetle
1. Few queen cells are present in the center of the comb. 2. The age of queen larvae in the cells is the same. 3. Scattered brood pattern is observed. 4. Drone brood often appears in worker cells.	Failing queen

• Unpleasant putrid "foul brood" odor can permeate the apiary if many colonies died over winter.

Any bad-smelling hive, especially if winter-killed, should be suspect, and if you look at the brood frames in good light, you should be able to see scales in the cell bottoms, or even some protruding tongues.

If a colony is suspected of being infected with AFB, follow these steps as soon as possible:

Step 1. Reduce entrance to minimize robbing.
Step 2. Distinguish infected colony from the rest by marking or painting it a different color, to reduce drifting.
Step 3. Either destroy diseased colonies (see below) or begin a medication (chemotherapeutic) program immediately; see "Foulbrood Disease Chemotherapy" below. If you do not want to use medications, you can also apply integrated pest management (IPM) practices outlined in the next chapter.

Call your state bee inspector or other officials, or your local beekeeping organization, for advice and to confirm diagnosis. If no one is available, first isolate the suspect colony so other bees won't rob it (move it or close the entrance to its smallest opening) and then collect and send a sample of the suspected brood comb. To do this, select a suspect frame and cut a sample of comb, about 4 to 5 inches square and free of honey, that contains the diseased brood. Wrap it in **newspaper** so it will not get moldy; do not use any other kind of wrapping. On a separate piece of paper, write your name and address and place it and the sample(s) in a sturdy cardboard box and mail it to your state bee lab or to the USDA-ARS Bee Research Lab in Beltsville, Maryland; see "Bee Laboratories (National and Provincial)" in the References. Visit their website for additional instructions. Send a letter, call, or email, letting the lab know that you are sending samples under separate cover, the problem you are having with the hive, and the following information:

- Name and address of beekeeper.
- Name and address of sender (if different).
- Location of samples and source.
- Number of samples sent (each labeled and numbered in a different package); indicate if the samples are from the same or different apiaries.

Testing

Use the "ropy test," described below, on larvae that have been dead for about three weeks. Since it is difficult to determine how long a larva has been dead, randomly test between 5 and 10 cells containing dead larvae from several frames. An accurate way of determining how long a larva has been dead is by checking for the presence or absence of its body segments or constrictions (like earthworm constrictions). If they are absent, the larvae have been dead for at least three weeks.

Insert a match, stem, or twig into a cell, stir the dead material, then slowly withdraw the testing stick (see the illustration on p. 200). If a portion of the decaying larva clings to the twig and can be drawn out about 1 inch (2.5 cm) or more while adhering to the other end (the dead larvae), its death was probably due to AFB. This method is not 100%, so to be safe, send samples to the lab. Be sure to burn the test stick. Scrub your smoker and hive tools with a soapy steel wool pad and wash your hands, gloves, and bee suit thoroughly in hot soapy water. Bleach does not kill AFB spores.

Treatment

It must be stressed that AFB is a very serious and contagious bee disease, and is the reason why bee inspection programs were started. Unfortunately, many states no longer have apiary inspections so it is up to you, the beekeeper, to be diligent and keep your bees healthy and disease free.

Some countries do not allow bees to be treated with antibiotics and only permit burning of all the infected equipment. This is a good option if you purchased old, diseased equipment, and it was the only option before the availability of chemotherapy and decontamination chambers. For chemotherapeutic treatments, see "Foulbrood Disease Chemotherapy" in this chapter. **Do not** use chemotherapy drugs as a prophylactic—treat only if you have a disease outbreak.

Now that drug-resistant strains of foulbrood are present, the only way to ensure healthy bees in the future is to use stock that is from queenlines resistant to diseases and mites.

Burning Hive Equipment. If you do not want to treat with antibiotics, an IPM approach is to burn diseased colonies, but first check to make sure you can legally burn contaminated equipment. To burn hive equipment, you can either (1) kill all the bees and burn the equipment or (2) save the bees and put them on new equipment:

1. Kill all adult bees by spraying at night all frames using a fast-acting insecticide. If you can spray bees on frames, use a 3 to 4 percent soapy water solution (or 1 cup liquid detergent per gallon of water). If possible, contain all bees and equipment in a tarp or plastic bags.
   - Remove entire colony with dead bees inside, to a field in which a pit has been dug at least 18 inches (45 cm) deep (deeper if a lot of equipment needs to be burned) and contains a hot burning fire. If you must carry the hive furniture separately, place each super on burlap or in cardboard boxes to keep dead bees and diseased honey from spreading the spores in the existing apiary.
   - Burn all brood, honey, bees, and wax from the infected hive in the deep pit. You can provide support for the larger hive bodies by laying tree limbs across the pit. Make sure all is con-

sumed and turns to ash. You can save dry wax in a container or plastic bag (to be sent to a rendering plant), as long as it is securely closed against robbing bees.

- Cover ashes and pit with fresh dirt.

2. Save the bees and put them on new equipment.
- Cage the queen, then shake the bees off all frames into a new body filled with foundation. Now is a good time to requeen, since this queen (and her workers) were not resistant to disease.
- Feed or spray with antibiotic-medicated syrup. Then use antibiotics according to the label directions. By such actions, bees will be forced to use up the (diseased) food in their honey stomachs to draw out the foundation, and thus not spread the disease. This practice is not advisable without the support of your apiary inspector, because many bees can drift to other colonies and thus spread the disease.
- If the bees still develop AFB after these precautions, kill the bees and burn or sterilize the hive furniture.

If there is a medical or veterinary incinerator in town, you could see if they will burn your (intact) equipment. Make sure to extract any honey from these diseased colonies and bottle it. AFB honey will be perfectly good to eat, but **do not** use it as bee food. Remember, spores can last up to 80 years!

Sterilization. You can save newer hive bodies if they are not too well coated with wax and propolis (which are loaded with AFB spores). Invert hive bodies, so rim edges and handholds are down, and stack them three or four bodies high. Scrap edges and inside free of all wax and propolis and then:

- Fill the inside of the stack with newspaper and ignite it; when the insides of the hive bodies are scorched, extinguish the fire.
- You can also paint the insides of the hive equipment with kerosene and light it with newspapers. This is a more thorough way to sterilize equipment.
- A propane torch can also be used for the tops and bottom boards as well as hive bodies; wood should be lightly browned and all edges and seams given special attention.

These methods, while satisfying, may not completely sterilize hive equipment. Before burning, again check with your state's bee inspection program to be sure the procedure is legal. Gamma radiation chambers or other fumigation facilities are the most effective method to decontaminate bee hive equipment, but many have been closed or are hard to locate. A few states may still have some chambers available, so check for information online or with your state's agriculture department.

Another way to sterilize equipment is to boil woodenware in lye baths or boiling paraffin. Of these methods, the paraffin bath is preferable, and can also be used to preserve new woodenware. Complete instructions, from Australia, can be found in the Rural Industries Research and Development Corporation (RIRDC) Publication No 01/051 (http://www.rirdc.gov.au).

Bacteria: European Foulbrood Disease

European foulbrood (EFB) is caused by a bacterium now called *Melissococcus plutius* (=*Streptococcus* or *Bacillus pluton*), although other bacteria may also infect larvae at the same time, producing similar symptoms. EFB is commonly found in weak colonies, or those stressed due to poor foraging resources, or trucked across country for pollination. The disease is usually prevalent in the spring, slowing the growth of the colony, but may disappear with the onset of a good honeyflow.

Larvae more than 48 hours old are at greatest risk, and thus those that die are usually **not** capped but are visible in the bottom of the cells. The bacterium is found in feces, in wax debris, and on the sides of cells of infected larvae. The dead larvae dry in the cells to form soft scales that are easily removed by bees. Not as serious as American foulbrood, EFB can be treated with Terramycin, and the colonies requeened or strengthened with additional bees.

EFB is transmitted from colony to colony in these ways:

- Cells in which larvae hatch may contain bacteria.
- Bacteria are present in honey and/or pollen and are passed on to larvae by nurse bees feeding them.
- As scales are removed by cleaning bees, bacteria are spread throughout the colony.

European Foulbrood Disease

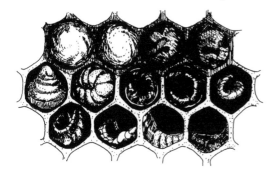

Symptoms:
Bacteria infect younger, uncapped larvae,
turning them brown

Test:
Larvae are twisted in the cells, no pupal
tongue is evident, no "foul" smell is present,
and the ropiness test does not apply

- Diseased robber bees enter a clean hive.
- Contaminated equipment is used.
- Bees from diseased hives drift to clean hives.

Symptoms

The symptoms of a colony infected with EFB are different from those for AFB. Learn to recognize these signs:

- Larvae die in coiled, twisted, or irregular positions in their cells (see the illustration on this page).
- Since most larvae die young, their cells are uncapped and you can see the discolored larvae clearly.
- Larval color may change from pearly white to light cream, then brown to grayish brown, darkening as the dead larvae dry up; normal healthy larvae stay a pearly white.
- Dry scales—the remainder of the larvae—are easily removed from their cells, unlike AFB scales, which are difficult to remove.
- Some larvae die in capped cells, scattered over the brood comb; cappings may be discolored, concave, and punctured.
- A sour odor may be present.

- Dead larvae are normally not ropy as in AFB.
- Drone and queen larvae are also affected.

Bees seriously infested with varroa mites may have EFB-like symptoms; if there is any question, send a sample of brood to the USDA-ARS Bee Research Lab in Beltsville (see "Bee Laboratories (National and Provincial)" in the References. Use the method previously described in the section on AFB testing. Read about varroa mites in Chapter 14.

Control

These methods have been used to control EFB:

- Requeen to break the brood cycle; this allows the bees to clean out dead and infected larvae.
- Use chemotherapeutic agents to treat the disease (see "Foulbrood Disease Chemotherapy," below).
- Feed with clean syrup and pollen supplement/substitute; make sure the pollen you purchased is irradiated, as it could contain EFB spores or other disease organisms.
- Restrict drifting between colonies, by relocating or redistributing hives (see the illustration on hive orientation in Chapter 4).

- Carefully inspect brood in frames before exchanging equipment.

Foulbrood Disease Chemotherapy

Drugs can be given to bees for both AFB and EFB once the disease has been diagnosed. Starting in the 1940s, beekeepers used the antibiotic oxytetracycline (sold as Terramycin) to control AFB. Unfortunately, in the 1990s this drug was found to be no longer effective for AFB. It is, however, used to treat EFB. Currently, the new antibiotic for AFB is used **only to treat** the disease, not to prevent it. This drug is tylosin tartrate, sold as Tylan soluble powder.

The antibiotic drugs **do not cure** the disease; they prevent the spores from germinating, allowing the bees to clear out diseased brood. The antibiotic must be present while the larvae are being fed, to prevent germination inside healthy larvae.

Never use any chemotherapeutic drug during a honeyflow. If drugs are used during a honeyflow, the honey must not be used for human consumption. **Follow the label directions before using any drugs on bees.** Otherwise, you may not be giving enough medication to treat the disease and are contaminating your honey.

Some beekeepers are deciding not to treat their bees with **any** drugs. If you decide to do this, check with your state bee inspector and determine where other beekeepers are within the flight limit of your colonies, to be sure that you are not picking up (or are the source of) pathogens or parasites from (or to) other bee colonies. A better alternative would be to practice an IPM scheme, outlined in the next chapter.

Fumigation Chambers

Special fumigation chambers or gamma radiation can decontaminate empty combs and equipment. This method kills the disease spores and allows the equipment to be reused. Most states no longer have these chambers. Check with the local bee inspectors or state agriculture extension offices for the licensed operators in your area.

Fungus: Chalkbrood Disease

Although common in Europe for decades, chalkbrood was first reported in the United States in 1968 on leafcutting bees (Megachilidae) and on honey bees in 1972; it is now spread throughout the country. The causative agent is the fungus *Ascophaera apis* (Maassen ex. Clausen), and it may reduce honey production but usually will not destroy a colony. Currently, chalkbrood disease has been reported to be on the rise and in some instances can kill colonies weakened by mites or other diseases or by stress (moving bees frequently for pollination) or exposure to pesticides (including fungicides). Some genetic lines, especially inbred bees, can be more susceptible than others, so one control may be to requeen a diseased colony with a new strain of bees.

There are about 20 other *Ascophaera* species, and some of them affect other pollinating insects. Infections are seen in the late larval stages, and drone brood can be especially vulnerable because they are on the cooler edges of the brood frame.

The symptoms are the appearance of white, mummified larvae, first in the cells and then on the bottom board. The infected larvae are usually removed from their cells by nurse bees. Dried mummies will eventually turn dark gray to black; eventually all these colors of mummies can be found in brood frames and on the bottom board (see the illustration of chalkbrood disease on p. 207).

The most susceptible larvae are four days old. The spores of the fungus are resistant to degradation and can be viable for 15 years. Spores are transmitted throughout an apiary by:

- Wind.
- Soil.
- Nectar, pollen, and water.
- Drifting bees or diseased robber bees.
- An infected queen.
- Equipment.
- During food exchange.

While chalkbrood is not a serious disease to part-time hobby beekeepers, in severe cases a colony can be reduced and honey crops lost. In some commercial operations, chalkbrood can be a serious problem. Pesticide exposure may be responsible for lowering bee immunity or killing beneficial fungi, but investigations in this area are still too premature. Staying current with bee research will help you become a better-informed beekeeper.

There is no chemical registered for use against this disease, but managing strong, healthy colonies keeps

Chalkbrood Disease

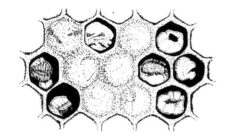

Symptoms:
Fungus infects young larvae, which appear chalky white in the cells, but turn gray and eventually black

Mummified larvae may resemble pollen packed in cells, but bees pull out these mummies, which will be clearly visible on the bottom board

many diseases under control. Additional methods that have worked are:

- Move hives to sunny location, with plenty of air ventilation and dry conditions.
- Remove infected combs and burn them; replace with new foundation.
- Add bees to strengthen weakened, diseased hives.
- Requeen if disease is severe, especially with hygienic queen lines.
- Make sure bottom board is dry throughout the year.
- Feed syrup and protein supplements to keep colony strong and healthy (see Chapter 7 and "Diseases and Pests" in the References).

While there are other diseases caused by fungi, such as stonebrood (*Aspergillus flavus* and *A. fumigatus*), they are rarely seen or go undiagnosed. If you have any questions about a disease, send samples to the USDA-ARS lab Beltsville.

VIRAL DISEASES

Viruses are different from bacteria—they are fragments of DNA or RNA (nucleic acids in a protein coat) that have become detached from the genomes (chromosomes) of bacteria. They are considered nonliving organisms because they lack all of the necessary features that allow them to reproduce on their own. They can only live, grow, and reproduce by altering the DNA of living host cells to manufacture more viruses and use the nutrients from host cells; after replicating themselves, the host cells are destroyed. Because of the small size (over 100 times smaller than bacteria), they cannot be seen with a conventional microscope; the only way to positively identify viruses is to use molecular technology.

Antibiotics, which kill only bacterial organisms, do **not** work on virus diseases. Many virus-prone bees may have a genetic predisposition to viral conditions, and the only reliable way to control such diseases is to requeen the colony with resistant stock. Unfortunately, investigators found a new virus (IAPV) while searching for the cause of CCD, and others have been recently identified (2006). As we have said before, especially in the area of pests and diseases, information changes very fast, so keep current with the new research.

Viruses are spread by mating of drones to queens (venereal transmission), through the egg to larvae and adults (vertical transmission), and between bees via food and fecal matter (horizontal transmission). Some viruses are also found in bee bread and wax. The prevalence of bee virus is still not known, but with the current CCD situation, better surveys and testing will help give us a clear picture of the spread and extent of these pathogens.

Prior to the introduction of parasitic mites, the only virus most beekeepers came in contact with was

Deformed Wing Virus

sacbrood virus. Now in the post-mite era, 18 viruses have been identified that cause bee diseases; the 7 outlined below are the most commonly encountered (in order of most seen to least). For more information see "Diseases and Pests" in the References.

Deformed Wing Virus (DWV)

This virus was first identified in 1991 and became noticeable only after varroa mites were discovered, with which it is associated. It is now distributed worldwide. Symptoms include young workers emerging with wings that are twisted, deformed, and wrinkled. These bees are underweight, discolored, and have a shortened life span since they will be dead within 48 to 67 hours. Drones will be the first to get this virus. Bees infected with DWV are usually evicted from the colony and can be seen crawling on the ground in the apiary; they are a sign that the colony may soon collapse. This virus is vectored by the varroa mite, and there is no known treatment, other than controlling the mites. Some reports suggest that requeening with an uninfected queen early enough in the season may help save the colony; however, to date, there is no way to test if queens are virus free.

Black Queen Cell Virus (BQCV)

First isolated in 1977, this virus kills capped queen larvae and prepupae and may be associated with *Nosema apis*. The diseased larvae inside the cell turn yellow and have a sacbrood-like covering. It is commonly found in commercial queen-rearing operations in the spring and early summer, but is also found in worker brood and adult bees. Infected queen larvae or prepupae sealed in the cells turn black, and the cell walls turn brown-black. Virus particles are transferred by nurse bees from the food glands, and bees with nosema can also be infected with BQCV. Infected queens may have a poor brood pattern, so requeening would certainly be advised. Further studies need to be done, and perhaps controlling nosema could reduce the incidence of the virus.

Israel Acute Paralysis Virus (IAPV)

This virus was discovered in 2004 and associated with dying colonies in the 2007 outbreak of CCD. It causes paralysis in bees, which then die outside of the hive, and is transmitted by varroa mites.

Acute Bee Paralysis Virus (ABPV)

This virus was first identified when bees were experimentally inoculated with chronic bee paralysis virus (CBPV) in 1963; the bees were seen trembling and crawling on the ground with dislocated wings. ABPV is found in brood and adult stages and is often seen during the summer when varroa populations are high. It appears that the mite also activates the virus in infected bees, but its replication in bees or mites is still not clearly understood.

Sacbrood Virus (SBV)

Sacbrood is globally distributed and was first identified in the United States in 1913. Both larvae and adults can get the disease, but it is most easily seen in larvae that are two days old or older. Nurse bees become infected when cleaning out diseased larvae, and the viral particles are found in the food or hypopharyngeal glands. Thus the virus is spread throughout the colony, including in foragers, who can infect the pollen loads when they regurgitate nectar onto their pollen baskets. Young bees can then become infected by feeding on virus-laden pollen. Sacbrood disease is easy to diagnose when seen in the larvae, but the adults with the disease may not live long. If you see a scattered brood pattern, examine the brood carefully; larval symptoms are clear and obvious:

- Larvae are darkened from white to yellow; eventually they will turn dark brown.
- Older larvae develop leathery skin and dark head regions.
- Black-headed larvae are bent toward cell center.
- Larvae fail to pupate and die with heads stretched out.
- Diseased larvae are easily removed in liquid-filled sacs (the larval skin).
- Scales are dry, brittle, and easily removed.
- Disease is often seen in the spring and early summer months.

Sometimes bees remove these diseased larvae quickly, but if there is any question as to the identification, send a sample to the USDA-ARS Bee Research Lab in Beltsville (see method in "American Foulbrood Disease" above). Sacbrood can also be found in the spring after many foragers have been killed by pesticides. A new strain of SBV from the Asian bee *A. cerana* has now been identified in India; it is called Thai SBV.

Strengthening a colony with clean food and more bees may help in its recovery. Because sacbrood is a viral disease, medication is ineffective; requeening the colony may remedy the situation.

Kashmir Bee Virus (KBV)

First isolated from Asian honey bees in 1977, KBV was subsequently found in Australia and New Zealand and later in North America. This virus attacks all stages of bees and brood with no clear symptoms; it is closely related to ABPV. While the virus could be present as an inapparent infection (no symptoms), KBV has been shown to be activated to a lethal level when varroa mites are present in high numbers; the mites can transfer the virus.

Chronic Bee Paralysis Virus (CBPV)

This virus was first identified while the tracheal mite outbreak was being studied in the United Kingdom. Later, virus particles were extracted from dying bees. This virus is now found worldwide, except South America. It has two forms: One form is represented by trembling bees crawling on the ground with abnormal wing posture and distended abdomens. The other form is called hairless black syndrome because the bees lose their hair, appearing shiny black or greasy; such bees are not allowed back into the hive, and therefore are found on the ground. Both forms can be present in the same colony. CBPV is less virulent than ABPV, which takes only one day to kill bees. It is found in overcrowded colonies where the hairs of bees can break, exposing the cuticle, so it is not transmitted by varroa. It is not usually seen anymore.

 Notes

CHAPTER 14

Pests of Honey Bees

MITES

Mites are arthropods belonging to the superorder Acari, which also contains ticks; all have eight legs (see the illustration of a female tracheal mite on p. 211). They differ from spiders in that spiders breathe with book lungs and most mites have a tracheal system. Mites represent an amazingly diverse and rich order of animals. Many of them are microscopic and so are not often seen; they range in size from smaller than 100 micrometers to over 2 mm (0.8 inch). In adults, there are two body regions, the gnathosoma (head) and idiosoma (abdomen and legs), which has four pairs of walking legs; larvae usually have six legs.

To date, over 50,000 named species of mites have been identified, but estimates of over one million species not yet described have been suggested. Their habitats are even more diverse than those for insects, and include in and on soil, water (fresh and salt), plants, arthropods, vertebrates, and invertebrates. Some more interesting habitats are sea snake nostrils, walrus lungs, sea urchin guts, and snail shells (the mites live in special pockets on the shell and eat snail slime). Many are parasitic, and some attack animals, such as the sarcoptic mange mites (family Sarcoptidae), which burrow under the skin. Demodex mites (family Demodicidae) live in the hair follicles of mammals (including humans). House dust mites (family Pyroglyphidae) originally inhabited rodent nests but moved in to our homes—a much bigger nest. Other mites live in food, such as cheese, flour, seeds, and fungi or mold. Because they have no wings, mites are dispersed by wind, water, insects, plant seeds, mold and fungal spores, and birds, to name a few modes. The scientific study of mites is called *acarology*. (Greek: *akari* means mites.)

Dr. G.C. Eickwort, in chapter 40 in *Africanized Honey Bees and Bee Mites* (1988), listed around 40 mites found inside a beehive. To date, the only two parasitic mites found inhabiting honey bees in the United States are listed below. The third one, *Tropilaelaps*, may still make its way out of Asia and inhabit new areas (see the information on other parasitic mites in this chapter).

Tracheal Mites (Acarine Disease)

Tracheal mites are the causative agent of acarine disease, which was originally called Isle of Wight disease, named after where it was first found in 1919. When the mites were discovered, all importation of honey bees into the United States was halted, and Congress passed the Honey Bee Act of 1922. However, the mite was found in Mexico in 1984, and from this first report, it quickly spread by way of migratory beekeepers and packages into northern states and Canada. Many colonies of bees were initially lost (owing to tracheal mite infestation) before the varroa mite appeared. Distribution of the tracheal mite is now worldwide, except in some Pacific islands and perhaps in some African countries.

This small mite (*Acarapis woodi* [Rennie]) lives inside the thoracic tracheae (breathing organs) of adult bees. A newly mated female tracheal mite emerges from an old host bee, and by crawling up on the bee's hair, quests to find a newly emerged or callow bee; see the illustrations of the female tracheal mite and the stages of tracheal mites on p. 211. Once the mite

Questing Female Tracheal Mite

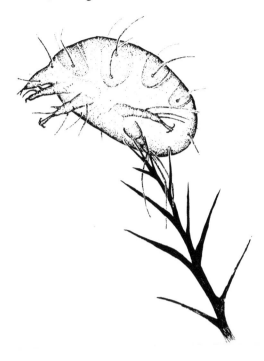

Stages of Tracheal Mite

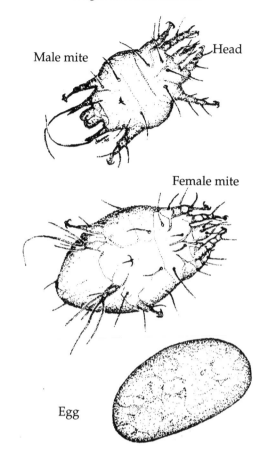

Male mite

Head

Female mite

Egg

Actual size　·

finds a suitable host, she enters the trachea by means of the spiracle opening, and can lay about one egg per day, for 8 to 12 days; see the sidebar on other parasitic bee mites. After the eggs hatch, the immature mites, or larvae, live as parasites inside all adult bees, feeding on bee hemolymph by piercing the walls of the tracheal tubes. New male mites emerge after 11 or 12 days while females emerge in 14 or 15 days.

The bees at risk are young, newly emerged bees (workers, queens, and drones) up to three days old, which are distinguished and selected by female mites over older bees (see the life cycle chart on p. 212). During the summer months, the number of tracheal mites can be quite low because of the short life of the workers. However, over winter when bees are clustered, mites can infest and re-infest bees until the mite population is high enough to cause the colony to perish. This is why the first signs of the tracheal mite are dead colonies in the early spring, hives empty of bees but full of honey.

These mites can cause severe losses in temperate climates but cause fewer problems in warmer climates. However, they may carry virus diseases or cause queen supersedure; purchased queens often have tracheal mites. Because of their small size, tracheal mites are often overlooked, and their impact is now overshadowed by the larger varroa mite. But

they have **not** disappeared. If your colonies are not thriving and have been checked for all other problems, you should look for tracheal mites.

Collecting Bees to Test for Tracheal Mites

External signs of tracheal mites are unreliable but include dwindling populations of bees, weak bees crawling on the ground with K-wings, and abandoned hives in the spring with plenty of honey stores. A positive diagnosis of the tracheal mite by gross examination of the colony or by the finding of bees walking around on the ground cannot be done. Some of the visible symptoms are not always reliable and are not necessarily due to the mite. Therefore, in order to determine if you have mites, you **must** dissect bees or send a sample of your bees to the USDA-ARS Bee Research Lab in Beltsville, Maryland (see "Bee Laboratories (National and Provincial)" in the Refer-

Life Cycle Chart

Tracheal mite (*Acarapis woodi* [R.])

AGE OF BEE   1 to 3 days old   3 days   8 days   12 days   Daughter mites exit old bee, quest on bee hairs, and transfer to a new, young bee host; enter trachea to lay eggs.

Female mite invades new bee 1 to 3 days old.

Mite feeds and lays about 1 egg per day.

Larvae hatch and feed on bee blood. Adult females hatch in 14 days, males in 12. Mating occurs in the trachea.

Varroa mite (*Varroa jacobsoni* Oud.)

AGE OF BEE   8 days old   10 days   12 days   18 days   21 days

Female mite, attracted to the brood pheromones, invades larva before it is capped. Mite will invade drone brood first.

Female foundress mite hides in the bee brood food until cell is capped over.

When bee larva has spun its cocoon, the foundress mite feeds on its blood and begins to lay eggs.

Mite lays up to five eggs, which damage developing bee by feeding on it, allowing pathogens to enter. Mating occurs inside the cell.

Daughter mites exit as injured bee emerges; mites disperse to nurse bees and invade new larvae. Male mite usually dies in the cell.

ences), as you would send brood samples for analysis for foulbrood. If you suspect that your apiary is infested with this mite, and you want to collect the specimens yourself, here is how to do so:

Step 1. Sample at least 50 percent of the colonies in any one apiary. The best time of year to collect bees for tracheal mites is early spring or late fall.

Step 2. Collect only "old" bees, including drones; old bees are most likely to have an infestation and are the easiest to diagnose (see below). They can be found on the inner cover, at the entrance, or out foraging, not near the broodnest.

Step 3. Place the collected bees in 70 percent ethanol (alcohol) or isopropyl (rubbing) alcohol or freeze them in a glass or plastic jar or bag.

Step 4. Send or deliver these specimens to the state bee inspector, state entomologist, or USDA lab, with the following information:
- Your name, address.
- Location of apiary tested (state, county, township).
- Number of colonies in the apiary.
- Source of bees (i.e., name of dealer).

If you are sending the samples through the mail, use bees stored in alcohol, or place the frozen bees in a very small amount of alcohol. Pour off as much alcohol as possible to reduce the weight if you are shipping by the postal service. Check the USDA-ARS Beltsville lab website for mailing instructions.

Dissecting Bees

If you want to dissect bees yourself, or for a science class project, follow the procedure outlined below. Patience and practice are the most important requirements for a successful dissection. Practice on drones first, especially those collected in the late summer or early fall; they are easy to hold and their tracheal tubes are larger. If you are requeening a colony, check your old queen, for she may have infested the entire colony. Finally, collect some old summer or early spring bees and dissect them. If you have used chemicals to control varroa mites, you may not

Dissenting Bees for Tracheal Mites

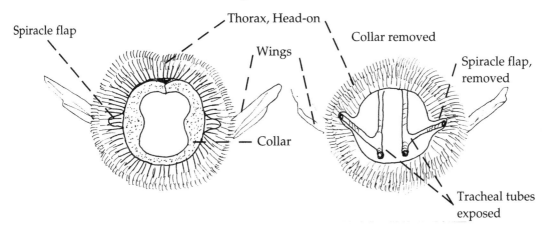

find a lot of tracheal mites, so collect bees before you treat.

A dissecting microscope (at about 40–60X) and a pair of fine jeweler's forceps will be needed. You can find older microscope models online or at school or university surplus departments. Now follow these steps:

Step 1. Soften a frozen bee by holding it in your hand for a few seconds. If the bee was stored in ethanol, it is already soft enough, but if it has been in alcohol for a few months, the tissues will be darkened and it may be difficult to see mites.

Step 2. Place the bee on its back and pin it through the thorax, between the second and third pairs of legs, to a piece of corkboard. You can also hold the bee in your fingers once you have become accomplished at this technique.

Step 3. While looking through the microscope, remove the head and pull off the collar surrounding the thoracic opening with the forceps (see the illustration on dissecting bees on this page).

The thoracic trachea will be exposed when this covering is removed. In a healthy bee the trachea looks like a pearly-white dryer hose. If mites are present, the trachea will have shadows or be spotted—the spots being mites of all ages. In severe infestations, the tube can be completely brown or black. Darkened tracheae will be visible to the naked eye, while healthy tracheae will be white and shiny. You can use this method to detect heavy infestations (spring and fall) but not light ones, such as in the summer. You should examine at least 25 bees per sample, as light infestations can be missed by sampling too few bees.

This is one of several ways to look for tracheal mites. Check the Internet, or your local beekeeping organization, for more information; for a video of tracheal mite dissection, go to the following USDA-ARS website: http://www.ars.usda.gov/pandp/docs. htm?docid=14370.

Controlling Tracheal Mites

Chemical. Menthol, from the plant *Mentha arvensis*, is sold in crystal form (98% active ingredient) at many bee supply companies; each two-story colony takes 1.8 ounces (50 g) of the menthol or one packet. The problem with menthol is that it is temperature dependent. Menthol vapors will sometimes cause the bees to leave the hive if the temperature is too hot. Conversely, the crystals will be ineffective if the temperature is too cold, because not enough vapors would be released. The crystals should remain in the colony at least two weeks. Remove all menthol at least one month before the surplus honeyflow, to keep honey from becoming contaminated. It is not used much anymore.

Other chemical control methods include formic acid and a commercial product (Apilife Var), which are the only chemicals approved for controlling tracheal mites in the United States. These compounds are also used for varroa mites. Check with state agriculture or apiary inspectors, bee suppliers, and bee journals for the current status of other chemical controls.

Oil Patties. An alternative method is to use oil patties; a vegetable shortening (**not lard**) and sugar patty kept in the colony over winter (October to April) has been shown to protect bees against these mites. Do not put these patties in the colony all the time because they will attract the small hive beetle.

To make the patties, use a 1:2 ratio of vegetable shortening to white sugar, or enough of each to have the patty hold its shape. Place a quarter-pound (93 g) patty, about the size of your hand, on the top bars of the broodnest in each colony in late fall. This will help protect emerging bees during the winter months. The patty should last about a month; after that, replace it with another one. Some colonies will remove the patty much more quickly; they may be displaying hygienic behavior, a good trait. Because young bees are continually emerging, it is important to have the patty present in the colony for an extended time before winter. The best time to treat is when mite levels are climbing—that is, in the fall and early spring (see the chart showing the sequence of suggested treatment times on p. 220).

Varroa Mite (Varroosis)

The varroa mite was first identified in 1904 as *Varroa jacobsoni* (Oudemans) on the Asian honey bee *A. cerana* in Indonesia, where it reproduced only in drone brood and otherwise caused little damage to the bees. Later, two other species have been described: *V. underwoodi* and *V. rindereri*, which are morphologically different from *V. jacobsoni* and are found on other bee species (see "Bee Mite Table," p. 219) but had not been reported on the European honey bee (*A. mellifera*).

Once the European bee became established in Asia, *V. jacobsoni* was found to reproduce on both drone and worker brood and also caused significant colony losses.

In the 1980s, differences in the *V. jacobsoni* mite were noted from the various regions where the mite was found. Variations in the shape and size of the adult female mite in their reproductive biology were recorded. Meanwhile, *A. mellifera* colonies in Europe, North America, and the Middle East were quickly succumbing to mite infestation, while bees in the tropics of South America were not. Finally, molecular techniques showed that varroa from Asia were genetically different from those mites in the United States. In 2000, *V. jacobsoni* inhabiting *A. mellifera* was found to be another species and was renamed *V. destructor* (Anderson and Trueman 2000). After further study, eighteen different haplotypes (mites with unique mitochondrial [mtDNA] sequences) were found, nine in *V. jacobsoni*, six in *V. destructor*, and three unresolved (see "Bee Mite Table," p. 219).

In 1986, varroa was first reported in the United States and is now one of the major killers of bee colonies. Adult female mites attach themselves onto adult bees and are thus inadvertently carried to other uninfested colonies or apiaries. This method of transportation is known as *phoresy*, and mites collected from adult bees are called phoretic mites. The movement of the African bees north from South and Central America has accelerated the mite's spread throughout the United States. Consequently, this mite, aptly named *V. destructor*, has since spread across most

Varroa Mite

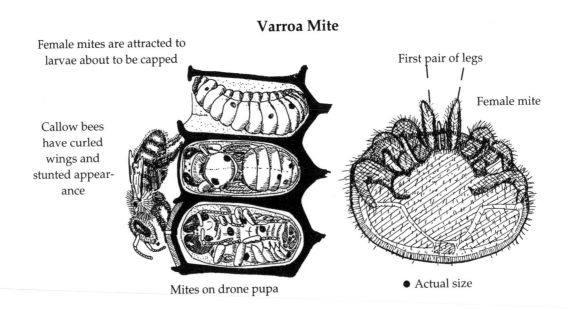

Female mites are attracted to larvae about to be capped

Callow bees have curled wings and stunted appearance

Mites on drone pupa

First pair of legs

Female mite

● Actual size

LIFE CYCLE OF VARROA MITES

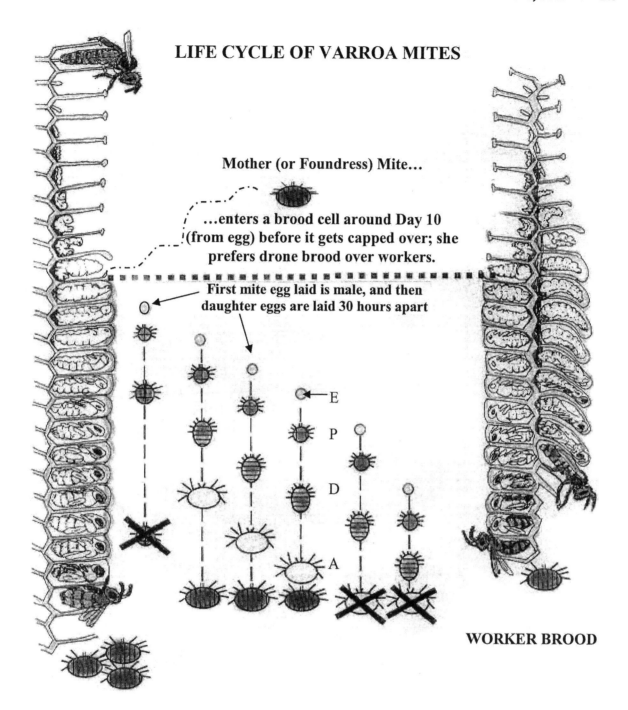

Mother (or Foundress) Mite...

...enters a brood cell around Day 10 (from egg) before it gets capped over; she prefers drone brood over workers.

First mite egg laid is male, and then daughter eggs are laid 30 hours apart

E
P
D
A

WORKER BROOD

DRONE BROOD

Up to three daughter mites can develop in drone brood because of the 24-day time frame it takes for drone adults to emerge. If the foundress mite infests worker brood, on average only one daughter has time to reach maturity; if in drone brood, 1 to 3 daughters can emerge. The foundress mother mite can re-infest at least one other brood cell. Males mate with their sisters in the capped cell and never leave the cell; they are killed by nurse bees cleaning the cell. Mites *very rarely* enter queen cells. E = egg, P = protonymph, D = Deutonymph, A = Adult. Schematic by D. Sammataro; see also http://www.ars.usda.gov/Services/docs.htm?docid=2744&page=14.

of the beekeeping world. (For a map of its historic spread, go to http://www.mylovedone.com/image/solstice/win10/SammataroandArlinghaus.) Most recently (2008–2010) varroa has been found in New Zealand, Hawaii, Madagascar, and parts of Africa.

Life Cycle of Varroa

Varroa is a big mite, with adult females measuring 1 mm long by 1.5 mm wide (0.04 x 0.06 in); it is easily seen with the naked eye, and is about the size of a large pinhead (see illustrations of the mite and its life cycle on p. 214). The life cycle of varroa mites is closely tied to the life cycle of bees, and since these mites don't have eyes or noticeable antennae, they navigate in the dark of a colony by touch, heat, and smell. Only adult female mites are found on adult bees, feeding on bee hemolymph by piercing the soft tissues between the abdominal segments or behind the head.

Female mites are attracted to the odor of the drone brood pheromone—but they will also invade worker brood if drone brood is lacking—and the prepupae, as the cells are about to be capped. When the bee larva is old enough (seven to eight days from the egg stage), the mother mite moves into the cell where the bee larva is developing, and hides at the bottom of the food fed to the growing bee. She is able to do this because she has two lateral *peritremes*, or breathing tubes, which help her breathe when she hides in the brood food. By hiding, she avoids detection by the nurse bees, which will soon cover the cell with a wax cap so the larva can spin a cocoon and finish the transformation into an adult bee.

Now the mother mite must act quickly to keep from becoming entangled in the cocoon's silk; she must crawl up onto the metamorphosing bee and hang on until the spinning process is complete. After the cocoon is formed, the mite then starts to feed, piercing a hole between the forming legs of the bee pupa. In about 30 hours she lays her first egg, an unfertilized egg that becomes a male, followed about every 30 hours or so by fertile daughter eggs. During this time, the mother carefully keeps the feeding wound open so her young ones can eat.

When mature (in five to six days), the male mite, who is small and whose chitinous shell never hardens, will mate with one or two sisters before he himself dies. He never leaves the natal cell. When the beleaguered bee finally completes metamorphosis and emerges as an adult, the mother and daughter mites also emerge to begin the cycle over again. The number of daughter mites produced depends on whether the mother mite invaded a worker bee or drone bee larva. Because workers take 21 days to develop from egg to adult, that leaves a maximum of 18 days for the mites to complete their own development. This means that only one (rarely two) daughter mites will have enough time to develop into adults. If the mother mite enters a drone cell, there will be one to three daughters capable of growing to adulthood because the drones' development time is 24 days (see the illustration of the life cycle of the varroa mites on p. 215).

Most times the infested young bees, if not killed outright by the feeding mites, will be weakened and soon die. The new female mites will live for a time outside on other bees, until they invade new brood to repeat the cycle. The damage varroa does to bees is subtle and still not clearly understood, but is probably responsible for:

- Reduced flight activity of foraging bees.
- Weight loss (6–25%).
- Reduced life span (by 34–68%).
- Reduced blood volume (by 15–50%) when fed upon by mites.
- External damage (chewed wings, legs, stunted growth) if more than five mites in one cell.
- Transmission of virus and other pathogens.

Clearly, detection and treatment are imperative to keep your colonies from perishing.

Symptoms of Varroosis

The symptoms of varroosis are many and can be confused with those of some other diseases or situations, such as pesticide poisoning. Bee parasitic mite syndrome, or BPMS, was first coined by researchers at the Beltsville Bee Research Lab, to explain why colonies with both tracheal and varroa mites were not thriving. The implication of BPMS was that bee mites vectored viruses, making bees susceptible to other pathogens and now may be included in CCD symptoms.

If your colonies have high levels of varroa mites, here are some common signs to look for:

- Infested capped drone or worker brood; cappings can be punctured, as in foulbrood.

- Disfigured, stunted adult bees with deformed wings, legs, or both; there can also be bees crawling on ground.
- Bees discarding infested or deformed larvae and pupae.
- Pale or dark reddish brown spots on otherwise white pupae.
- Spotty brood pattern and the presence of diseases.
- Dead colonies in the late summer, right after honey has been harvested.
- Queens superseded more than normal.
- Foulbrood and sacbrood symptoms present.
- American foulbrood disease symptoms existing, but no ropiness, odor, or brittle scales present.
- No predominant bacterial disease found.

Detecting Varroa Mites

There are four basic techniques you can use to detect varroa mites. It is important to be able to test your colonies periodically to determine the levels of mites so you can choose which treatment will be the most efficacious. Because mites are now resistant to acaricides, you must choose treatments that will control mites without contaminating hive products. Check at least 10 to 20 percent of the colonies in each of your apiaries.

Cappings Scratcher

You can observe mites inside capped bee cells by using a cappings scratcher (with forklike tines), a tool for scraping the wax cappings off of honey frames. Use this tool to pull up capped drone pupae. Here's how:

Step 1. Pick a frame of drone brood or find a large patch of drone brood on several frames.
Step 2. Hold the forked tines parallel to the comb and insert the tines into the top third of the cappings, into the pupae below.
Step 3. Pull the capped drone pupae straight up or lift up the handle end of the fork, leaving the tines on the comb, until the drone pupae are pulled out of their cells.
Step 4. Examine the pupae carefully. A heavy infestation is at least two mites per cell; a moderate level is five mites per 100 pupae. The mites are clearly visible: females are reddish brown and look like ticks on the white pupae. Immature female mites are white or light brown.

Testing for Varroa
Sticky Board and Strips or
Tobacco Smoke

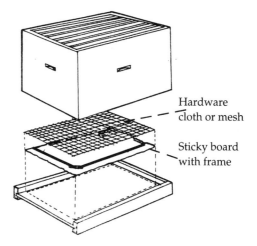

Hardware cloth or mesh

Sticky board with frame

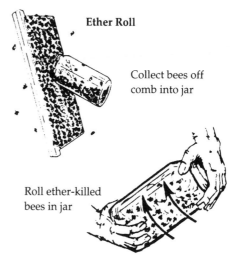

Ether Roll

Collect bees off comb into jar

Roll ether-killed bees in jar

Courtesy of M. Burgett and A.I. Root Company.

Ether Roll

This method should be used by experienced beekeepers, as it can be unreliable (not all the mites are dislodged, giving you a false reading).

Step 1. Collect approximately 300 bees in a wide-mouthed jar with lid.
Step 2. Scrape bees (do **not** get the queen) into the jar; you can also use a car vacuum modified to collect bees in a jar.
Step 3. Knock bees to bottom of jar with sharp blow; there should be about a 1-inch (2.5 cm) layer of bees on bottom.
Step 4. Remove lid and spray 2-second burst of

ether starter fluid into the jar. Alternatively, you can add enough 70 percent alcohol, or soapy water to cover the bees. (Ether and alcohol are flammable—be careful.)

Step 5. If you used ether, replace the lid and agitate or roll jar for about 10 seconds to dislodge the mites from the bees. The mites should stick to walls (see the illustration on testing for varroa on p. 217).

Step 6. If you used soapy water or alcohol, shake the jar vigorously for three to five minutes and strain out bees with a coarse hardware cloth strainer. The mites will be in the liquid, which can later be strained through a coffee filter and the mites counted. Numbers of mites will vary, depending on the time of year.

Sugar Roll

This method will not kill the bees and is used by researchers to collect live mites for laboratory research.

Step 1. Collect bees in a jar, as above for the ether roll. This time, use a canning jar (such as a Mason jar) with a removable lid, which has been replaced with a piece of 8-mesh hardware cloth, cut to fit inside the metal rim.

Step 2. When you have about 300 bees, add about 1 to 3 teaspoons of powdered sugar to the jar, to coat them. You may need more if there are more bees; the bees should be well coated.

Step 3. Shake the jar vigorously and invert it over a piece of white paper to catch the falling mites. Shake and rest for one to two minutes, then repeat. You can also put the jar in a sunny spot for a minute and then shake the jar again. Remember, the number of mites you find will depend on the time of year and your region.

Dr. Marla Spivak has devised a method to calculate the number of mites in your colony: If you know how many bees were in your sample, you can estimate the number of mites per 100 bees. If there is brood in the colony when you sample, you should double this number to factor in the number of mites in worker brood. For example, if there are 5 mites/100 bees, the total infestation is probably 10 mites/100 bees. If your colony has more than 10% infestation, you should consider treatment. See also the website: http://www.extension.umn.edu/honeybees/components/freebees.htm and check Appendix H and "Mites" in the References for more information.

Sticky Board

A sticky board is a sheet of paper or other material coated with some glue or other sticky substance and placed on (or under) a bottom board in your colony. Varroa mites will occasionally fall off and will stick onto the board; when placed on the bottom board for three days, the sticky board is a reliable way to get an idea how infested a colony is.

You can purchase ready-made boards from companies that make traps to monitor insect pest populations (see "Diseases and Pests" in the References), or you can make a board using card stock or other stiff paper coated with petroleum jelly (use Tanglefoot if you live in a hot climate). Place these boards into the hives in your apiary.

Step 1. Place sticky board on bottom board of the colony. The paper should fit inside the bottom board. Cover this paper with a sheet of 8-mesh hardware cloth with the edges turned under to keep it off the sticky paper. You can also staple the mesh to ¼-inch-high (0.6 cm) wooden lath strips, to keep the bees from contacting the sticky board.

Step 2. Leave the board in for at least three days.

Step 3. Pull out sticky board and count mites. If you divide the number of mites by the number of days the board was in the colony, you will get an average number of mites dropping per day. High numbers of mites (more than 50 per day) may indicate the apiary needs to be treated.

These techniques will tell you, with varying degrees of accuracy, the number of mites in a colony. The most accurate is the sticky board, which is a passive and generally noninvasive way to determine mite loads. The mite populations are always changing, depending on the time of year, amount of brood, size of the bee population, and race of bees. Stay current with the literature on economic injury level for varroa mites, the number for which treatment is recommended.

Many bee supply companies sell bottom boards that can accommodate sticky boards, as well as those that have a permanent screen, to allow mites to natu-

Parasitic Bee Mites and Their Honey Bee Hosts

Bee Host	Mite Species									
	Varroa destructor	V. jacobsoni	V. underwoodi	V. rindereri	Euvarroa sinhai	E. wongsirii	Tropilaelaps clareae	T. koengerum	T. mercedesae n.sp.	T. thaii n.sp.
Apis florea			X Nepal, S. Korea		X					
A. andreniformis										
A. cerana	X	X	X			X				
A. koschenikovi				X Sumatra						
A. nuluensis Borneo		X	X?							
A. nigrocincta Sulawesi		X?	X							
A. dorsata dorsata Asia, Indonesia, Palawan	## Korea							X Sri Lanka	X Palawan, Sri Lanka	
A. d. breviligula							X Philippines (not Palawan)			
A. d. binghami Sulawesi									X**	
A. laborisoa Nepal								X	X Vietnam	X Vietnam
A. mellifera	X Japan and Korean haplotypes	X Papua N.G, Irian jaya	##				X Philippines		X	
A. m. scutellata Africa	X									

Notes: X = Positive identification; ** = Currently unresolved; ## = Incidental visitor. *Mesostigmatic mites parasitizing honey bees, arranged according to host bee species. Sources:* Compiled by D. Sammataro and D.L. Anderson (from Anderson and Morgan 2007 and Navajas et al. 2010), published in *Honey Bee Colony Health: Challenges and Sustainable Solutions,* edited by D. Sammataro and J. A. Yoder (Taylor and Francis, 2012).

Sequence of Suggested Treatment Times for Bee Mites

Untreated Populations of Both Mites and Bee Brood

Tracheal mites

Day-old bee larvae

Varroa

WINTER	EARLY SPRING	SPRING EARLY SUMMER	SUMMER	LATE SUMMER	FALL
Use oil patties over winter to reduce tracheal mites and pathogens.	Feed (pollen, syrup) and medicate (fumagillin, TM-25); add an oil patty.	Monitor mite levels, and if needed, treat for varroa in the spring. Only one treatment is necessary if colonies are not reinfested later on.	Monitor mite levels; high varroa here may kill colony. Treat again if needed with strips, or to save honey, use a cultural control. Use strips if mite load is high.	Monitor mite levels; treat with oil patties as fall approaches. If not already treated, insert strips, in time to remove before winter. Feed fumagillin.	Remove all strips; feed oil patty, and prepare for winter.

Typical Colony Management Sequence

| Wrap hive (optional); provide clean and adequate honey stores. | Reverse, feed, and medicate as needed. Test for mites and treat if necessary. | Reverse as needed. Before flow starts, take out all varroa strips; feed oil/sugar patties. | Monitor varroa levels; add super for honey as needed. If varroa levels are high, use a cultural control. | Remove honey, and treat with strips if necessary; apply oil patties. | Requeen colony if queen is more than one year old. Reduce forager and drone levels. Use oil patties. |

Mite Treatment   **Broodnest**   **Dry comb**   **Honey**   **Foundation**

rally drop out. Check the catalogs to see what is available. To make counting easier, a company in Michigan (Great Lakes IPM) makes a board with a third of the grid blackened for easier counting; they also come in nuc sizes.

Treatment for Varroosis

Choose the treatment appropriate to the season. If high mite populations are found during the summer honeyflow (most common), you cannot put in chemical acaricides, as this will contaminate honey and beeswax.

To date there are only a few chemical control products registered for treating varroa: Apistan, a plastic strip (like a flea collar) impregnated with the pesticide fluvalinate. Apistan is now mostly replaced with CheckMite+ strips (active ingredient is coumaphos) because the mites, in many areas, are resistant to fluvalinate. Other chemical acaricides include formic acid and a commercial product (Apilife Var); these two are the only approved control chemicals in the United States. Sucrocide is another product, but

some researchers have not found this to be very effective. Acaricides are generally placed in colonies in early fall, after the honey supers have been removed (see the chart of the sequence of suggested treatments on this page) and again in the spring. Due to the ineffectiveness of some of these chemicals, check out other options now available.

Other products that are coming are Hivastan and ApiGuard. Check to see if these are permitted (some still have an EPA Section 18 Emergency Exemption) or are not yet approved. All registered chemicals are available from bee supply companies, and it is important to follow the label directions carefully. Do not be tempted to use other, home-made chemical cocktails, as some of these are now being detected in the honey and especially the wax, making them unfit for human consumption, use, or sale. Timing of treatment is very important.

Formic acid is effective but its liquid form is very dangerous to use. It acts as a fumigant, killing both types of parasitic mites, but it can be toxic to bees too if not applied correctly. It is **extremely caustic to** humans, so respirators must be worn. The same is true for oxalic acid, which is also used but not yet registered for use as a miticide.

If you don't want to use chemical controls, an integrated pest management (or IPM) approach presents a multiple of options, the last of which is chemicals. See more about IPM in this chapter. Recent research has found that the pesticides found in the beeswax comb can interact with one another or with contaminants that bees bring in from their foraging trips; the end result appears to be that some combinations (e.g., fungicides and miticides) make the compounds toxic to bees. Keep current with the ongoing studies; see "Pesticides" in the References.

MAJOR INSECT ENEMIES

Wax Moth

First reported in the United States in 1806, this pest was probably introduced with imported bees. The female greater wax moth (*Galleria melonella* L.) is about ½ to ¾ inch (1.3–1.9 cm) long and is gray-brown (color varies somewhat). This moth holds the wings tentlike over the body instead of outstretched, or upright, like a butterfly. The wax moth is thought to have evolved with honey bees from Asia and commonly inhabits nests of all honey bee species.

This moth deposits eggs in cracks between hive parts or in any other suitable place inside the hive. After hatching, the larvae are quite active, moving up to 10 feet (3 m) to infest other hives, where they tunnel into the wax combs, hiding at the midrib to keep from being discovered by house bees. The dark wax of brood combs contains the shed exoskeletons of bee larvae and some pollen, both of which are highly attractive to wax moth larvae. The larvae can grow to 1 inch (2.5 cm) long in 18 days to 3 months, depending on the temperature. As these larvae tunnel along, silk strands mark their trails through the combs (see the illustration of the wax moth on p. 222). Before pupating, the larvae fasten themselves to the comb face, on the wooden frames, or inside the walls, inner covers, or bottom boards of the hive and spin a large silk cocoon. The moth larvae can damage the hive furniture by chewing into the wooden parts. Left untended, wax moths can destroy weak hives within one season.

Symptoms of wax moth damage are:

Other Parasitic Bee Mites

There are two other parasitic mites but they are not yet found in the United States: Euvarroa and Tropilaelaps. Euvarroa was first identified from *Apis florea*, the dwarf honey bee from India, in 1974 and is reported to parasitize only drone brood. These mites are smaller than Varroa and currently two species have been identified: *Euvarroa sinhai* on *A. florea* throughout its natural range, and *E. wongsirii* on *A. adreniformis* from Malaysia and Thailand (see table on p. 219). Euvarroa generally have long setae or hairs on the posterior edge of the pear-shaped body shield and so far are not a threat to our bees.

The Tropilaelaps mite is the newest threat to global apiculture but for now is confined to its home range in Asia where four species have been identified (see References in the "Mites" section and the table on p. 219). This mite is an important pest of the introduced European honey bee *A. mellifera* wherever it has been introduced throughout Asia. Of importance is that the adult mites, both male and female, are fast, active, and will invade bee brood before it is capped. They are reported to be phoretic for only a few days, however, and may not survive for long in a broodless colony. One method of control is to cage the queen for a week or so to keep the colony broodless. These mites carry some of the same bee viruses as do Varroa. Be alert when monitoring mite levels in your colonies to watch for this new mite (see illustration comparing Varroa and Tropilaelaps). If you do find it, contact your local apiary inspector or the National Bee Lab in Beltsville (see References).

Note: The USDA plans to conduct a survey in 2011 to look for *Tropilaelaps* and for *Apis cerana*.

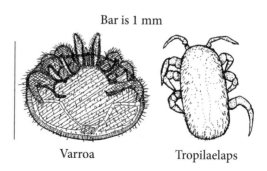

Bar is 1 mm

Varroa   Tropilaelaps

Comparing Varroa Mite and Tropilaelaps

Wax Moth

Cocoons on frame

Remains of comb eaten by wax moth larvae become silken tunnels, frass droppings, and wax debris

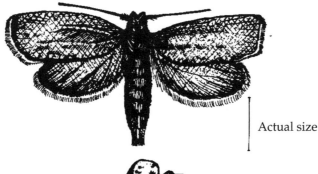

Adult female
(wings extended)

Actual size

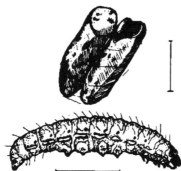

Pupal cocoons

Larva or caterpillar
(destructive stage)

- Tunnels in combs.
- Silk trails, crisscrossing one another over combs.
- Small dark specks (excrement of wax moth larvae) on bottom board or in the silk trails in a hive.
- Silk cocoons attached to wooden parts.
- Destroyed comb, piles of debris, and moth larvae on bottom board.

Control

To control wax moths, use these methods:

- Maintain strong colonies (the best defense against wax moths).

- Store empty combs in cold places; cold temperatures will slow the rate of growth and deter the adult moths from laying their eggs.
- Freeze empty combs or comb honey at 20°F (–7°C) for 5 hours; at 10°F (–12°C) for 3 hours; or at 5°F (–15°C) for 2 hours. If you are treating a lot of comb honey, it is best to freeze for at least 24 hours to kill any eggs.
- Extract supers quickly and store properly.
- Burn frames and/or hive bodies that are heavily infested.
- Keep bottom board (or Varroa boards) clean of debris.

- Brood combs can be stored if they are exposed to light 24 hours a day; female wax moths cannot lay eggs if light is present.
- Fumigate dry combs with carbon dioxide (CO_2) composed of 74 percent CO_2 and 21 percent nitrogen (N), at 50 percent relative humidity (RH), at 100°F (37.8°C) for 4 hours, 115°F (46°C) for 80 minutes, or 120°F (49°C) for 40 minutes. Be careful, as beeswax starts to melt at 148°F (64°C).

Some other chemicals can be used to fumigate combs, but their permitted use varies from state to state; the use of moth crystals in stored comb is no longer recommended. The state bee inspector or extension entomologist should be consulted before using chemicals.

Other controls include using the bacteria *Bacillus thuringiensis*, or BT, which was produced commercially to control moth and butterfly caterpillars. BT was impregnated in wax foundation to control wax moths, and sold separately. The formulation most effective against wax moths is called Certan; however, this has not been available in recent years.

A cultural practice is to put combs in the sun; in areas that have fire ants (*Solenopsis invicta* Bunen), you can leave moth-infested combs near an active ant nest—the ants will kill wax moth larvae effectively. But use caution because these ants also kill bee colonies and can bother beekeepers.

Wax moths are naturally beneficial because they destroy diseased wax combs of feral or varroa-killed colonies, thus eliminating diseased comb and providing a new clean nesting space for bees. In addition, they are a valuable commodity, sold as fish bait and pet food for reptiles and other exotic animals. They can be reared off beeswax, using baby cereal, glycerin, and honey, and are often a secondary business to many beekeepers (see Appendix F).

The lesser wax moth (*Achroia grisella* Fabricius) does similar damage to wax comb, but unless the infestation is great, the damage is minor compared to that of the greater wax moth. But if left unchecked, this moth can get into dry goods (flour), seeds and grains, and stored pollen and pollen substitutes.

Small Hive Beetle

The small hive beetle *Aethina tumida* (Coleoptera: Nitidulidae), our newest bee pest, was first identified in a Florida apiary in the spring of 1998. Before its discovery in the United States, the beetle was known to exist only in South Africa. How it found its way to North America is still not understood. It is not considered to be much of a pest in its native home, but it has been a major problem in the United States. The beetle is now found in all of the states and parts of Canada, but it is especially a problem in the southeastern states, where the winter weather is moderate. In most northern areas, populations do not seem to build up to high numbers, as they do in southern states. As of 2002, the beetle was found in Australia.

The small hive beetle (SHB) is in the Nitidulidae family, which includes picnic beetles, known for their attraction to fermenting fruit. SHB has been found in traps containing fermenting pollen and bee bread, cantaloupe, pineapple, and bananas, and may use fruit when bee colonies are not available.

Description

The adult beetle is about one-third the size of a bee, around ¼ inch (5.5–5.9 mm) long, ⅛ inch (3.1–3.3 mm) wide, weighing 12 to 15 mg, reddish brown or black, and covered with very fine hair (see the illustration of the small hive beetle on p. 224). The larvae are cream colored and similar in appearance to young wax moth larvae. You can differentiate the beetle larvae from wax moth larvae by examining their legs. Beetle larvae have three sets of legs just behind the head. Wax moth larvae, like all moth and butterfly larvae, have **three sets of legs** behind the head and a series of paired *prolegs* that run the length of the body. Prolegs are absent in beetle larvae. The following is a quick check list to identify the SHB:

Larvae
- Color: tan.
- Spines along the back.
- Size of mature larva, ready to pupate: ¾ inch (10 mm).
- Three pairs of legs, behind the head.
- No webbing or tunnels with black droppings (wax moth frass).

Adult
- Size: around 1/4 inch (5–6 mm); see the illustration comparing sizes on p. 224.
- Antennae: clubbed.
- Color: reddish brown to black.

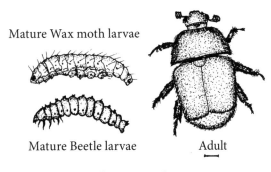

Mature Wax moth larvae

Mature Beetle larvae Adult

Small Hive Beetle

Size Comparison of Small Hive Beetle with Worker Bees

- Very short hairs.
- Runs from light, hides on the bottom board.

Life Cycle

Adult females are strong flyers and invade colonies to lay their egg masses in the cracks and small crevices that bees are not able to reach. There have also been reports of eggs laid under wax cappings of brood and in empty cells. A single female beetle is capable of laying up to 1000 eggs in her lifetime. The eggs hatch in one to six days, producing a great number of small larvae, which consume pollen, honey, bee brood, and wax. They complete their larval stage in 10 to 16 days and then drop to the bottom board, where they crawl outside in order to pupate in the soil up to 100 feet (30 m) from the hive entrance, preferring light, sandy soils. Adult beetles emerge from the soil in approximately three to four weeks and are sexually mature approximately one week later. They are good flyers and easily disperse to new colonies where they mate inside the hive and deposit eggs to begin a new generation.

Beetles live 30 to 60 days depending of food resources, temperatures, and moisture; they do best in hot, humid climates. During the winter months, adult beetles will be found in the bee cluster next to food and warmth, and will die if they leave the cluster. The beetles completely shut down reproduction during winter.

Damage

In the southeastern states of the United States, SHBs thrive and are of significant economic importance. The beetles readily take over even strong colonies with little resistance by the bees. A few female beetles can produce masses of larvae, which can soon overwhelm a bee colony. In addition to consuming the resources of the colony (according to a 1940 study by Dr. A.E. Lundie, Union of South Africa), the adult beetles defecate in the honey, causing it to ferment and run out of the combs. Larvae also tunnel through comb, damaging it, killing brood, and eating stores of honey and pollen. In addition, beetles can be in harvested honey supers that are taken off the hives. When these supers are stored in the honey house, the beetles will contaminate the honey by their feeding, rendering the honey not only unsalable but unpalatable to bees, to the extent that the bees will not even eat it. "Rotten oranges" is how some observers have described the smell of this fermenting mass.

Thus, full honey supers stored in the honey house or on hives above bee escapes, and weak colonies with honey but few bees, are the most vulnerable to attack by SHBs (these include nucs and small queen mating colonies). When SHB infestations are heavy, beetle larvae by the thousands have been seen crawling out of the colony entrance; even in strong colonies, queens will stop laying eggs and the bees may abscond.

Detection

All hive inspections should be done with an awareness for this pest. When a hive containing beetles is opened, they can be seen running across the combs to find hiding places. Adults may also be detected under top covers or on bottom boards. If an infestation is heavy, both adults and masses of larvae may be seen on the combs and bottom board. Combs can be full of holes, but these larvae do not produce silken tunnels, webbing, or cocoons in the hive (as wax moth larvae do); wax moth could also be present. It is important to be able to identify the beetle larvae.

If beetles are suspected in a honey super, firmly shake it over an upturned outer cover; beetles can be dislodged and seen running to find a hiding place. Fermented honey exuding from full supers in storage, waiting to be extracted, or on active colonies, is a sign that hive beetles may be present; the "decaying orange" odor will be detectable.

Another place to check is syrup feeders, grease patties, and pollen patties; the latter two can attract many SHBs. Corrugated cardboard, with the paper removed from one side and placed on the bottom board at the rear of the hive, has been successfully used in detecting adult beetles. Plastic corrugated "cardboard" is preferred because the bees will chew up regular cardboard.

Control

Strong colonies are the best defense against serious infestations. However, if too many beetles enter, they can overwhelm even a strong colony by the masses of eggs they lay in a short time. Workers can be seen trying to sting or pull at the beetles, but the beetle's hard exoskeleton and rounded body make this impossible. Beetles that are harassed will try to hide from the bees, and have been found entrapped in "corrals" made of propolis. This entrapment behavior may be encouraged by selective breeding of queens.

Reducing the hive entrance lets the guard bees do their job better, but will not work in apiaries that are heavily infested. In the honey house, fans and a dehumidifier help keep beetles under control; freezing the supers for 24 hours at 23°F (–12°C) is reported to kill the beetles at all life stages. Recent progress on SHB traps offers promise of control when the infestation is light to moderate.

If SHBs are detected, and you are in an area where they may thrive (warm winters), the following safety measures are recommended:

1. Keep your honey house clean; do not store full supers for long or leave honey-filled wax cappings exposed.
2. When supering or removing honey supers, be aware that SHBs can invade and thus be spread through your apiaries.
3. Moving infested colonies to another apiary site

may reduce the buildup of pupae in a particular area. Some locations may not be as suitable for the beetles as others. Fire ants prey on pupating beetles, and chickens love the larvae.
4. If beetles have contaminated honey and it has started to ferment, stimulate bees to clean it up by power washing out as much of the honey as possible.
5. Experiment with trapping beetles or other cultural controls. Check websites and journals for SHB traps.
6. Some colonies may be able to keep SHB populations lower, so breed from those queens.

To reduce the threat of this pest in your apiary, take the following precautions:

- Maintain only strong, healthy colonies and don't mix infested equipment with uninfested colonies.
- Keep apiaries clean of **all** unused equipment; do not store empty supers on colonies.
- Store honey in a cool, dry place.
- Extract honey as soon as it is removed from colonies.
- Destroy beetles in stored honey supers as soon as they are detected.
- Wash honey drums to reduce their attraction to SHB.
- Melt stored cappings wax.

Check "Diseases and Pests" in the References for more information and websites.

Hive Treatment

Some chemicals used to control the SHB have become less effective and also contaminate hive furniture and wax. A better management strategy is to keep colonies strong and use in-hive bait traps. Check bee journals for current information on SHB traps.

Soil Drench

If you have too many beetles in an apiary, you could use a soil drench. GuardStar is a liquid soil treatment (40% permethrin—see note below) and has been approved in controlling the SHB around honey bee colonies. Hive beetles must pupate in the soil to complete their life cycle, and this insecticide will kill them in the soil. This pesticide provides treatment for the beetles, while minimizing contact with bees and honey. **Read, understand, and follow label directions.**

Note: Permethrin is highly toxic to bees, and extreme caution must be taken to avoid contact by spray or spray drift with the bees, hive equipment, or any other surfaces that bees may contact. Do not contaminate any water or food source that may be in the area or apply during windy conditions. For better soil penetration and improved efficacy, cut the grass around the hive prior to application.

As with varroa, keep current with the literature and check with your state's apiary inspectors or cooperative extension office.

Africanized Honey Bees

In 1956, a researcher in Brazil (W.E. Kerr) imported over 50 queens of *Apis mellifera scutellata* from South Africa into South America to improve the bee stock in tropical regions. The volatile, defensive nature of the Africanized honey bee (AHB), however, is a problem, a serious one in some urban areas. First reported in Texas in 1990, swarms of AHBs have now spread into many southern states. The most recent advance has been into Florida in 2005. If you live in an area that is subject to this invasion, make sure you can differentiate European honey bees (EHBs) from the AHBs (for a map of the latest locations where AHBs have been found, go to http://www.ars.usda.gov/Research/docs.htm?docid=11059&page=6). In general, AHBs are smaller than EHBs, faster moving, more defensive, and more apt to abscond and swarm (see Appendix E).

AHBs can usurp weak colonies, such as a mating nuc, a colony with a caged queen, or a colony recently stressed by beekeeper manipulation. A small cluster or swarm, about the size of a grapefruit, may contain one or more queens; it will land on the outside of a weakened colony. The workers will slowly infiltrate the weaker colony, killing guard bees and eventually the resident queen. Once she is dead, the AHB queen, protected by a "ball" of bees, will enter and resume her duties.

AHBs are extraordinarily successful, and many beekeepers in AHB areas who have converted to keeping them (out of necessity to stay in the bee business) admire them for their low disease and mite incidence, and their adaptability. They work hard and build large populations quickly, then cast many swarms (or abscond if food is scarce) to take advantage of new resources. This behavior is one of the mechanisms responsible for the AHB advancing into North America. Beekeepers keeping AHBs in areas where this bee is endemic have learned to manipulate them enough to produce extra honey and, in some areas, collect pollen to sell.

EHB queens that mate with AHB drones produce an Africanized colony which can be fine for a season, but it should be requeened before the defensiveness becomes too extreme. By keeping good records, you can quickly tell if your colonies have been overtaken; it is **imperative** to mark all your queens in each hive and requeen with non-AHB stock when the marked queen is missing. If the behavior of a colony suddenly changes and the queen has no marking, suspect a takeover and requeen with a new, marked European queen. You may have to kill over one-half of the adult bees (pest control specialists often use 3% soapy water solution) and introduce the new queen on the emerging brood. Cover the hive entrance with a piece of queen excluder to keep the new, marked queen in and the AHB queens out. Having volatile AHBs in your apiary is not to your advantage—legally, socially, or economically. For further information, see "Africanized Bees" in the References.

MINOR INSECT ENEMIES

Although bees are often preyed on by other insects and spiders, these predators usually do not have any appreciable effect on a colony's well-being. In some areas, however, any of these minor predators might become a serious problem. Spiders (Araneae) do catch adult bees; some species even wait for bees to arrive at a flower before attacking them. The most common types of spiders that would catch a bee are the orb weaver, grass, and house spiders.

In some southern areas, brown recluse, jumping, and black widow spiders are commonly found under inner covers or on the sides of hive furniture, especially in equipment that is stored outside. Scorpions can also be found in these regions, so be careful.

While not much of a problem in the temperate climates, in the subtropical areas ants (Hymenoptera: Formicidae) are a serious pest, and hives have to be placed on top of greased posts or oiled cans to keep out these marauders. The more harmful ones in North America include Argentine ants (*Iridomyrmex humi-*

lis), fire ants (*Solenopsis invicta* Bunen), and carpenter ants (*Camponotus* spp.). Ants can be controlled by keeping the apiary free of weeds, debris, and rotting wood and by placing hives on stands. For more serious infestations, ant baits can be used (those containing boric acid and corncobs are not toxic to bees and pets); contact your local extension agents on what ants are in your area and what control measure to use that won't affect your bees.

Other ants (sweet-attracted species), earwigs, and cockroaches may use various hive parts, especially the inner cover, as a shelter or nest. Earwigs (Dermaptera), found on top of the inner cover, may annoy bees. Keep tall vegetation mowed around hives. Nematodes are small worms and some also live on bees, but they are not really serious threats. Termites (Isoptera) can damage hive parts, especially if they are resting on the ground, and in some states (southeastern and southwestern states) can be a serious problem by eating wooden hive parts. As before, keep your apiary mowed and hives on stands, and many of these problems will disappear.

Antlions (Myrmeleontidae) larvae (also called doodle bugs) live in sandy soil and dig the typical cone-shaped pits for trapping ants or other insects that fall into them. They dig their pit traps by throwing out sand by means of upward jerks of the head; the long mandibles serve as a shovel. The small pits they form are about 1 inch deep (2.5 cm) with sloping sides (as steep as the soil will allow). At the bottom of the pit, the antlion will hide in the soil, with its head just below the bottom of the crater, waiting for some hapless insect to fall in. While not a serious problem, if there are many in an apiary the larvae can feast on crawling bees (and queens) and the ants they attract (which may be beneficial).

Some insect predators eat bees but do not typically pose a serious threat to strong colonies, and no control measures will be needed. Your field bees may be caught by:

- True bugs (Hemiptera), such as assassin bugs (Reduviidae); ambush bugs (Phymatidae) are particularly voracious.
- Robber flies (Diptera: Asilidae).
- Mantids (Mantodea).
- Hornets and wasps (Hymenoptera: Vespidae). These may be a problem in the fall or if colonies have died from pesticides or mite predation.

Wasps, hornets, and yellowjackets will clean out dead hives, feeding on dead insects, brood, pollen, honey, and even wax moths.

- Dragonflies (Odonata: Antisoptera) and damselflies (Zygoptera).

Other insects prey on the stored products in a colony or on the insects that eat the stored products; they can be a problem if there are many dead colonies. There are also non-insect invaders, including the pollen mite (Acari), which is found in deserted comb or in hive debris. The most common insects found in a hive, or living in stored equipment include:

- Moths—dried fruit moth (*Vitula edmandsae*), and Indianmeal moth (*Plodia interpunctella*).
- Beetles (Coleoptera), which may live inside eating hive debris and litter found there. The most common ones are dermestid beetles (Dermestidae), weevils (Curculionoidea), sap beetles (Nitidulidae), and scarab beetles (Scarabaeidae); the last two eat stored pollen. Some beetles are predators, such as ground beetles (Carabidae), or parasites, like the blister beetles (Meloidae), which eat or parasitize live bees.

Certain flies (Diptera) bother bees at times but are mostly considered a minor nuisance unless their natural prey is unavailable. Some flies are predators, but others are opportunists, found in colonies that died of other causes. Others will parasitize bees, but these are found mainly in tropical climates.

The following have been noted in the literature as being pests of bees:

- Humpbacked flies (Phoridae), blow flies (Calliphoridae), thick-headed flies (Conopidae), flesh flies (Sarcophagidae), and tachinid flies (Tachinidae).
- The bee louse (*Braula coeca*). This fly, which eats food at the bee's mouth and looks like a varroa mite (except that it has six instead of eight legs), may reach damaging levels in some regions. However, because bees are treated with acaricides for mite control, these flies are difficult to find in regions where varroa mites are treated.

To control these insect pests, store your empty equipment in cold or freezing temperatures. **Never** use insecticides or pest strips in stored hive bodies and comb: these will also kill bees and can be absorbed into the wax.

ANIMAL PESTS

Skunks and Raccoons

Skunks (family Mustelidae), which include weasels and badgers, and raccoons (family Procyonidae) are serious pests to bees, often visiting hives in the early evening as well as during the day. They can cause damage to both equipment and bees and dig up the beeyard looking for food to eat. By scratching at the entrance, a skunk entices bees to come out of the hive, and as the bees crawl out, the skunk eats them. Skunks even teach their young that beehives are a good place to get some tasty snacks, and can deplete hive populations drastically. Your apiary could become decimated quickly if you do not take some countermeasures to protect colonies. They also feed on bumble bee colonies.

Raccoons often take and scatter anything loose in the apiary, including feeder jars and frames of brood or honey that have been left out. Some raccoons are strong enough to lift off the covers of hives and feed on bees, honey, or brood. Depending on the severity of winter in your area, these pests could be nearly a year-round problem, especially in queen-rearing yards where the colonies are smaller. In the Southwest and South and Central America, the coati (*Nasua* and *Nasuella* spp.), a member of the raccoon family, can also be a problem. Signs of mammalian visitors are:

- Defensive bees.
- Grass near hive entrance is torn up.
- Scratch marks on the hive front or on earth at hive entrance.
- Outer covers are off or skewed.
- Weak colony for no other apparent reason.
- Area near entrance is muddy after a rain, and tracks and scat can be seen.

Discouraging and eliminating these pests may be accomplished by:

- Using hive stands, at least 18 inches (46 cm) high, to keep bees out of reach (see "Hive Stands" in Chapter 4). This is the best and easiest way to eliminate mammal predation, especially skunks.
- Sprinkling rock salt crystals on the ground around the hive. Although this method may deter these pests (until it rains), it will also kill the vegetation around the hives.
- Trapping skunks (may be illegal in your area) will cause the skunks to discharge, which will not make you popular with your neighbors. Alternatively, live trapping skunks and raccoons can be done successfully (use cat food or marshmallows). If you cover the trap with an old blanket, skunks will generally not discharge; practice makes perfect.
- Using poison baits; this method is not recommended because it is not selective enough and can harm other animals and pets. Before killing or using poison bait traps, contact your state game and wildlife departments and comply with regulations for controlling fur bearers.
- Placing a strip of carpet tacking, nail side up, on landing board; this does not always discourage the skunks, and many times they pull it out.
- Extending a piece of hardware cloth in front of entrance, which will allow bees to sting the skunk's belly; this is a good temporary measure. Make sure it is fastened to the bottom board, or the skunk will tear it off.

Bears

Bears (Ursidae) eat brood and honey and do extensive damage to equipment, especially in Canada, where large bear populations exist. However, bears are now found in almost all states in the continental United States, and they are capable of destroying apiaries. Signs of bear damage are overturned hives, smashed hive bodies, frames scattered over the apiary, and entire supers that have been removed from the apiary and scattered 30 to 50 yards (27–46 m) away.

An electric fence around the apiary is probably the only effective control against this animal. Kits for putting up fences are available in most rural feed or farm supply stores and are getting easy to set up. Most kits now come with solar panels to recharge the batteries that supply the electric charge.

Raised platforms are used in Alaska where grizzly bears are a problem, but these are extremely difficult and expensive to build. Locating apiaries away from bear routes may help, because these animals keep to knolls, forest edges, and stream banks. Do not leave combs or hive debris around an apiary, as this will attract not only bears but also other pests. Paint hives to blend into background. Alternative ways to reduce bear damage include moving bees to a new location and seeking the assistance of local conservation departments, who may trap bears that are frequent visitors.

Mice

Mice (genera *Mus, Micromys, Clethrionomys, Peromyscus*) are the most damaging animals to bee hives, next to vandals. They enter hives in the fall and winter and, although they appear not to harm the bees, can cause extensive damage to comb and woodenware. They may destroy weak colonies by feeding on pollen, honey, wax moth larvae and cocoons, bee brood, and bees. Their droppings and urine are another irritation that often disrupts cluster behavior, especially if the colony is weak. If you wrap or otherwise winterize your hives, mice can invade and make nests in the winter packing. Because mice carry viruses (such as hantavirus) and vermin (fleas) that may affect humans, keeping these pests out of bee hives can be important to your health. In the desert Southwest, packrats can be a problem if they build their nests in abandoned equipment; their nests also contain kissing bugs (subfamily Triatominae), which can invade homes; they can vector Chagas disease.

Signs of mouse damage are:

- Chewed combs or wood.
- Droppings on the bottom board.
- Holes chewed in entrance reducers, enlarging the opening to allow mice to enter.
- Nesting materials (grass, paper, straw, cloth, or such) in hives, usually in between frames.

Bee hives placed along forest edges and in fields of tall grasses are especially at risk. The following measures may help to control damage from mice:

- Place hives on stands (although mice can climb).
- Use entrance reducers; some beekeepers line theirs with metal to keep mice from chewing through them.
- In the fall, close the entrance with ½-inch mesh

hardware cloth or metal mouse guards (can also be purchased from bee suppliers). Mice can squeeze through a space that measures 3/8 × 3 inches (1 × 8 cm).
- Keep weeds down around hives.
- Place mouse bait on bottom boards or around the base or under the hive. This measure is not recommended because it is not selective. The only place to use poison bait and traps is in an enclosed space where extra equipment is stored, such as your honey house.

VANDALS *(HOMO SAPIENS)*

There has been an increase in the number of hives stolen or otherwise vandalized in recent years, which makes vandals the number one vertebrate pest of bees. The increasing demands for equipment, honey, bees, and hives for pollination services have all contributed to the prevalence of thieves. Furthermore, colonies are also vandalized by the curious, who think they will be able to obtain some free honey simply by opening up a colony. Those bent on mischief can overturn or otherwise damage hive furniture.

Vandals can be discouraged if apiaries are placed near year-round dwellings. If it is not possible to place them near your residence, you can often rent land from a homeowner in exchange for a few pounds of honey a year. Branding your hive bodies and frames is good protection. If your hives are stolen, for example, and the bee inspector finds your brand on hives in some other yard, the person responsible for the act is more likely to be apprehended and your equipment returned to you. See Chapter 3 for other ideas on personalizing your hive furniture.

Instead of painting your hives white and placing them in open, highly visible areas, try going to the paint stores and getting cans of premixed colors that other consumers returned. Mixing these together often results in a nice, mud-colored paint that makes your hive boxes disappear into the background. Tree hedges or judiciously placed shrubs to create screens around the yard also help to discourage would-be thieves. So does fencing with a locked gate!

MISCELLANEOUS MINOR PESTS

Although many birds are insectivorous, few, if any, eat bees in large quantities in North America. Bee

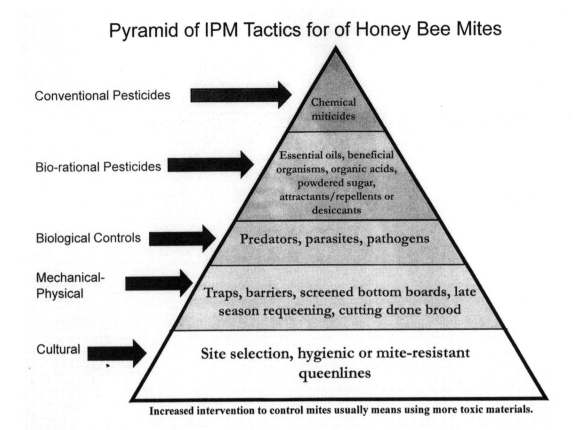

Pyramid of IPM Tactics for of Honey Bee Mites

Conventional Pesticides → Chemical miticides

Bio-rational Pesticides → Essential oils, beneficial organisms, organic acids, powdered sugar, attractants/repellents or desiccants

Biological Controls → Predators, parasites, pathogens

Mechanical-Physical → Traps, barriers, screened bottom boards, late season requeening, cutting drone brood

Cultural → Site selection, hygienic or mite-resistant queenlines

Increased intervention to control mites usually means using more toxic materials.

eaters (family Meropidae), common in Asia, Africa, and Europe, can decimate apiaries and can eat many virgin queens on mating flights. In North America, flycatchers and kingbirds (family Tyrannidae) feed on bees, and woodpeckers (family Picidae) can damage old abandoned or weak hives. But you should make no attempt to control birds by poisoning or shooting them, which is illegal.

Other minor pests, which could be major in some areas, include frogs, toads, lizards, squirrels, opossums, rats, and shrews. Livestock will knock over hives if they are not otherwise protected in pastures. Bees could also bother livestock at watering containers or ponds, especially in desert regions. Additionally, coyotes (*Canis latrans*) have been known to eat brood and honey if they can get into a hive.

Snakes can be problematic if you live in areas where venomous snakes are common. These reptiles, while beneficial in keeping down populations of mice and other rodents, like to nest in the warmer inner covers in cold weather, or underneath stored equipment. While most snakes are harmless, be careful and mindful of them when opening abandoned equipment.

INTEGRATED PEST MANAGEMENT

The discovery of mites, small hive beetles, and now a new Nosema pathogen has altered beekeeping practices forever. Gone are the days of the laissez-faire beekeeping of our grandfathers, when all they had to do was put bees in a hive and harvest honey. Now beekeepers face many challenges, which have changed beekeeping from a hobby into an occupation with risks. The discovery of these threats has also focused attention on how important bees, in fact all pollinators, are to the survival of our world. Without pollinating insects we would be eating rice and wheat.

That said, there are additional problems with keeping our bees healthy; before mites, we only needed to use a few chemical antibiotics. Now, not only are we putting many different kinds of chemicals into bee colonies (not all of them are legal), but also we are stressing bees by moving them around to pollinate crops and thus exposing them to other bees that have more serious problems (resistant mites and diseases). Bees are not easily domesticated, and we don't really need to domesticate them; they work with us but should not be totally dependent on us. This total reliance on keeping mites and disease at bay artificially has led to weakened bees, not to mention the arms race of keeping ahead of resistance. Studies have shown that intensive reliance on chemicals to control pests or diseases lasts for only about a decade. Because there are now chemical-resistant mites and diseases, as part-time beekeepers we need to develop an alternative action plan. We need to rely less on chemical crutches and more on augmenting the natural survival tactics and instincts of bees, such as breeding survivor bees.

Integrated pest management (IPM) is a strategy that is environmentally sensitive and uses **multiple tactics** such as requeening, biological "controls," and cultural or other soft "controls" to "manage" the mites and diseases, instead of killing all of them, provided they do not damage the colony (see the pyramid of IPM tactics on p. 230). Chemical acaricides or other products should be used as a last resort. The key words here are "pest management" not "eradication." Remember, it is not possible to kill all mites in a colony or all mites in a region. The presence of some mites in the colony does not necessarily detract from hive health, provided the colony is strong, is well fed, and has nutritious food, and the mite numbers do not get out of hand. (See "Integrated Pest Management (IPM)" in the References; also see www.extension.umn.edu/honey-bees, http://maarec.cas.pus.edu, www.viginiabeekeepers.org, and other websites).

IPM also implies certain facts:

- Know your pest/disease; sample if needed for positive identification.
- Monitor levels of mites.
- Prevent the spread of pests/disease.
- Use novel controls first:
  - Biopesticides or bio-rational pesticides (essential oils, organic acids, powdered sugar, repellents, etc.).
  - Biological controls (fungus, predators, etc., still being studied).
  - Cultural controls (site selection, resistant queens).
  - Mechanical controls (sticky boards, screened bottom boards, requeening, drone trapping).
- Apply conventional pesticides/controls only when mite levels are too high.

Use the resources available to you (your county or state agriculture department, your apiary inspectors) or take courses and workshops to learn about the pests and diseases that might affect your apiary. Memorize the life cycles of pests and learn to recognize the different disease symptoms. For example, look at the chart (p. 220) showing the sequence of estimated treatment times to see when the mite populations are the lowest (summer for tracheal mites, winter/spring for varroa mites) and time your treatments accordingly. Sample suspect colonies to get a positive identification so you can treat correctly.

Good practices to keep bees healthy include:

- Make sure all your colonies have ample, even large stores of uncontaminated pollen and nectar. Starving colonies can succumb more easily.
- Replace all brood comb in the colony every two or three years. If you do not want to use foundation (because it might contain pesticides), consider letting the bees draw out their own comb (such as leaving frames with strips of foundation), or converting to top-bar hives.
- Sample colonies for nosema, mites, and other pests or diseases. If you treat colonies without knowing if they have a disease or parasite, you not only are wasting money but also may be helping them become resistant to treatment.
- Keep colonies headed by queenlines that are mite or disease resistant; you can also select those survivor queens in your own apiary.
- Use cultural, mechanical, or other nonchemical control techniques.

Reducing Varroa Mites

Here are some different techniques that you can "mix and match" to fit your particular situation for reducing varroa mites:

- Restrict brood rearing by caging the queen, removing capped brood.
- Treat adults (shake them into a screened box and dust with powdered sugar) to knock off the mites.
- Trap mites in drone brood (using drone foundation) and freeze.
- Scrape off mites with pollen traps.
- Use botanical oils (Apilife Var) or organic acids.

Management Strategies for Both Mites

If you want to reduce the levels of both mites, try this:

- Split colonies, kill older foragers, requeen, and treat with bio-rational remedies.
- Keep bees healthy with plenty of pollen and honey stores.
- Provide food supplements (protein and syrup) if needed.
- Use drone brood to trap varroa; cut out or freeze frames after cells are capped over.
- Use oil patties for tracheal mites.
- If nosema is present, treat with fumagillin.

In the fall:

- Harvest last honey supers.
- Check mite levels.
- Reduce numbers of old foragers by moving the colonies a few feet and collecting the older foragers, then destroy them.
- Requeen colonies in fall by using resistant bee stock, for example, Russian for mites. Hygienic lines are beneficial for varroa mites and some diseases.
- Place shortening-sugar patties over winter for tracheal mites. For varroa, use essential oils or other measures.
- Treat for nosema if needed.
- If colony is highly infested, split the colony, kill older foragers, requeen, and treat with bio-rational controls (such as powdered sugar).
- Keep bees healthy with plenty of pollen and honey stores; provide food supplements if needed.
- Make sure there is enough food for bees to store going into the winter.
- If mite levels are too high, use chemical controls.

Other Treatment Options for Mites

Powdered sugar or flour sprinkled on bees knocks off mites. Use this method if you have honey supers on the colony and can't use other chemicals. An alternative is to smoke the colony heavily to knock off mites. Insert sticky boards to catch the dislodged mites and keep them from crawling back up into the colony. You can also install a screened bottom board so mites will fall outside the colony entirely (check bee supply catalogs). Dusting can be repeated weekly, but remember, it only knocks the phoretic mites off the bees, not the mites still in brood cells. This technique may not keep varroa populations from spiking in the fall, so be prepared to use another control measure.

Cultural Practices for Disease

If a colony has been diagnosed with American foulbrood or even European foulbrood, shake all the adult bees off the combs. Then:

- Place bees into new equipment, including all new foundation, and requeen with a resistant queen.
- Feed sugar syrup and pollen supplement until the disease spores carried by the adult bees are flushed.

If this colony still develops the same disease, destroy it. As a last resort, use antibiotics or acaricides if needed.

To prevent diseases from developing or spreading, use these techniques:

- Rotate combs every three years (date the frames).
- Clean all tools (hive tools, smokers, suits) regularly.
- Learn to recognize and identify bee diseases; send in samples if in doubt.
- Requeen with resistant queenlines.

We all need to keep our bees healthy and able to develop natural immunities and resistance mechanisms. Let's be part of the solution and not part of the problem.

CHAPTER 15

Pollination

It is important for beekeepers to know how flowers, ranging from those on apple trees to almond trees, are pollinated, set seed, and bear fruit, and to recognize the importance of adequate pollination from bees and other pollinators. Both growers and beekeepers should know that wild or native bee pollinators help honey bees with the pollination tasks. Many beekeepers are supplementing the pollination workforce by rearing native bees such as *Osmia* bees, bumble bees, and leafcutter bees, to name a few (see "Non-*Apis* Bee Pollinators" in the References).

SOME DEFINITIONS

Flowers are the reproductive parts of angiosperm plants where seeds are formed and from which fruits and vegetables develop. The earliest angiosperm in the fossil record found so far dates back to 125 million years. The flower is the action site for pollinators, offering not only food (nectar and pollen) but a landing platform (petals), landing lights (flower color), and directional signals (nectar guides) to point the way to the food. Some petals even have special conical cells to help bees get a firm foothold while pollinating the flower. A few flower species provide not food but wax and resins for nest material, and some even mimic certain female bees, to attract "pollinating" males.

The male part of the flower, called the *anther*, on the tip of the stamen produces a powdery substance called the *pollen grains*, which are the sperm cells. The female structure is called the *pistil* and contains the *stigma* (usually sticky) and the *ovule* (where the seeds are formed). Each species of flowering plants has its unique shape and form of pollen, which en-

ables paleobotanists (those who study ancient plants) to identify 10,000-year-old pollen from the mud of bogs and lake bottoms.

Most flowers that we recognize are "perfect"; that is, the flower has both female and male parts. *Fertilization* takes place when pollen unites with a female ovule of the same flower species in the ovary (see the illustration of fertilization of a flower on p. 234) to form the seed and fruit. The transfer of pollen from male to female sex organs is called *pollination*. Other plant species have "imperfect" flowers; in other words, they have either the male or the female flower on each plant; these are called *dioecious*. The holly (*Ilex* spp.) is a good example of this; you must have a male plant to produce pollen and a female plant to receive the pollen so the ovary can produce those familiar red berries. *Monoecious* plants possess both the male and female flowers separately, but on the same plant. A good example of this is the cucumber. A large and familiar family that has many florets on a single flower head is the sunflower; they are called *composite* flowers.

All plants must be pollinated before seed (or fruit) will set. Pollen is transferred from the anthers to the stigma either abiotically (by wind, water, gravity) or biotically (by mammals, birds, humans, and insects). If the transfer takes place on the same blossom or on another blossom on the same plant, it is called *self-pollination*. Beans, for example, are self-pollinating. Though many kinds of beans and other plants do not need insect visitors, they do benefit from the extra pollen carried by them and may even set better or more fruit. Other examples are soybeans, peaches, and lima beans.

Fertilization of a Flower

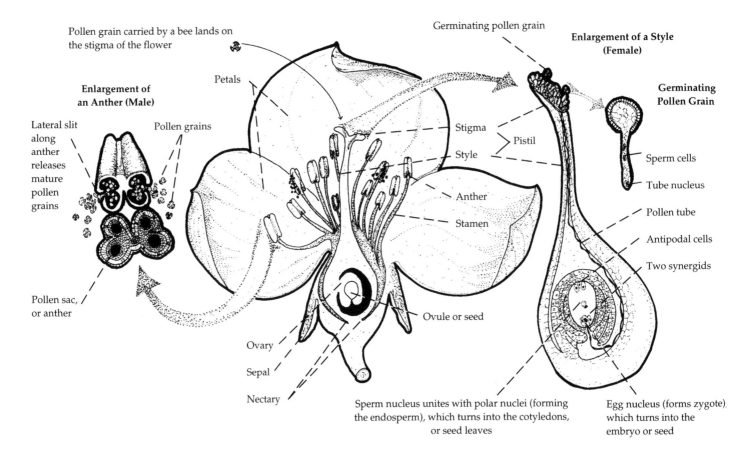

Enlargement of an Anther (Male)

Lateral slit along anther releases mature pollen grains

Pollen grains

Pollen sac, or anther

Pollen grain carried by a bee lands on the stigma of the flower

Petals

Stigma

Style

Anther

Stamen

Ovule or seed

Ovary

Sepal

Nectary

Sperm nucleus unites with polar nuclei (forming the endosperm), which turns into the cotyledons, or seed leaves

Germinating pollen grain

Pistil

Enlargement of a Style (Female)

Germinating Pollen Grain

Sperm cells

Tube nucleus

Pollen tube

Antipodal cells

Two synergids

Egg nucleus (forms zygote), which turns into the embryo or seed

But if the pollen goes from a Red Delicious apple tree flower to a Granny Smith apple tree blossom, this is called *cross-pollination*. Apples and many fruits have a further complication—they are *self-sterile* or *self-incompatible*, which means that the pollen from the Red Delicious apple will **not** pollinate itself or flowers from other Red Delicious trees. It must have another variety of apple to set fruit. The placement of apple varieties, size of the blocks of *pollinizers*, and length of rows may be factors in getting good fruit set in the orchard. It is important not to have too big a block of any one variety in any single area of the orchard. To correct this, some growers graft a limb of an appropriate pollinizer (usually a variety of crab apple, which are excellent pollen producers) onto every six trees. They must choose the correct variety that will bloom at the same time as the variety of apples in the orchard block.

Most angiosperm plant pollination is biotic (80%), with insects being the most common carriers of pollen. In plants that are pollinated abiotically, about 98

percent of them are pollinated by wind and 2 percent by water. Wind-pollinated plants are mostly the gymnosperms, including all the grasses (and their cultivated cousins corn, oats, wheat, and rice), poplars, most nut trees (except almonds), ragweed, and evergreen coniferous or cone-bearing trees (pine and spruce). In these cases, the flowers are generally small and inconspicuous. Such pollen is light and is produced in enormous quantities; this is the cause of allergic reactions or hay fever for many people. It is the pollen from ragweed (*Artemisia* spp.) that most people are allergic to, not the goldenrod (*Solidago* spp.) flowers, which bloom at the same time.

THE MECHANICS OF FERTILIZATION

When a bee visits an apple tree flower, she picks up pollen grains from the anther and pokes her head into the flower to get a nectar reward. As she moves on to another apple bloom, her body may brush up against the stigma of the second flower, where she

Fruit Proportion

Fruit	Proportion (honey bees)[a]
Almond	1.0 (100%)
Apple	1.0 (0.9)
Cherry	0.9 (0.9)
Citrus	0.8 (0.9)
Cranberry	1.0 (0.9)
Kiwi	0.9 (0.9)
Peach	0.6 (0.8)
Pear	0.7 (0.9)
Plum/prune	0.7 (0.9)
Strawberry	0.2 (0.1)
Field crops	
Alfalfa	1.0 (0.6)
Canola	1.0 (0.9)
Sunflower	1.0 (0.9)
Vegetables and melons	
Asparagus	1.0 (0.9)
Cucumber	0.9 (0.9)
Melons	0.8 (0.9)
Squash	0.9 (0.1)
Vegetable seed	1.0 (0.9)
Watermelon	0.7 (0.9)

Source: R.A. Morse and N.W. Calderone. 2000. The value of honey bee pollination in the United States. Bee Culture 128:1–15.
[a]The first number is the proportion dependent on insects for pollination; the number in parentheses is the percentage for which honey bees are the principle pollinator.

seed. The other sperm nucleus goes to the center of the ovule to unite with the polar nuclei; this develops into tissue called the endosperm, which nourishes the developing embryo. The endosperm becomes the seed leaves or *cotyledons* of the new plant; these are the first leaves that come up out the ground. The quantity of the pollen that a plant produces (as a result of favorable growing conditions) may influence the pollinated plant to produce more or less seeds, thus stabilizing the next generation of plants produced; see "Pollination" in the References.

After fertilization, the ovules secrete hormones that stimulate the wall of the ovary to thicken into the surrounding fruit tissue. From this complex, double fertilization, almost all flowering plants on earth are pollinated. Even "seedless" varieties of some crops need to be pollinated for fruit development; as the crop starts to form a seed, it is aborted early in its development, and thus the crop is not truly seedless but instead has undeveloped seeds.

BEES AS POLLINATORS

The most efficient pollinators—highly motile, small, and plentiful—are the insects. Major insect pollinators include beetles, flies, butterflies, moths, and bees. Bees are probably the principal pollinating agents of plants whose flowers have colors within the bee's visual range of blue, yellow, green, and ultraviolet.

Although honey bees are the insect of commerce, for some plants, such as alfalfa, bees are not very efficient. They do not like to work the flowers because of the unique tripping mechanism in the flower, which hits the bee's body while she works (see the illustration on tripping an alfalfa flower on p. 236). Remember, over **$15 billion** worth of crops are pollinated by bees in the United States each year; growers need bees.

For the most part, honey bees are the best pollinators. It takes several trips by many bees to adequately pollinate one apple or one cucumber, because there are many seeds in each fruit. Each seed (ovule) needs a pollen grain to fertilize it. If the ovule does not get adequate pollination, the fruit could be lopsided or the cucumber curly, making them less valuable. For example, in cucumbers, complete pollination takes over 20 visits by bees, and the female blossom lasts only **one day**. Pollen must be moved from the male flower to the female flower before it closes up for the

again seeks a pollen and nectar reward. The pollen grains adhere to the bee's body by means of static electricity and her plumose body (covered with fine branched hairs). (See Appendix A.) The grains of pollen sticking to the hairs are thus rubbed onto the moist tip of the stigma, after which incredible things begin to happen.

The pollen grains start to grow a root, called a *pollen tube*, down the pistil, to deliver two sperm nuclei to the female ovule (embryo seed). This tube has to grow down the entire length of the stigmatic tissue to reach the ovary. Once it has found an unfertilized ovule, the two sperm nuclei are released. When one nucleus cell fuses with an egg nucleus, it becomes the

A Honey Bee Tripping an Alfalfa Flower

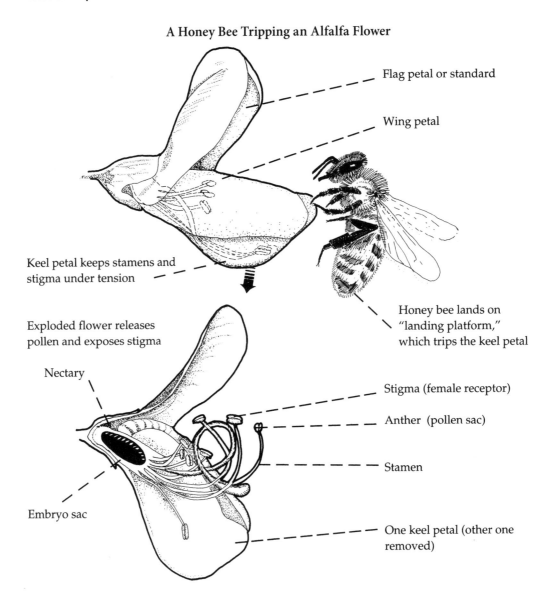

Flag petal or standard

Wing petal

Keel petal keeps stamens and stigma under tension

Exploded flower releases pollen and exposes stigma

Honey bee lands on "landing platform," which trips the keel petal

Nectary

Stigma (female receptor)

Anther (pollen sac)

Stamen

Embryo sac

One keel petal (other one removed)

night; bees are the most cost-effective way of doing this. That is one reason why bees are needed in the pollination of large cucurbit fields (pumpkin, cucumbers, melons, and the like).

Colonies of bees can be moved to crops needing pollination (see "Moving an Established Colony" in Chapter 11). Honey bees are desirable pollinating agents for these reasons:

- Each colony contains large populations of foragers to work crops within a narrow pollinating window.
- Bees will usually work only one type of flower on each trip (flower fidelity), not mixing pollen types. For example, when the honey bee flies out to gather nectar and pollen from an apple blos-

som, this is the only type of flower she will visit on this flight.
- Crops can be sprayed with certain attraction pheromones, which ensure that the bees will work only the target crop.

RECOMMENDATIONS FOR GROWERS

Many growers and orchardists have planted large blocks of crops that require migratory honey bee pollinators. A good rule of thumb is to place one or two colonies per acre of crop, and more would be better, especially if the weather turns wet or cold. This is especially true in early spring: the more bees in the orchard, the closer they will be to the target crops.

Commercial Crops Needing Pollination

Common name	Scientific name	Pollinator
Okra	*Abelmoschus esculentus*	Honey bees (including *Apis cerana*), solitary bees (*Halictus* spp.)
Kiwifruit	*Actinidia deliciosa*	Honey bees, bumble bees, solitary bees
Onion	*Allium cepa* and other *Alliums* spp. (chives, garlic, leek)	Honey bees, solitary bees
Cashew	*Anacardium occidentale*	Honey bees, stingless bees, bumble bees, solitary bees (*Centris tarsata*), butterflies, flies, hummingbirds
Celery	*Apium graveolens*	Honey bees, solitary bees, flies
Strawberry tree	*Arbutus unedo*	Honey bees, bumble bees
Carambola, starfruit	*Averrhoa carambola*	Honey bees, stingless bees
Beet	*Beta vulgaris*	Hover flies, honey bees, solitary bees
Mustard	*Brassica alba, B. hirta, B. nigra*	Honey bees, solitary bees (*Osmia cornifrons, O. lignaria*)
Rapeseed	*Brassica napus*	Honey bees, solitary bees
Broccoli	*Brassica oleracea* cultivar	Honey bees, solitary bees
Cauliflower	*Brassica oleracea* Botrytis Group	Honey bees, solitary bees
Cabbage	*Brassica oleracea* Capitata Group	Honey bees, solitary bees
Brussels sprouts	*Brassica oleracea* Gemmifera Group	Honey bees, solitary bees
Chinese cabbage	*Brassica rapa* var. *chinensis* and others	Honey bees, solitary bees
Turnip, canola	*Brassica rapa* var. *rapa*	Honey bees, solitary bees (*Andrena ilerda, Osmia cornifrons, O. lignaria, Halictus* spp.), flies
Pigeon pea, cajan pea, congo bean	*Cajanus cajan*	Honey bees, solitary bees (*Megachile* spp.), carpenter bees
Chile pepper, red pepper, bell pepper, green pepper	*Capsicum annuum, C. frutescens*	Honey bees, stingless bees (*Melipona* spp.), bumble bees, solitary bees, hover flies
Papaya	*Carica papaya*	Honey bees, thrips, large sphinx moths, moths, butterflies
Safflower	*Carthamus tinctorius*	Honey bees, solitary bees
Caraway	*Carum carvi*	Honey bees, solitary bees, flies
Chestnut	*Castanea sativa*	Honey bees, solitary bees
Watermelon	*Citrullus lanatus*	Honey bees, bumble bees, solitary bees
Tangelo	*Citrus* spp.	Honey bees, bumble bees
Tangerine	*Citrus reticulata*	Honey bees, bumble bees
Coconut	*Cocos nucifera*	Honey bees, stingless bees
Coffee	*Coffea* spp., *C. arabica, C. canephora*	Honey bees, stingless bees, solitary bees
Coriander	*Coriandrum sativum*	Honey bees, solitary bees
Hazelnut	*Corylus cornuta* var. *californica*	Honey bees, solitary bees
Azarole	*Crataegus azarolus*	Honey bees, solitary bees
Cantaloupe, melon	*Cucumis melo* L.	Honey bees, squash bees, bumble bees, solitary bees (*Ceratina* spp.)
Cucumber	*Cucumis sativus*	Honey bees, squash bees, bumble bees
Squash (plant), pumpkin, gourd, marrow, zucchini	*Cucurbita* spp.	Honey bees, squash bees, bumble bees, solitary bees

Commercial Crops Needing Pollination *(continued)*

Guar bean, goa bean	*Cyamopsis tetragonoloba*	Honey bees
Quince	*Cydonia oblonga* Mill.	Honey bees
Hyacinth bean	*Dolichos* spp.	Honey bees, solitary bees
Longan	*Dimocarpus longan*	Honey bees, stingless bees
Persimmon	*Diospyros kaki, D. virginiana*	Honey bees, bumble bees, solitary bees
Cardamom	*Elettaria cardamomum*	Honey bees, solitary bees
Loquat	*Eriobotrya japonica*	Honey bees, bumble bees
Buckwheat	*Fagopyrum esculentum*	Honey bees, solitary bees
Feijoa	*Feijoa sellowiana*	Honey bees, solitary bees
Fennel	*Foeniculum vulgare*	Honey bees, solitary bees, flies
Strawberry	*Fragaria* spp.	Honey bees, stingless bees, bumble bees, solitary bees (*Halictus* spp.), hover flies
Soybean	*Glycine max, G. soja*	Honey bees, bumble bees, solitary bees
Cotton	*Gossypium* spp.	Honey bees, bumble bees, solitary bees
Sunflower	*Helianthus annuus*	Honey bees, bumble bees, solitary bees
Walnut	*Juglans* spp.	Honey bees, solitary bees
Flax	*Linum usitatissimum*	Honey bees, bumble bees, solitary bees
Lychee	*Litchi chinensis*	Honey bees, flies
Lupine	*Lupinus angustifolius* L.	Honey bees, bumble bees, solitary bees
Macadamia	*Macadamia ternifolia*	Honey bees, stingless bees (*Trigona carbonaria*), solitary bees (*Homalictus* spp.), wasps, butterflies
Acerola	*Malpighia glabra*	Honey bees, solitary bees
Apple	*Malus domestica,* or *M. sylvestris*	Honey bees, orchard mason bee, bumble bees, solitary bees (*Andrena* spp., *Halictus* spp., *Osmia* spp., *Anthophora* spp.), hover flies (*Eristalis cerealis, Eristalis tenax*)
Mango	*Mangifera indica*	Honey bees, stingless bees, flies, ants, wasps
Alfalfa	*Medicago sativa*	Alfalfa leafcutter bee, alkali bee, honey bees
Rambutan	*Nephelium lappaceum*	Honey bees, stingless bees, flies
Sainfoin	*Onobrychis* spp.	Honey bees, solitary bees
Avocado	*Persea americana*	Honey bees, stingless bees, solitary bees
Lima bean, kidney bean, mungo bean, string bean	*Phaseolus* spp.	Honey bees, solitary bees
Scarlet runner bean	*Phaseolus coccineus* L.	Bumble bees, honey bees, solitary bees, thrips
Allspice	*Pimenta dioica*	Honey bees, solitary bees (*Halictus* spp., *Exomalopsis* spp., *Ceratina* spp.)
Apricot	*Prunus armeniaca*	Honey bees, bumble bees, solitary bees, flies
Sweet cherry	*Prunus avium* spp.	Honey bees, bumble bees, solitary bees, flies
Sour cherry	*Prunus cerasus*	Honey bees, bumble bees, solitary bees, flies
Plum, greengage, mirabelle, sloe	*Prunus domestica, P. spinosa*	Honey bees, bumble bees, solitary bees, flies
Almond	*Prunus dulcis, P. amygdalus,* or *Amygdalus communis*	Honey bees, bumble bees, solitary bees (*Osmia cornuta*), flies

Peach, Nectarine	*Prunus persica*	Honey bees, bumble bees, solitary bees, flies
Guava	*Psidium guajava*	Honey bees, stingless bees, bumble bees, solitary bees (*Lasioglossum* spp.)
Pomegranate	*Punica granatum*	Honey bees, solitary bees, beetles
Pear	*Pyrus communis*	Honey bees, bumble bees, solitary bees, hover flies (*Eristalis* spp.)
Black currant, red currant	*Ribes nigrum, R. rubrum*	Honey bees, bumble bees, solitary bees
Rose hips, dogroses	*Rosa* spp.	Honey bees, bumble bees, carpenter bees, solitary bees, hover flies
Boysenberry	*Rubus* spp.	Honey bees, bumble bees, solitary bees
Blackberry	*Rubus fruticosus*	Honey bees, bumble bees, solitary bees, hover flies (*Eristalis* spp.)
Raspberry	*Rubus idaeus*	Honey bees, bumble bees, solitary bees, hover flies (*Eristalis* spp.)
Elderberry	*Sambucus nigra*	Honey bees, solitary bees, flies, longhorn beetles
Sesame	*Sesamum indicum*	Honey bees, solitary bees, wasps, flies
Eggplant	*Solanum melongena*	Honey bees, bumble bees, solitary bees
Naranjillo	*Solanum quitoense*	Bumble bees, solitary bees
Rowanberry	*Sorbus aucuparia*	Honey bees, solitary bees, bumble bees, hover flies
Hog plum	*Spondias* spp.	Honey bees, stingless bees (*Melipona* spp.)
Tamarind	*Tamarindus indica*	Honey bees (incuding *Apis dorsata*)
Clover (not all species)	*Trifolium* spp.	Honey bees, bumble bees, solitary bees
White clover, also alsike, red, and crimson	*Trifolium alba, Trifolium* spp.	Honey bees, bumble bees, solitary bees
Blueberry	*Vaccinium* spp.	Honey bees, alfalfa leafcutter bees, southeastern blueberry bee, bumble bees (*Bombus impatiens*), solitary bees (*Anthophora pilipes, Colletes* spp., *Osmia ribifloris, O. lignaria*)
Cranberry	*Vaccinium oxycoccus, V. macrocarpon*	Honey bees, bumble bees (*Bombus affinis*), solitary bees (*Megachile addenda, Alfalfa leafcutter bees*)
Tung tree	*Vernicia fordii*	Honey bees
Vetch	*Vicia* spp.	Honey bees, bumble bees, solitary bees
Broad bean	*Vicia faba*	Honey bees, bumble bees, solitary bees
Cowpea, black-eyed pea, blackeye bean	*Vigna unguiculata*	Honey bees, bumble bees, solitary bees
Karite	*Vitellaria paradoxa*	Honey bees
Jujube	*Zizyphus jujuba*	Honey bees, solitary bees, flies, beetles, wasps

Sources: Information from Wikipedia; K.S. Delaplane and D.F. Mayer. 2000. Crop pollination by bees. New York: CABI. S.E. McGregor. 1976. Insect pollination of crops. Washington, DC: USDA-ARS. See "Pollination" in the References.

If the weather turns bad, more bees in the field will help ensure adequate pollination. Also, colonies in the spring may be weaker than colonies in the summer, and to compensate for this, make sure there are four or more frames of brood covered with bees per colony.

If growers are not diligent, bees can be killed by pesticides sprayed on the crops or on nontarget plants and weeds or as it drifts over hives or in the water supply. Work closely with the grower to time spraying when bees are **not** in the field. Make it a clear part of the lease contract (see below). Pollen (and nectar) can become contaminated by pesticides, and many crops are sprayed during bloom with fungicides, which are considered nontoxic to bees. Recent research is showing that these compounds can cause problems in the colony, so keep current with the findings. There are many published lists on bee-toxic pesticides (see "Pesticides" in the References and the sidebar on types of pesticides in Chapter 13). In areas where so many pesticides have been used and all the pollinators are dead, some plants must be pollinated by hand.

In many cases, native pollinators have been killed by destructive farming practices or pesticide use. Many non–honey bee pollinators are valuable to growers and need to be protected and cultivated. You, as beekeepers, can also raise alternative pollinators as a sideline (see "Non-*Apis* Bee Pollinators" in the References and the website for the USDA-ARS lab in Logan, Utah). These native bees may be more important to future growers as the number of feral honey bee colonies succumb to mite infestation. But both these non-*Apis* bees and honey bees need flowers for an entire season to stay alive and rear offspring. By placing your colonies near uncultivated areas (or by planting certain forage crops or native wildflowers), you will have better success in keeping and establishing all kinds of bees (see list of common bee plants).

LEASING BEES

Many beekeepers lease their hives to fruit and vegetable growers whose crops benefit from or require bees for pollination. The demand for bees is increasing, due in part to declining bee populations (especially feral or wild honey bee nests), caused by urbanization of natural foraging land, pesticide use, mites, and pollution (and colony collapse disorder).

Some factors to consider when leasing or renting bees are:

- Number of hives: If other factors are favorable, count on one colony per acre of fruit crops, more for other crops.
- Weather: Optimum flying conditions for bees include temperatures between 60° and 90°F (15.6° and 32.2°C), winds of less than 15 mph (24 km/hr), and fair, sunny days.
- Colony strength: Each colony should have at least four frames of brood and bees and a laying queen.
- Timing: Set out bees just as the crop comes into bloom (about 10% bloom); if set out too early, bees may work other blooming plants and may not switch to target crop.
- Leasing fees: Although there is no flat fee for leasing bees, some factors that may affect the price include time of year; pesticide hazard; loss of queen, bees, and/or honey; and the difficulty of getting into and out of the field.

Some beekeepers remove frames of pollen from colonies to stimulate bees to collect more pollen. Others use pollen traps for the same reason, or install pollen-hoarding strains of queens, or put in queenless colonies that have queen cells. All of these techniques require close attention to the condition of the bees and the crop, as well as the weather.

POLLINATION CONTRACTS

To be fair, the beekeeper and the grower should draw up a pollination contract or agreement. This document will help prevent misunderstandings while at the same time detailing the expectations of all parties. Key points include:

- Location of crop.
- Date of placement of bees into the crop and their removal (relative to bloom time and condition).
- Number and strength of colonies.
- Pattern of colony placement.
- Rental fee and date on which it is paid.
- Agreement by grower not to apply bee-toxic pesticides while bees are in the crop, or to give the beekeeper 48 hours notice.
- Agreement by grower to warn beekeeper of other spraying in the area.

Poisonous Plants

Name	Toxic part
Abies alba, silver fir	Honeydew reported
Aconitum spp., monkshood[a]	Honey?/Pollen
Aesculus californica, California buckeye	Nectar/Honey/Pollen
Andromeda spp., andromeda	Honey/nectar?
Arbutus unedo, strawberry tree	Nectar
Astragalus spp., locoweed, tragacanth	Nectar
A. miser v. *serotinus*, timber milk vetch	Nectar
Caltha vulgaris, march marigold	Pollen occasionally
Camellia reticulate, netvein camellia	Nectar
Coriaria arborea, New Zealand tutu[a]	Honeydew from passion vine hopper *Scolypupa australis* feeding on tutu
C. japonica[a]	Honeydew
Corynocarpus laevigatus, New Zealand laurel or karaka, summer titi	Nectar
Cuscuta spp., dodder	Nectar
Cyrilla racemiflora, southern leatherwood, titi	Nectar
Datura stramonium, jimsonweed[a]	Honey toxic to human
D. metel, Egyptian henbane	Honey
Delphinium consoida, forking larkspur	Plant toxic to bees, mammals
Digitalis purpurea, foxglove[a]	Pollen?
Euphorbia sequeirana, spurge	Honey/pollen
E. marginata, snow on the mountain	Plant poison to bees
Gelsemium sempervirens, yellow jessamine[a]	Nectar/pollen?
Helenium hoopwaii, orange sneezewood	Toxic to mammals/bees
Hyoscyamus niger, black henbane	Nectar/pollen, plant toxic to bees
Kalmia latifolia, mountain laurel[a]	Nectar/plant toxic to mammals
Ledum palustre, wild rosemary	Honey/pollen?
Macadamia integrifolia, macadamia	Cyanide gas from bloom
Nerium oleander, oleander[a]	Honey
Papaver somniferum, opium poppy	Pollen
Ranunculus spp., buttercup	Nectar/pollen?
Rhododendron spp., azalea, and *R. anthopogon, R. lutea, R. occidentalis, R. ponticum, R. prattii, R. thomsonii*[a]	Nectar/pollen
Sapindus spp., soapberry	Nectar
Senecio jacobaea, tansy ragwort	Nectar bitter, plant toxic to mammals
Solanum nigrum, black nightshade	Toxic plant
Sophora microphylla, yellow kowhai	Nectar
Stachys arvensis, nettle betony, staggerweed	Nectar or yeasts in nectar
Taxus spp., yew	Pollen
Tilia spp., basswood	Occasional nectar toxicity under drought conditions
Tulipa spp., tulip	Nectar, occasional toxicity
Tripetaleia paniculata, an azalea	Nectar
Tulipa spp., tulips	Stigmatic nectar?
Veratrum spp., Western false hellebore	Nectar
Zygadensis (= *Zygadenus*=*Zigadenus*) *venenosus*, death camas	Nectar/pollen

Sources: J. Skinner. 1997. In: Honey bee pests, predators and diseases, 3rd edn. Medina, OH: A.I. Root. J. Ramsay. 1987. Plants for beekeeping in Canada and northern USA. Cardiff, UK: IBRA.
[a] Plant parts also toxic to humans and animals.

Common Bee Plants

Common name	Latin name
Acacias, esp. catclaw acacia	*Acacia* spp., *A. greggii*
Alfalfa	*Medicago sativa*
Almond, apricot	*Prunus dulcis, A. armeniaca*
Alsike clover	*Trifolium hybridum*
American holly	*Ilex opaca*
Anise hyssop	*Agastache foeniculum*
Appalachian Mountain mint	*Pycnanthemum flexuosum*
Apple	*Malus domestica*
Asparagus	*Asparagus officinalis*
Aster spp.	Asteraceae family
Basswood, American	*Tilia americana* and other *Tilia* spp.
Bee bee tree	*Evodia danelli (=Tetradium)*
Birdsfoot trefoil	*Lotus corniculatus*
Black cherry	*Prunus serotina*
Black chokeberry	*Aronia melanocarpa*
Black gum or tupelo	*Nyssa* spp.
Black locust	*Robinia pseudoacacia*
Blackberry	*Rubus* spp.
Blackhaw	*Viburnum prunifolium*
Bladderpod	*Lesquerella gordonii*
Blue curls	*Trichostema* spp.
Blue thistle, viper's bugloss, blue weed	*Echium vulgare*
Blue vervain	*Verbena hastata* L.
Blue vine	*Cynanchum leave (Gonolobus laevis)*
Blueberry	*Vaccinium* spp.
Bonset	*Eupatorium* spp.
Borage	*Borago officinalis*
Brazilian pepper tree	*Schinus terebinthfolius*
Buckeye or horsechestnut	*Aesculus* spp.
Buckthorn	*Rhamnus* spp.
Buckwheat	*Fagopyrum esculentum*
Butterfly weed	*Asclepias tuberosa*
Buttonbush	*Cephalanthus occidentalis*
Cactus (giant saguaro)	*Carnegiea gigantea*
Cajeput, tea tree	*Melaleuca quinquenervia*
Camphorweed	*Heterotheca subaxillaris*
Canola, rapeseed	*Brassica rapa* and other *Brassica* spp.
Catalpa	*Catalpa speciosa*
Catnip	*Nepeta mussinii*
Chaste tree	*Vitex agnus-casteum*
Cherry	*Prunus cerasus*
Chickweed	*Stellaria media*
Chives	*Allium schoenoprasum*
Citrus (orange, lemon, etc.)	*Citrus* spp.
Cleome or spider plant	*Cleome* spp.
Clethra summersweet	*Clethra alnifolia*
Clover, sweet and white Dutch	*Melilotus* spp. and *Trifolium* spp.
Common buckthorn	*Rhamnus cathartica*
Common hackberry	*Celtis occidentalis*
Common vetch	*Vicia sativa*
Cotton	*Gossypium* spp. Extrafloral nectaries
Crab apple	*Malus sylvestris*
Cranberry	*Vaccinium macrocarpon*
Creeping thistle	*Cirsium arvense*
Creosote bush	*Larrea tridenta*
Crimson clover	*Trifolium incarnatum*
Currant	*Ribes* spp.
Dandelion	*Taraxicum officinale*
Dead nettle	*Lamium* spp.
Devil's walking stick	*Aralia spinosa*
Dogbane	*Apocynum* spp.
Dogwood	*Cornus* spp.
Elm	*Ulmus* spp.
Eucalyptus	*Eucalyptus* spp.
Fairy duster	*Calliandra eriophylla*
Figwort	*Scrophularia californica* and other species
Fireweed	*Epilobium angustifolium*
Gallberry	*Ilex glabra*
Garlic chives	*Allium tuberosa*
Germander, thyme	*Teucrium canadense*
Globe thistle	*Echinops ritro*
Golden honey plant, wingstem	*Verbesina alternifolia (=Actinomeris)*
Goldenrod	*Solidago* spp.
Hawthorn	*Crataegus* spp.
Heath	*Erica* spp.
Heathers	*Calluna vulgaris*
Henbit deadnettle	*Lamium* sp.
Holly	*Ilex* and various species
Honey locust	*Gleditsia triancanthos*
Honeysuckle	*Lonicera* spp.
Horehound	*Marrubium vulgare*
Horsemints	*Monarda* spp.
Joe-Pye weed (=Eupatorium)	*Eutrochium* spp.
Knapweed	*Centaurea* spp.
Knotweed or smartweed	*Polygonum* spp.
Korean evodia, bee bee tree	*Tetradium daniellii (=Evodia)*
Lavender	*Lavandula angustifolia*
Leadwort	*Amorpha fruticosa*
Leopardsbane	*Doronicum cordatum*
Lespedeza	*Lespedeza* spp.
Lima bean	*Phaseolus limensis*
Linden, basswood, lime tree	*Tilia* spp.
Loosestrife, purple	*Lythrum salicaria*
Lungwort	*Pulmonaria* spp.

Common name	Latin name	Common name	Latin name
Mangrove, black	*Avicennia nitida, A. germinans*, and other spp.	Saw palmetto	*Serenoa repens* and other *Serenoa* spp.
Manzanita	*Arctostaphylos* spp.	Selfheal	*Prunella vulgaris*
Maple	*Acer* spp.	Shad, shadblow or serviceberry	*Amelanchier arborea* and other spp.
Marigold	*Calendula officinalis*	Smartweed	*Polygonum* spp.
Mesquite	*Prosopis* spp.	Snakeweed	*Gutierrezia* spp.
Milk vetch	*Astragalus* spp.	Sourwood	*Oxydendrum arboreum*
Milkweed, common, swamp, whorled	*Asclepias* spp.	Soybean	*Glycine max*
Mints	*Mentha* spp.	Spanish needles, beggar's ticks	*Bidens* spp.
Mountain bluet	*Centaurea montana* (Knapweed)	Speedwell	*Veronica spicata*
		Spurge	*Euphorbia* spp.
Mountain mint	*Pycnanthemum flexuosum*	Star thistle	*Centaurea* spp.
Mustard	*Brassica arvenisi* (L.) and others	Sumac; African sumac	*Rhus* spp., *R. lancia*
N.J. tea	*Ceanothus* spp.	Sunflower	*Helianthus annuus* and other spp.
Ohio buckeye	*Aesculus glabra*	Sweet autumn clematis	*Clematis terniflora*
Oilseed rape (canola)	*Brassica napus* L., *Brassica rapa*	Sweet corn	*Zea mays* (pollen only)
Oregano	*Origanum vulgare*	Tall ironweed	*Vernonia altissima*
Palms	*Sabal* spp.	Tallow tree	*Sapium sebiferum*
Palo verde	*Cercidium* spp.	Thistles	*Cirsium* spp. and other genera
Plum	*Prunus*	Thyme	*Thymus* spp.
Privet	*Ligustrum* spp.	Tufted vetch	*Vicia cracca* and other *Vicia* spp.
Pussy willow	*Salix discolor*		
Rabbitbrush	*Chrysothamnus* spp.	Tulip-tree or tulip poplar	*Liriodendron tulipifera*
Raspberry	*Rubus* spp.	Tupelo	*Nyssa* spp.
Red chokeberry	*Aronia arbutifolia, Photinia pyrifolia*	Vetch	*Vicia* spp.
		Virgin's bower	*Clematis virginiana*
Red clover	*Trifolium pratense*	White clover, lawn clover	*Trifolium repens*
Red maple	*Acer rubrum*	White sweet clover	*Melilotus alba*
Redbud	*Cercis canadensis*	Wild buckwheat	*Eriogonum* spp.
Red-flowering thyme	*Thymus praecox*	Wild carrot	*Daucus carota*
Russian olive, autumn olive	*Eleagnus angustifolia, E. umbellata*	Willow	*Salix* spp.
		Woundwort	*Stachys byzantina*
Russian sage	*Perovskia atriplicifolia*	Yellow sweet clover	*Melilotus officinalis*
Sage	*Salvia* spp.	Yellow rocket	Cruciferae family (Mustards)
Salt cedar	*Tamarix* spp.		

Source: J.M. Graham (ed.). 1992. Hive and the honey bee. Hamilton, IL: Dadant and Sons.
Notes: Some species of honey plants can be invasive weeds. Check your state agriculture department and websites.
Some good websites are http://ag.arizona.edu/pima/gardening/aridplants/; http://www.aces.edu/department/ipm/npt1.htm, http://www.aces.edu/department/ipm/npt2.htm; http://uvalde.tamu.edu/herbarium//index.html; http://www.vftn.org/projects/bryant/navbar_pages/texas_regions.htm; and http://MAAREC.cas.psu.edu.
Also see References under "Honey Plants."

- Agreement by grower to reimburse beekeeper for any additional movement of colonies in, out, or around the crop.
- Agreement by grower to provide right-of-entry to beekeeper for management of bee colonies.

Sample contracts can be obtained from *Bee Culture* magazine http://www.beeculture.com/storycms/index.cfm?cat=Story&recordID=634, the Department of Entomology at Penn State University in University Park, or websites.

PROBLEM PLANTS

Some plants are harmful, even poisonous to bees. For example, the sundew (Droseraceae), Venus fly-trap (*Dionaea musipula*), and pitcher plants (Sarraceniaceae) are insect-eating plants that attract their victims by secreting a sweet sap, odor, or both. These plants grow in wet areas and are not usually attractive to bees; the number of bees lost to them is minimal. Other plants produce nectar that either is toxic to some insects (*Rhododendron* spp.) or make a toxic honey. While this is a rare problem, if your apiary is near an area where these plants are growing, you may want to get suspect honey tested or be alert for bee kills. Certain environmental conditions, such as abnormally cold or dry weather, may cause otherwise nontoxic plants, such as linden trees (*Tilia* spp.), to yield toxic nectar or pollen. On some instances, wild bees, such as bumble bees, can collect these substances that are toxic to honey bees without adverse effects. The list in the sidebar summarizes information on toxic plants.

HONEY PLANTS

If you want to secure a good honey crop, you need to have your apiaries adjacent to or nearby good bee forage. Although you cannot control the weather, you can look over an area before locating an apiary, to see if there will be ample and diverse forage at all times of the year. In general, highly developed (industrial), residential, or intensely agricultural land (even tree farms) will not be good for bees, and may indeed expose them to additional contaminants, such as pesticides or industrial or commercial waste. For example, if you see honey of a strange color (or flavor) you may want to investigate further for a nearby factory or waste site (e.g., bright orange honey was seen in a colony located near a soda factory that was throwing sugary orange waste in a landfill).

Lists of honey plants are available from local bee groups, newsletters, books, and websites, so please look them up. Major honey plants are those that will give a surplus crop of honey, whereas minor honey plants provide the day-to-day source of food used for brood rearing or for building up a colony coming out of winter or during a dearth period. Remember that soil type, moisture, sunlight, temperature, and humidity will affect the amount of food (nectar and pollen) plants produce. In some instances, you can sow seeds of crops (like alfalfa or sunflowers) or locate your apiary near honey-yielding plants (including trees) if you have enough land. With the current interest in the decline in pollinators, planting a "pollinator garden" in your area may be a good way to change fallow or other land into a field used by all kinds of pollinating insects. Check the list of honey plants in the sidebar, "Pollination" in the References, and other web sources for honey plants in your area.

 **Notes**

Anatomy of Honey Bees

The anatomy of the honey bee is similar to that of other insects except for the specialization of certain organs and structures needed by bees to carry out functions peculiar to them. Parts common to other insects include the three basic insect parts—head, thorax, and abdomen; the exoskeleton (the hard, waxy protein [chitin] covering); the free respiratory system (no lungs); the ventral or bottom spinal cord; and the free circulatory system (no veins).

EXTERNAL ANATOMY OF DRONE

Drones are larger than both the queen and the workers and can be recognized by their large eyes, which meet at the top of the head, and the blunt abdomen. The reproductive organs are enclosed within the body and remain inside the body until the mature drone flies to mate with a virgin queen. Drones mature after emerging from the cell in about 12 days and can be seen flying back to the colony in the afternoon after orientation flights. During the copulatory act, the organs are everted except for the testes (which remain inside); after coupling, the entire organ will break off and the drone dies. Queen breeders who instrumentally inseminate queens collect semen from mature drones by squeezing the thorax, which causes a drone to evert its sex organs.

External Anatomy of the Drone Honey Bee

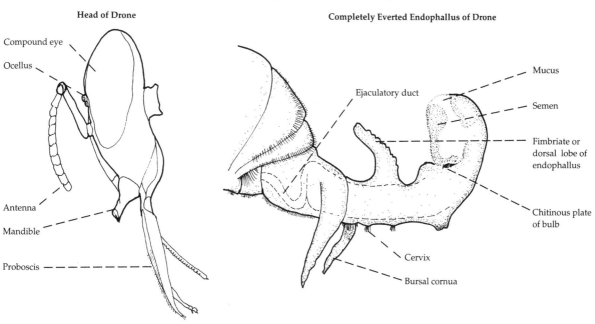

Head of Drone

Compound eye
Ocellus
Antenna
Mandible
Proboscis

Completely Everted Endophallus of Drone

Ejaculatory duct
Mucus
Semen
Fimbriate or dorsal lobe of endophallus
Chitinous plate of bulb
Cervix
Bursal cornua

What Makes Bees Exceptional Pollinators

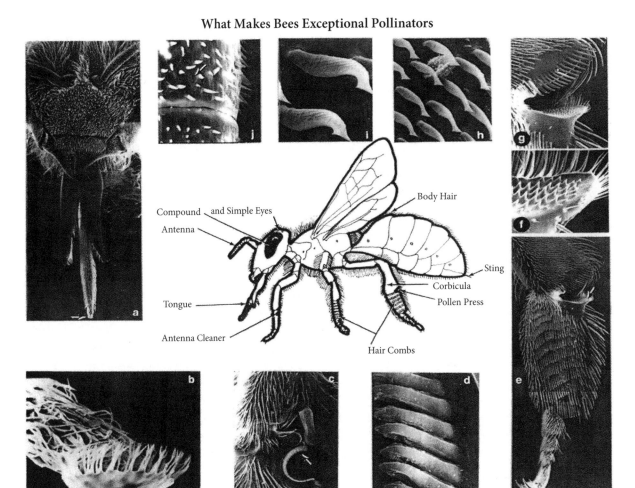

WORKER: WHAT MAKES BEES EXCEPTIONAL POLLINATORS

As mentioned in Chapter 2, the diagram showing the worker anatomy gives the names of the parts of a worker honey bee. In the diagram here, the parts of the worker are blown up using a scanning electron microscope (SEM) to give the fascinating details of how this pollinating machine is put together. Diagram includes:

a. the head of a honey bee, showing the mouthparts.
b. The tip of the proboscis which has sensory hairs to help guide the proboscis to the nectary as well as to provide information on the food being ingested.
c. Close-up of antenna cleaner, showing modified hairs that act as a comb. The antennae are important sensory organs of the bee and must be kept clean and free of debris.

d. Close-up of the modified hair comb of antenna cleaner in *c*.
e. Hind leg showing the hair combs (corbicula) that help make the pollen pellet. The hairs are aligned in one direction to keep the pollen sorted. Pollen press is on the top.
f. Close-up of base of pollen press.
g. Pollen press showing modified hairs to collect pollen grains.
h. Close-up of modified hairs on hind leg of bee with a captured pollen grain from a sunflower.
i. Modified hair in *h*, at higher magnification; the peg-like hair aid in collecting pollen grains.
j. Close-up of antenna showing the sensory hairs and pegs that enable the bee to smell.

For more information and photos, see Erickson et al. 1986, in References under "Bees and Beekeeping."

Internal Organs of a Worker Honey Bee

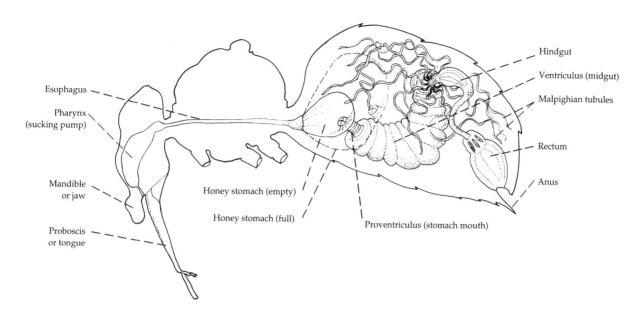

Internal Organs of a Queen Honey Bee

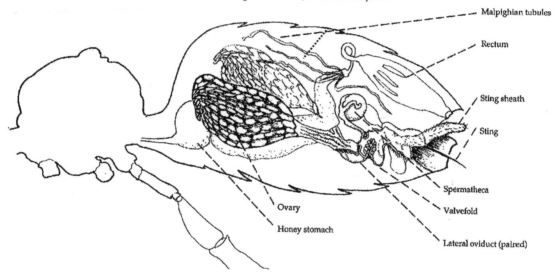

INTERNAL ANATOMY

Compare the internal anatomy of the queen and drone to that of the worker. The queen has larger ovaries than the worker, a sting, but no wax glands or pollen-carrying structures on her hind leg. See the figures showing the external anatomy (pp. 245 and 248) and the table on morphological differences between honey bee castes and drones.

Sting Anatomy

The sting structure of bees is a modified egg-laying device, or *ovipositor*, developed into a defensive structure. Made up of two barbed lancets, these are supported by the quadrate plates and extensive musculature in the last abdominal segment (called VIIT, for the seventh tergite; see the illustration showing the sting anatomy of the worker on p. 249). When the bee stings, the two lancets move into the skin with a sawing action and anchor themselves by their barbs.

Morphological Differences between Honey Bee Castes and Drones

Item	Queen	Drone	Worker
Larval diet	Royal jelly	Drone jelly?	Worker jelly
Cell orientation	Vertical	Horizontal	Horizontal
Development time	16 days	24 days	21 days
Eyes	3920–4920 facets	13,000 facets	400–6300 facets
Brain	Small	Large	Large
Mandibular gland products[a]	9-Oxodec-*trans*-2-enoic acid[b]	Absent	10-Hydroxydecenoic acid
	9-Hydroxydec-*trans*-2-enoic acid[c]		
Mandibles	Unmodified, cutting only		Small and notched, modified for comb building, flat center ridge
Tongue	Short	Short	Short
Honey stomach	Small	Small	Well developed
Wings	Appear reduced in length	Extend over abdomen	Extend over abdomen
Legs	Middle tibia = no spine	Middle tibia = no spine	Middle tibia = spine Hind leg = pollen- collecting apparatus
Wax glands	Absent	Absent	Present, 4 pairs
Nasonov gland	Absent	Absent	Present
Ovaries (n = ovarioles)	Well developed; n = 300	Absent	Undeveloped; n = 2–12
Spermatheca	Present	Absent	Rudimentary
Sting	Lightly barbed, curved, waxy	Absent	Strongly barbed, straight, no wax
Antennae	1600 chemoreceptors	?	2400 chemoreceptors
Reproductive role	Egg laying	Mating	Maternal care
Number per colony	1 or 2	100–2500	30,000–60,000
Food	Fed royal jelly	Begs/eats honey	Honey, pollen
Life span	1–8 years	Up to 60 days	Up to 4–6 months

[a] Brood food comes from the mandibular (white liquid) and the hypopharyngeal (clear) glands of young, nurse bees.
[b] (9-ODA).
[c] (9-HDA).

Once implanted, the sting sac can pump venom into its victim unless it is scraped off. Queens rarely sting and have no barbs on the lancet; thus they can sting rival queens repeatedly. Drones have no sting.

The poison gland secretes venom into an attached poison sac, which holds the venom until it is pumped through the sting. This gland was originally labeled the acid gland and is found as such in the older bee literature. Another smaller gland, the Dufour (or alkali) gland, also opens in the sting chamber and may secrete lubricants for the sting apparatus or for the eggs.

Internal Organs of the Drone Honey Bee

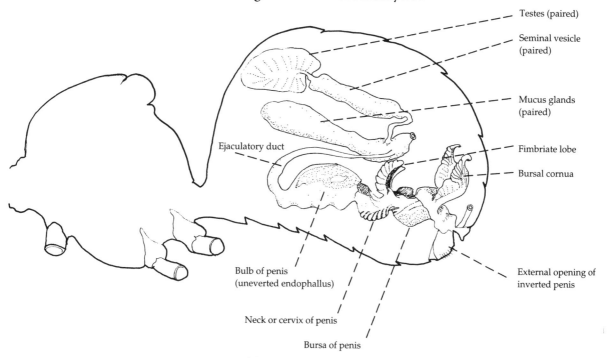

Testes (paired)

Seminal vesicle (paired)

Mucus glands (paired)

Fimbriate lobe

Bursal cornua

Ejaculatory duct

Bulb of penis (uneverted endophallus)

External opening of inverted penis

Neck or cervix of penis

Bursa of penis

Sting Anatomy of a Worker (Honey Bee)
(Ventral View)

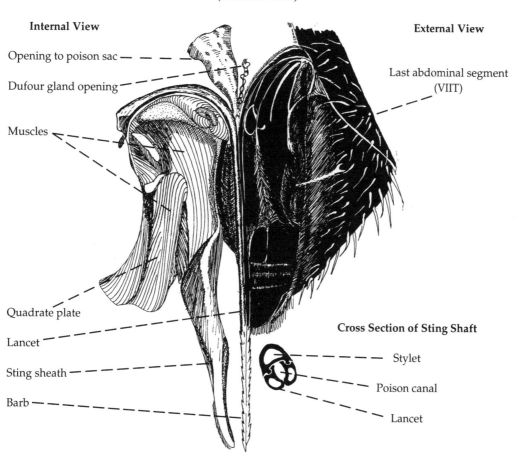

Internal View

Opening to poison sac

Dufour gland opening

Muscles

Quadrate plate

Lancet

Sting sheath

Barb

External View

Last abdominal segment (VIIT)

Cross Section of Sting Shaft

Stylet

Poison canal

Lancet

APPENDIX B

Pheromones

Honey bee behavior both inside and outside the hive is regulated to a large extent by chemical substances called pheromones. Pheromones are secreted by glands and trigger certain behavioral responses or physiological activities. Refer to the table on glands of honey bees for more information.

QUEEN PHEROMONES

Located in the mandibles of a queen's head are the mandibular glands, which produce and secrete the pheromones called queen substance, which include the chemicals *(E)*-9-hydroxydec-*trans*-2-enoic acid (10-HDA), (E)-9-oxo-2-decenoic acid (9-ODA), methyl *p*-hydroxybenzoate (HOB), and 4-hydroxy-3-methoxyphenylethanol diacetate (HVA). These compounds elicit various responses in worker and drone honey bees (see the table on the glands of honey bees). Inside the hive these substances have been shown to inhibit ovary development in workers and deter them from making new queens. Their absence invokes the opposite response: queen cup construction (which leads to queen cells) is undertaken.

A swarm—either flying out of the hive to a new

Glands of Honey Bees

Gland location	Number	Substance secreted	Worker	Queen	Drone
Head					
Hypopharyngeal	2	Brood food and royal jelly, enzymes, glucose oxidase	X		
Salivary	2	Fatty substance, enzymes	X	X	X
Mandibular	2	Lipid component of larval food + 10-HDA	X		
		Alarm, alerts other workers (2-Heptanone)	X		
		Enzymes	X		
		Queen substance[a] contains over 20 pheromones including 9-HDA,[b] 9-ODA,[c] HOB,[d] and HDA + HVA		X	
		Congregating pheromone			X
Thorax					
Salivary	2	Saliva, enzymes (derived from larval silk gland)	X	X	X
Legs					
Arnhart, tarsal "gland"	?	Worker pheromone: attract workers to nest or to forage	X		
		Footprint + pheromone + HDA mandibular from queen inhibits queen cup construction		X	
		Drone pheromone: ?			X

Gland location	Number	Substance secreted	Worker	Queen	Drone
Abdomen					
Wax	4 × 2	Beeswax: nest construction	X		
Nasonov		Scent pheromone[e]: to attract other workers	X		
Abdominal tergites (Ab. Terg. 3–5)		Pheromone: recognition of queen by worker; stabilize queen retinue, inhibit worker ovary; attract drone to queen (precopulation)		X	
Rectal	6	Enzyme: catalase to break down starch	X	?	?
Spermathecal		Polysaccharide + proteins		X	
Drone accessory	2	Peptides + mucus			X
Sting apparatus		Venom	X	X	
Dufour		Egg coating, distinguishes eggs laid by queen and those laid by workers		X	
Kozhevnikov	2	Pheromone: queen attracts and recognized by workers		X	
Sting sheath	2	Alarm pheromone (acetates, isopentyl, etc.)	X		
Alkaline		Lubricant for sting?	?	X	
Sting shaft		Waxy esters? (lubricant?)	?	X	
Larvae and pupae					
Larval silk	2	Silk: spun into cocoon	X	X	X
?		Brood pheromones: incubation, attracts and recognized by workers	X	X	X
?		Inhibits queen rearing, workers' ovary development		X	
?		Stimulates foraging, especially for pollen	?	?	?
? Diploid drone larva		Cannibalism pheromone		?	

Sources: E. Crane. 1990. Bees and beekeeping: Science, practice, and world resources. Ithaca, NY: Comstock. M.L. Winston. 1987. Biology of the honey bee. Cambridge, MA: Harvard University Press.

Note: 2-Hep = 2-heptanone; 9-HDA = 9-hydroxy-2-enoic acid; 10-HDA = 10-hydroxy-2-decenoic acid; 9-ODA = (E)-9-oxodec-2-enoic acid; HOB = methyl p-hydroxybenzoate; HVA = 4-hydroxy-3-methoxy phenylethanol.

[a] There are many pheromones in queen mandibular glands, one or more of which (1) inhibit worker ovary development, (2) stimulate workers to release Nasonov pheromone, (3) stimulate workers to forage, (4) regulate workers clustering in swarms, and (5) attract workers.

[b] 9-HDA inhibits queen cup construction, allows workers to recognize queen (virgin), and stabilizes a swarm.

[c] 9-ODA attracts drone to queen, inhibits queen rearing and queen recognition by workers, and suppresses ovary development in workers.

[d] HOB + 9-ODA + two isomers of 9-HDA + HVA induces attendant behavior of workers to the queen.

[e] Scent pheromones: citral, nerol, geraniol, (E,E)-farnesol; 9-HDA, 9-hydroxy-2-decenoic acid; HOB, methyl p-hydroxybenzoate; HVA, 4-hydroxy-3-methyoxyphenylethanol; 9-ODA, 9-oxo-2-decenoic acid.

homesite or settled in a cluster—is aware of its queen's presence by means of these and other substances. Queen substance also guides drones toward queens who are on mating flights. Researchers are still investigating other pheromones or other volatile compounds, and with the more sophisticated instrumentation, new compounds are being discovered, such as brood pheromones and volatiles from mites.

WORKER PHEROMONES

Several different chemical pheromones are produced by workers. Two of these are alarm pheromones. One identified alarm odor (isopentyl acetate or isoamyl acetate) is released from a membrane at the base of the sting. It smells like banana oil and stimulates bees to sting or fly at intruders. Other alarm com-

Glands of a Worker Honey Bee

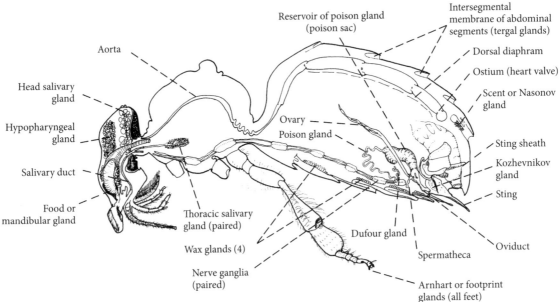

pounds from the sting have recently been isolated and are currently being studied. Another alarm odor (2-heptanone) is released by the mandibular glands of workers. Items anointed with this odor are attacked by bees; it currently is being investigated as a mite and wax moth deterrent.

A third pheromone is the scenting or orientation odor and is composed mainly of four chemicals (nerolic, geranic, and citrol acids, and geraniol). This chemical complex is produced by the Nasonov or scent gland near the dorsal tip of the abdomen. Dispersal of these substances is aided by the fanning action of bees secreting it. Upon smelling these chemicals, bees move toward the source (as in a swarm). Honey bees produce many other pheromones, some have yet to be identified or the functions understood. Researchers are continuing to study drone and brood pheromones.

Honey bee pheromones are divided into two groups, the *primer* and the *releaser*. Releaser pheromones (short-term effect) trigger a rapid, immediate response from the receiving bees, usually outside the colony; primer pheromones (long-term effect) trigger or suppress developmental events inside the hive. The pheromones are liquids but may change into gas, solids, or remain liquid after secretion. Nasonov gland produces a releaser pheromone; queen substance is both a primer and a releaser pheromone. In the latter case, the queen substance suppresses queen cup (and queen cell) construction, which is a long-term effect; but on her nuptial flight, the pheromone has a short-term effect on drones. Our knowledge of pheromones is increasing thanks to new technological tools used today; we can now identify and trace these chemicals within the bee colony as well as from individual bees.

Bee Sting Reaction Physiology

LOCAL REACTION

What happens in your body when you are stung by a bee? As with any invader, the body's natural defenses are called on to help. Basically, bee venom is a foreign protein (called an *antigen*) that stimulates the production of the body's defense proteins (called *antibodies*). Antibodies belong to a family of proteins known as gamma globulin, also called immunoglobulins. The bee sting antigens appear to stimulate a specific class of immunoglobulins known as immunoglobulin E, or IgE (see illustration on bee sting reaction physiology on p. 255).

Because the bee venom antigen causes the production of specific antibodies, in this case IgE, people not otherwise exposed to honey bee proteins must be stung more than once before any type of reaction will occur. After the initial injection of venom, the body seems to "remember" that particular antigen and will likely react faster to subsequent stings, with further antibody production.

In a *local reaction*, the antigen of the bee venom appears to cause the production of the IgE bodies, which are attached to tissue cells (called *mast cells*). Mast cells contain numerous vesicles filled with histamine and other substances, which when released promote inflammation. The action of the antigen with the IgE/mast cell complex causes the histamine-filled vesicles to empty. Once released into the body, histamine has several effects, which include the expansion of blood vessels (vasodilation), the increased permeability of capillary cell walls to proteins and fluids, and the constriction of the respiratory passages. The first two may be responsible for the in-

flammation, swelling, and itching associated with bee stings. Most beekeepers are reported to have this kind of local reaction. With repeated stings, the body may become immune to the venom thereby reducing the localized swelling (except it always hurts).

If you are being stung repeatedly by bees, cover your eyes and mouth (carbon dioxide [CO_2] that you exhale attracts bees) and run as fast and as far as possible from the stinging insects (or go into a car or shed). The lethal dose reported by researchers is 5 to 10 bee stings per pound of body weight, which means an average male, weighing 160 pounds (72.6 kg), can receive more than 1500 stings and still live; the critical factor is that the kidneys may shut down, due to the overload of venom in the bloodstream.

SYSTEMIC REACTION

In a systemic reaction, the same mechanisms come into play as in the local reaction, with one big difference: the antigen/IgE/mast cell complex reaction can cause death. This allergic reaction, called hypersensitivity, appears to be a result of the large amounts of histamine released from the mast cells.

In some people, the second bee sting may be enough to be fatal. Because the body "remembers" the bee venom antigen, the subsequent inoculations usually cause a faster reaction, which means more histamine is released each time a person is stung. Usually, a systemic reaction builds up gradually, with the victim showing greater distress, such as difficulty breathing, after each stinging incident.

An antihistamine and adrenaline (epinephrine) should be immediately administered to people experiencing a systemic reaction, to counteract the effects of the released histamine and give relief for breathlessness. In these cases, medical advice must be sought immediately to treat allergies.

Desensitization or Immunity

People who develop hypersensitivity to bee stings can become desensitized to bee venom. Most beekeepers become immune to venom after repeated exposure. Desensitization can also be undertaken by an allergist. What an allergist does is to incrementally increase the amount of venom that the victim receives, thus allowing the body to form enough of these blocking antibodies to combat the allergic reaction. In either case, the immune processes or desensitization are probably the same; frequent injections of the specific venom induce the body to manufacture a "blocking" antibody, IgG. The IgG competes with the IgE in its reaction activities to bee venom antigens. Because the IgG antibodies are not fixed to the mast cells, but float freely, they are better able to combine with the venom antigens. Less histamine is therefore released, and the discomfort or allergic response is prevented.

Some beekeepers can, over long periods of time and exposure, become allergic to bee venom as well as beeswax, honey, propolis, bee debris, and bee bodies. The percentage of beekeepers who do become allergic to bee stings, although hard to assess because individual body chemistry, allergy history, exposure, and genetic predisposition are so varied, is small. Most people will probably already know if they are allergic, but those who do not can contact local allergy clinics or hospital outpatient facilities for testing. See "Venom" in the References for more information.

Bee Sting Reaction Physiology

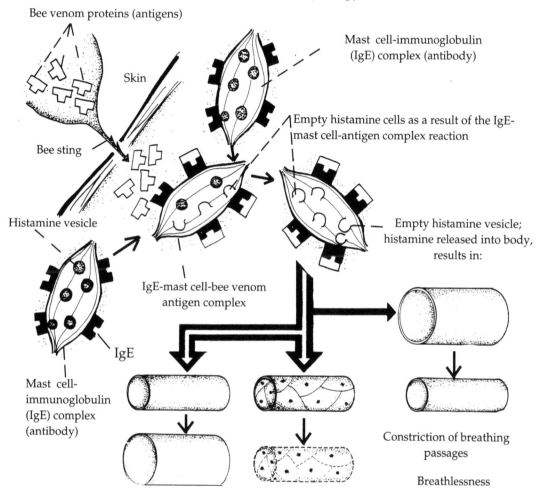

Bee venom proteins (antigens)

Mast cell-immunoglobulin (IgE) complex (antibody)

Skin

Bee sting

Empty histamine cells as a result of the IgE-mast cell-antigen complex reaction

Histamine vesicle

Empty histamine vesicle; histamine released into body, results in:

IgE-mast cell-bee venom antigen complex

IgE

Mast cell-immunoglobulin (IgE) complex (antibody)

Constriction of breathing passages

Breathlessness

Blood vessels dilate

Capillaries become more permeable

Swelling, inflammation

Paraffin Dipping

Today, some wood preservatives are contaminating the beeswax comb in bee colonies, and beekeepers are reluctant to use them to extend the life of wooden hive furniture. A better option to preserve hive woodenware is to dip it in boiling paraffin. This process is used in many countries and is very inexpensive if there is a lot of equipment to treat. It will clean up older boxes or can be used on new, fresh wood. It is also used to sterilize diseased equipment. Use a vat for dipping that has these features:

- Made of metal.
- An enclosure to provide a "double skin" around the vat.
- An electric heat source.
- Deep enough to hold two shallow supers or one deep, with some wiggle room.
- All joints welded from both sides to reduce the risk of bursting when in use; hot wax is dangerous.

You can dip: outer covers (without metal tops), supers and bottom boards, nucs.

Paraffin wax blocks can be purchased from commercial warehouses or online. The grade with a melting point of 140°F (60°C) is best. Heat the wax to 320°F (160°C); this can take up to 2 hours. Use a thermometer attached to a stick—**do not** rely on guesswork here; a thermostat would be better, one that can regulate the heat automatically. Some beekeepers add linseed oil and beeswax to the mixture.

Clean old equipment first by pressure washing or sanding to remove old paint and wax. If water is used, **make sure** wood is dry before immersing; otherwise

you could cause the wax to boil over. Additionally, the woodenware must be **dry**, as dry as possible, as the wax will seal any water in the wood and could lead to rot from the inside out.

Immerse the boxes into the boiling wax for about 2 minutes, using a heavy weight to keep them below the surface. Diseased boxes must be treated for 10 to 15 minutes. Then lift the supers from the wax onto a rack or other place to drain. Boxes previously painted with oil-based paints blister and will need to be scraped.

Boxes can be used straight after dipping without being painted; paraffin wax is a good preservative on its own. You can still paint the boxes while they are hot, with water-based latex. Two coats are best; the paint should be pulled into the wood by the drying wax.

The secret is to immerse the boxes long enough in the hot wax so the wood is thoroughly heated and good penetration by the wax can be obtained. The boxes will be very hot when they come out of the dipper, so use thick fireproof gloves and tongs.

Finally, some commonsense safety precautions:

- Locate the wax dipper away from anything flammable; outdoors is best.
- Wear eye protection in case the wax splashes.
- Have a metal cover ready to put over the vat in case the wax boils over.
- Wear protective footwear.
- Wear an apron and face mask.
- Have on hand a first aid kit, for burns.
- Have available a fire extinguisher rated for oil/wax fires (dry chemical powder).

For more information, go to these publications:

Goodwin, M., and H. Haine. 1998. Using paraffin wax and steam chests to sterilise hive parts that have been in contact with colonies with American foulbrood disease. New Zealand Beekeeper 5:4, 21.

Griffiths, G.L. 1992. Protocol for wax dipping bee equipment. In: Preservation of wooden hive equipment. Miscellaneous Publication 4/92. March 1992. Compiled by L. Allan, Senior Apiculturalist, Department of Agriculture, Western Australia.

Hot wax dipping of beehive components for preservation and sterilization. 2001. Publication No. 01/051, Project No. DAV-167A. Kingston, Australia: Rural Industries Research and Development Corporation and the Department of Natural Resources and Environment; http://www.rirdc.gov.au.

Kalnins, M.A., and B.F. Detroy. 1984. Effects of wood preservative treatment on honey bees and hive products. J. Agricultural Food Chemistry 32:1176–1180.

Matheson, A. 1980. Easily constructed paraffin wax dipper. New Zealand Beekeeper 41 (4): 11–12.

Robinson, P.J., and J.R.J. French. 1986. Beekeeping and wood preservation in Australia. Australian Bee Journal 67:1, 8–10.

Tew, J. 1984. Paraffin wax dipping of hive equipment. Gleanings in Bee Culture 112 (8): 422, 426.

Weatherhead, T.F., and M.J. Kennedy. 1980. Adhesives and wood preservatives for the beekeeper in the 1980s. Australasian Beekeeper 82 (4): 88–94.

Williams, D. 1980. Preserving beehive timber. New Zealand Beekeeper 41 (4): 13.

Differences between European (EHB) and Africanized Honey Bees (AHB)

Differences in Physical, Behavioral, and Developmental Characteristics

	AHB	EHB
Physical characteristics		
Worker length	12.73 mm	13.89 mm
Cell size[a]	4.6–5.0 mm	5.2–5.4 mm
Wing beat[b]	290 Hz	243–270 Hz
Wing position	Wings elevated off abdomen	On abdomen
Fresh weight[c]	2 g	2.5 g
Nest size, volume	22 liters	45 liters
Comb area (1000 cm²)	8–11	23.4
Honey area	0.9	2.8
Queen's egg laying		
Maximum/day	4000, 2–4 times brood area	2500
Total/year	105,000	58,000
Behavioral differences		
Swarming	3–16 times/year	3–6 times/year
Absconding	30% of colonies	Not common
Supersedure	> 40% in tropics	?
Pollen	Much more collected	Normal
Mating time	Late afternoon	Early afternoon
Defensiveness/stinging	Bounce off veil, sting in vicinity	Sting in apiary
Time before first sting	14 seconds	229 seconds
No. stings/minute	35	< 2
Pursuit distance	160 m	22 m
Recovery time	28 minutes	3 minutes
Entrance	Attack other bees at entrance	Not so prevalent
Drone rearing	Stimulated, more drones, mate more frequently with EHB queens	Suppressed
Returning foragers	Returning bees fly into entrance	Land on landing board
Foraging		
Unassisted spread/year	200–500 km	4 km
Maximum distance to new nest site	75 km	5 km

	AHB	EHB
Running on comb	Excessive, runs off comb, festoons at lower edge; parades around inside box or inner cover	Normal, not usual to run much
Reaction to smoke	More bees leave colony, more bees eat honey	Fans to expel smoke, some eat honey
Cell caps	Thin, hard to uncap	Normal
Usurpation rates	20–30%/year in fall (Arizona)	
Requeening	Can be very difficult, if requeening with EHB about 40% loss	Normally 10% loss
Life expectancy	Shorter than EHB, 12–18 days as foragers	Longer than AHB
Developmental stage		
Egg to adult worker	18.5 days	21 days
Egg to adult queen	15 days	16 days
Worker's first flight	3 days old	10–14 days old
Queen's mating flight	5–6 days	7–10 days
Drone's mating flight	7.5 days	13 days
Brood production	Almost 2 times as much when resources good	Slower

Sources: D.M. Caron. 2001. Africanized honey bees in the Americas. Medina, OH: A.I. Root. E. Crane. 1990. Bees and beekeeping: Science, practice, and world resources. Ithaca, NY: Comstock. S.S. Schneider, G. DeGrandi-Hoffman, and D.R. Smith. 2004. The African honey bee: Factors contributing to a successful biological invasion. Annu. Rev. Entomol. 49:351–376.

[a] Cell size: measure 10 cells three times; if average ≥ 49 mm, then bee is AHB; if 52 mm, then bee is EHB.
[b] Wingbeat: measure free-flying, undisturbed bees.
[c] Fresh weight of 30 callow bees, average.

Weight Differences between All Developmental Stages

Worker bee stage	Age (days)	EHB weight (mg)	AHB weight (mg)
Egg[a]	1	0.1435	0.1064
	2	0.1196	0.0723
	3	0.0961	0.1122
Larva	4	0.1692	0.2218
	5	1.6493	1.1640
	6	9.8635	4.9440
	7	54.547	30.060
Capped over	8	109.10	93.16
Capped and spinning	9	160.60	113.66
Spinning	10	148.35	116.24
Prepupa (immobile)	11	142.74	113.48
White pupa	12	145.69	114.10
	13	144.48	104.88
Light pink compound eye	14	142.89	111.20
Dark pink compound eye	15	136.30	109.88
Dark pink-purple compound eye and ocelli	16	141.05	107.26
Dark purple compound eye, light brown wing bases	17	135.34	107.78

Weight Differences between All Developmental Stages *(continued)*

Worker bee stage	Age (days)	EHB weight (mg)	AHB weight (mg)
Dark purple compound eye, light brown head, and thorax	18	136.11	101.38
Black compound eyes, light gray wing pad, dark brown head, and thorax	19	133.12	92.34 Worker emerges
Black antennae, medium gray wing pads	20	122.37	
Imago/emerging	21	116.11	

Source: D. Sammataro and J. Finley. 2007. The developmental changes of immature African and four lines of European honey bee workers. Apiacta 42:64–72.

[a] Egg dimensions: eggs are 1.3–1.8 mm long and take 48–144 hours to hatch, the average being 72 hours.

Rearing Wax Moth *(Galleria mellonela)*

Wax moth larvae, while a pest in the beeyard, can be a lucrative sideline operation, selling them as bait for fishermen or as pet food for exotics, such as turtles, lizards, snakes, and even birds. Look around for potential markets in your area.

DIETS FOR REARING WAX MOTH LARVAE

There are two basic diets you can use. Here they are:

1. Boil together for several minutes ⅓ cup sugar (or honey), ⅓ cup glycerol, and ⅓ cup water. Cool, and then add ¼ teaspoon of a vitamin supplement, such as you would use to feed to your chickens. (Do not add antibiotics.) Then stir in 5 cups dried cereal, such as a mixed-grain baby cereal. Mix together and place wax moth eggs on top of the mixture in a glass jar.
2. A better recipe is to warm 200 ml glycerin and 200 ml (6.76 fl oz) honey. Add to 2 boxes complete or fortified baby cereal. Mix together and store in the refrigerator until used (or add wax moth eggs right away). This will be a dry mix.

BEGINNING YOUR CULTURE

Select a glass (or plastic) jar with a wide top. A peanut butter or 2-quart fruit or Mason jar works fine. Fill container 2 inches (5 cm) deep with prepared diet. Cover the jar with a paper towel, a 20-mesh wire screen disk, and a jar lid with a large hole cut in the center (or you could use a Mason jar rim). The paper towel is to keep foreign material from entering and the young wax moth larvae from leaving; the larger larvae will eat through the paper if not protected by the wire.

Now add eggs, larvae, pupae, or adults to your culture. Whichever you add, they should all be at the same stage (i.e., don't mix eggs and pupae). Add 10 to 50 larvae, or whatever stage you are starting with, to each jar, depending on the size. Do not crowd too many larvae into one jar, as you will get smaller worms. It is best to start with eggs or adults that will lay eggs, since field-collected larvae may have parasites.

Now place your culture jars in an enclosed room, out of the sun, where it is warm, around 80° to 93°F (27–34°C). The larvae can tolerate a range of temperatures, from 77° to 99°F (25–37°C). But remember, the warmer the room, the faster they will grow; the opposite is also true. Write the date you started the culture on the outside of the jar, so you can time when the wax moth larvae will mature.

HARVESTING AND INCREASING

If you plan on harvesting larvae to eat, collect them as they begin to spin their cocoons, since they void all fecal matter just before spinning. Try not to disturb the growing medium (diet) when collecting them. If you wish to hold the larvae for a long period, you can store them at 59°F (15°C) and 60 percent humidity for up to a year. This will keep them from pupating.

To increase your colony, you will need to rear some larvae to adult moths. Either place some adults, or around 50 cocoons, in a separate jar (you do not need food in these jars, since adults do not feed). Now place a piece of paper, folded like an accordion, into the jar to give the female moths a place to lay eggs. Cover the jar with paper towels tied in place and keep the moths at room temperature.

By putting in 50 cocoons, you will get a good mix of males and females. Within 10 days, the adult moths will emerge. Each female moth will mate in the jar and begin to lay eggs in the folds of the paper. Each female can lay more than 100 eggs.

Once you see the eggs (laid in strips, they are round and slightly tan in color), place a 2-inch strip of eggs in separate jars on top of fresh food. Don't wait too long to transfer the eggs, or the larvae could crawl out and starve to death (they are very tiny and mobile at this stage).

BAIT WORMS

If you are using your worms just for bait, harvest them before they spin their cocoons. Put them in cold storage for 4 to 5 hours at 32° to 33°F (0–0.6°C). To ensure an even temperature of 32°F, put ice cubes and some water in a large bowl. Now put your larvae in a smaller jar and place the jar in the ice bath. Keep the jar in the bath until all the ice has melted.

After that the larvae will not spin cocoons, and you can store them at 40° to 45°F (4.4–7.2°C) until they are ready to sell as bait. Place the larvae in a separate container and keep them moist with paper towels, wood shavings, or other absorbent material.

HELPFUL TIPS

If your worms start dying suddenly, are attacked by fungi or virus, or change to a dark color, you could have a disease problem. Burn that batch and any other batch that may have been contaminated by it, and wash the jars in hot, soapy water and pour boiling water in them. Then start with a fresh batch.

TASTY INSECT TREATS

It has been suggested that the growing larvae can be delicately flavored by incorporating small amounts of various spices into their diet. Wax worms are great as snack items. Fried in hot oil, they pop like popcorn and, if lightly salted, are reported to have good or better flavor than potato chips or corn puffs.

REFERENCES

Gojmerac, W. 1978. Grow your own wax worms. Fin and Feathers (Jan.).Hocking, B., and F. Matsumura. 1960. Bee brood as food. Bee World 41 (5): 113–120.

Keeney, G. Director, Ohio State University Insectary, Columbus, OH.

Taylor, R.L. 1976. Butterflies in my stomach. Santa Barbara, CA: Woodbridge Press.

Taylor, R.L., and B.J. Carter. 1977. Entertaining with insects: Original guide to insect cookery. Santa Barbara, CA: Woodbridge Press.

Pointers for Extreme Urban Beekeeping (NYC), by Jim Fischer

1. Urban "Location" Measured in Blocks, Not Miles: Despite the well-known "3-mile radius" of a bee colony's immediate foraging area, distances seem more daunting to urban bees. New packages established on rooftops and installed on foundation within what should be "easy flying range" of large areas of parkland (such as Central Park in Manhattan and Prospect Park in Brooklyn) had trouble drawing out even 2 mediums of foundation from mid-April to late July and required extensive feeding. Hives closer to major parks did much better.

2. Urban Winds Are Not the Beekeeper's Friend: Rectilinear street grids tend to increase wind velocities where wide avenues are lined with tall buildings. This effect is even more pronounced along rivers. The result is that some locations have winds that make it difficult for bees returning to the hive. Worker bees weigh only a tenth of a gram, so careful site evaluation using a recording anemometer and a bee counter might be required to determine why some hives do not seem to prosper. Hives above 15 stories might not be able to make it back, due to winds.

3. Rooftop Hives Are Often Warmed by the Rooftops over Winter: Heat rises, and many roofs are black, so hives on older roofs may be kept unexpectedly warm, fail to cluster, and eat themselves into starvation when other hives have clustered and enjoy reduced consumption of stores. Putting one's hives on bricks, cinderblocks, or a hive stand is a very good idea.

4. Mouse Guards Are Not Just For Country Bees: While mice are rare in cities, and rats tend to be far too large to fit into a hive entrance, there are many species of roaches attracted to a hive. For example, the Bronx Zoo is famous for having a significant number of species of roaches who live well off bird seed and other animal feed living on bottom boards and above inner covers in the Bronx Zoo hives. Blocking the roach's access to the hive was solved with a 4-inch wide strip of ¼-inch hardware cloth or mesh, cut to the width of the hive entrance. Fold the mesh into a "V" shape and push into the entrance with a hive tool; this should stay in place by friction alone.

5. Urban Swarming and Supersedure Can Result in Unmated Queens: Because there are a smaller number of honey bee colonies in urban areas and fewer feral colonies, hives that swarm or replace their queen by means of supersedure may result in unmated queens. In New York City, we have seen naturally mated queens only in the Greenpoint and the Red Hook neighborhoods of Brooklyn. No other area within city limits has yet produced its own mated queens; but as more members of the NYC Beekeeping and the Gotham City Honey Co-Operative start hives, we expect improvement.

6. Trees Are the Primary Urban Nectar Sources: New York City has 5.2 million trees, nearly 600,000 of them lining streets, according to the Million Trees NYC organization. More than 44,509 acres of the city are shaded by trees, 24% of the total land area. This means that urban beekeepers should "Super Early, Harvest Often."™ ** Maples begin blooming in mid-February and one cannot super early enough to exploit all the blooms. The question becomes: How early will the bees start flying?

7. What the Neighbors Don't Know Won't Scare Them (or, Keep It Quiet): While beekeepers expect their neighbors to share a deep, abiding concern for the well-publicized problems facing bees, and to comprehend that one cannot have local food without also having local bees, many neighbors seem to expect locally grown, biodynamically managed, organically certified veggies to appear by magic. A generation has grown up without ever being stung, and they are now raising their own children, so there are many otherwise rational people who are absolutely certain that an insect sting is dangerous. Outreach can only go so far, and one cannot give away 300 bottles of honey to residents in each of the high-rises that surround one's hives. The Gotham City Honey Co-Op purchases paint in colors that match the most common brands of rooftop air-conditioning equipment to use on the outside of the hives, and we suggest that urban beekeepers wear workmen's coveralls and a veil rather than the usual bee suit if beehives are exposed to view. If the coveralls say "Joe's Air Conditioning" in large letters on the back, so much the better. Subtlety is an art form.

8. That being said, swarm control is crucial to continued tolerance of beekeeping in urban areas. Swarms are still treated by police, fire, and emergency management as if they were unexploded bombs. The prudent beekeeper utilizes splits, swarm traps, and constant vigilance in spring to control swarming, and those interested in swarm rescue are trained to respond to the unique challenges of swarms in the city. Our group has a quarantine yard set up to give rescued swarms and feral colonies a temporary home where Varroa can be eliminated while the swarms are broodless, and queen performance can be checked prior to being colony adopted.

9. Decontaminate Packages and Quarantine Nucs Entering Your City: With only a small number of feral hives in the larger parks prior to urban beekeeping becoming popular, much of New York City was essentially disease- and pest-free since at least 2006. Both the Hudson and East rivers are too wide and too windy for bees to fly across, so we are keeping bees on small islands off the coast of North America. Other urban areas may have similar barriers in the form of a solid ring of auto dealers, big-box retailers, and other suburban sprawl surrounding the urban core. Our group obtained the use of a remote corner of a NYC Parks' work yard for quarantine to hold hived swarms (see above) and imported nucs. All our packages are decontaminated by spraying with oxalic acid at pickup locations, and purchased nucs are treated with formic acid before being hived. The result has been a majority of hives without detectable varroa, a real pleasure. Sadly, some beekeepers refuse to decontaminate packages, which has resulted in some areas having varroa infestations and higher nosema counts. Outreach and education are ongoing in this regard, as one's bees can be no healthier than one's neighbors' bees. Although you may not be your brother's keeper, in cities, you are your neighbor's beekeeper, like it or not.

10. Welcome Back, My Friends, to the Class That Never Ends: Apologies to Emerson, Lake, and Palmer, but education is the best solution to most beekeeping problems, and with more than 950 members, we may be the largest beekeeping entity in the United States. Our completely free beekeeping class starts in October, and simply does not end. The 15-session winter lectures are standing-room only, and our nearly weekly hands-on apiary workshops are often fully booked within a few hours of when they are announced on our website. None of this would be possible if not for the generosity of the NYC Parks Department and others who allow us to use lecture halls and other spaces without charge. Not everyone can make it to every class or workshop, so bootleg recordings are passed around. Asking novices to mentor other novices is also unusual, but the ratio of novices to experienced beekeepers is high. This will likely change over time.

11. Logistics Are a Pain in the "Beehind": When most members of an organization live in one-bedroom apartments and do not own cars, simple things like assembling and painting wooden-ware, extracting honey, and storing honey supers in winter become daunting logistical challenges. Shared facilities and organized work days are the only possible way to meet the needs of our members. We store supers in an unused airplane hanger, and we extract honey in state-inspected

commercial kitchens located in private buildings, churches, and synagogues. Equipment is assembled in the same warehouse space in which we sort and distribute our cooperative bulk purchases of everything from foundation pins to bottom boards. Our ultra-local artisanal honey is labeled with hive zip codes and sells as quickly as we can bottle it. Even mundane things like glassware must be planned, ordered, and provisioned in advance of need. Plan ahead. . . .

Notes: Jim Fischer kept bees and pollinated apples in the mountains of Virginia for more than a decade, and when he moved to New York City in 2006, he found that beekeeping was prohibited by Health Department regulations dating from the late 1990s. "Legalizing" beekeeping required supporters, so he taught free beekeeping classes and workshops in Manhattan's Central Park. The classes turned enough city dwellers into bee enthusiasts to support a campaign to decriminalize beekeeping, which began in early 2008 and succeeded in 2010. Fischer invented Fischer's Bee-Quick (http://bee-quick.com/), a nontoxic repellent to clear bees from honey supers. All profits from the sale of Bee-Quick go to the *Eastern Apicultural Society* Bee Research Fund.

NYC Beekeeping is an association of more than 960 beekeepers and bee-lovers, operating on a not-for-profit basis. We offer a free course in beekeeping together with our sister group, the Gotham City Honey Co-Op. Members work with urban farms, community gardens, the NYC Parks Department, schools, and other organizations to foster a strong community of responsible urban beekeepers and to bring training and economic opportunity through beekeeping to underserved areas. NYC Beekeeping, http://www.nycbeekeeping.org, 917-740-4BEE (4233), nycbeekeeping@gmail.com.

APPENDIX H

Varroa Mite Infestation

Mite Density Chart

Number of Mites per 300 Adult Bees	Colony Infestation (%)	Number of Mites per 8 Samples of 300 Adult Bees	Apiary Infestation (%)
1	1	8	1
2	1	16	1
3	2	24	2
4	3	32	3
5	3	40	3
6	4	48	4
7	5	56	5
8	5	64	5
9	6	72	6
10	7	80	7
11	7	88	7
12	8	96	8
13	9	104	9
14	9	112	9
15	10	120	10
16	11	128	11
17	11	136	11
18	12	144	12

Notes: Count the number of mites in a sample of 300 adult bees. Use that number and this simple table to determine percentage of infestation. For example, if 12 mites were counted on 300 bees, check the number 12 on the table: that would be an 8% colony infestation. If the colony (or apiary) infestation is more than 10 to 12%, treatment of some sort is recommended. After Lee et al. 2010 and by permission of the author.

Chemotherapy for Control of Varroa Mites

A list of some of the current chemical miticides (acaracides) used to control Varroa; some of these may also work for *Tropilaelaps*.

Product Trade Name ®	Active Ingredient	Chemical Class
Apiguard	thymol	essential oil
Apilife VAR	thymol, eucalyptol, menthol, camphor	essential oils
Apistan **	fluvalinate	synthetic pyrethroid
Amitraz, Miticur, Api-warol (tablets)	formamidine	Formetanate, Methanimidamide
Apitol	cymiazole	iminophenyl thiazolidine derivative
Apivar **	amitraz	amadine
Bayvarol **	flumethrin	synthetic pyrethroid
Check-Mite+ **	Perizin, coumaphos	organophosphate
Folbex	bromopropylate	chlorinated hydrocarbon
Sucrocide	sucrose octanoate	sugar esters
Hivestan	fenpyroximate	pyrazole (alkaloid)
generic	formic acid	organic acid
generic	lactic acid	organic acid
generic	oxalic acid	organic acid

** No longer effective in some areas

Sources: Chemical controls for varroa mites compiled from Rosenkranz et al. 2010 and http://www.maf.govt.nz/ biosecurity/pests-diseases/ animals/varroa/guidelines/control-of-varroa-guide.pdf. In: *Recent Developments Focused on the Problems of Honey Bee Pollinators*, edited by D. Sammataro and J.A. Yoder, in press.

Glossary

abdomen The posterior or third region of the body of the bee that encloses the honey stomach, stomach proper, intestines, sting, and reproductive organs.

abdomen of a drone bee The posterior or third region of the body that contains the stomach, intestine and reproductive organs.

abdomen of a queen bee The posterior or third region of the body that contains the stomach, intestine, reproductive organs, sting, and Dufour gland.

abdomen of a worker bee The posterior or third region of the body that contains the honey stomach, stomach proper, intestines, sting, reproductive organs, wax glands, and pheromone gland (Nasonov gland).

abscond Bees abandoning a hive because of wax moth, excessive heat or water, mites, lack of food, or other unfavorable conditions.

acarine disease The name of the disease later known to be caused by the mite *Acarapis woodi* (Rennie) infesting the thoracic trachea of adult bees; see **tracheal mite**.

acarology The study of mites.

Africanized honey bee The common name for *Apis mellifera scutellata* hybrids inhabiting the Americas that have retained much of their defensive characteristics.

afterswarms Swarms that leave the hive after the first, or prime, swarm has departed, accompanied by a virgin queen. Such swarms in the order of their departure are often referred to as secondary, tertiary, etc., swarms.

alarm odor or pheromone A chemical substance released from the vicinity of a worker bee's sting that alerts other bees to danger. Isopentyl acetate is the principle alarm odor; 2-heptanone, found in the worker bee's mandibular glands, is a secondary alarm odor.

allergic reaction A systemic or general reaction to bee venom.

American foulbrood (AFB) disease A brood disease of bees that affects the late larval and prepupal stages and is caused by the bacterium *Paenibacillus* (=*Bacillus*) *larvae* spp. *larvae.*

antenna (*pl.* antennae) One of two long, segmented, sensory filaments extending from the head of insects; includes taste and odor receptors.

apiary The location and sum total of honey bee colonies in a single location; a beeyard.

apiculture The science and art of raising honey bees for economic benefit.

Apis cerana The scientific name for Indian or Asian honey bees. This species is found throughout Asia, and in certain countries it is the honey bee of commerce. It is the natural host of the varroa mite.

Apis dorsata The scientific name for the largest of the species of honey bees, often referred to as the giant bee. This species is found only in Southeast Asia.

Apis florea The scientific name for the smallest of the species of honey bees, often called the dwarf bee. This species is found in Asia, though it is more western in distribution than the other Asian species.

Apis mellifera The scientific name for the European honey bee, which is found throughout the Western World though originally it is thought to be Near Eastern in origin. This bee has been carried from Europe to all areas of the world except the Arctic and Antarctica.

Apistan strips Plastic strips impregnated with the pesticide tau-fluvalinate, used to control varroa mites; however, it is currently not effective in controlling varroa in most areas because the mites have become resistant to this pesticide.

***Bacillus* larvae (now *Paenibacillus larvae* spp. *larvae*)** The old scientific name of the organism (bacterium) that causes American foulbrood disease.

ball a queen   An attack on a queen by a number of worker bees intent on killing her by pulling at her legs and wings and then suffocating and overheating her. In this process the bees form a small cluster or ball of bees around the queen.

bee bread   Pollen gathered by bees, mixed with various bacteria, enzymes, and fungi, is converted into a nutritious substance. It is an essential food for larvae and young bees. Bee bread is stored in honeycomb and is often covered with a layer of honey for preservation.

bee cellar   A portion of a dwelling or building made especially for overwintering bee colonies in climates where winters are severe.

bee escape   A device that permits bees to pass only one way, preventing their return. It is used to remove bees from honey supers.

Bee Go   A chemical (a butyrate) repellent used with a fume board to remove bees from honey supers.

bee louse   The common name for *Braula coeca* (Diptera: Braulidae), a flightless fly found only on honey bees. The fly steals food from a bee's mouth but is not considered a serious problem. Since the introduction of the varroa mite and the chemicals used to control it, the bee louse is rarely seen.

bee paralysis   The condition which affects adult bees that are unable to fly or work normally. It is now known to be induced by a virus or carried by parasite bee mites, such as the varroa mite.

bee space   A space that permits safe passage for bees on opposite sides of facing combs but is too narrow to encourage comb building and too large to induce propolization by bees. It measures from ¼ to ⅜ inches (6.4–9.5 mm).

bee suit   Coveralls, usually white, that fit over normal clothing with or without an attached veil. This apparel is worn when working with bees.

bee veil   A cloth or wire netting worn with a hat or helmet for protecting the head and neck from stinging insects.

beehive   A box or receptacle for housing a colony of bees. Modern beehives are made of wood or plastic and adhere to the bee space dimensions.

beeswax   A complex mixture of organic compounds secreted by four pairs of glands on the ventral or underside of the worker bee's abdomen and used by bees for building comb. Its melting point is from 143.6° to 147.2°F (62–64°C).

bottom board   The floor of a beehive.

brace comb or burr comb   Small pieces of comb made as connecting links between two frames or between a comb and the hive itself. Burr comb often does not connect to any other part but is an extension of the comb, often built during a honeyflow.

brood   Immature stages of the bees not yet emerged from their cells (eggs, larvae, and pupae).

brood chamber   The part of the hive interior in which brood is reared; the brood chamber includes one or more hive bodies and the combs within.

brood foundation   Heavier wax foundation sheets, usually wired with vertical wire and used in broodnest frames; designed for frames to be placed in the broodnest.

brood rearing   The raising of young bees from eggs to adults.

broodnest   The part of the hive interior in which brood is reared; this is the warmest part of a colony, since the eggs and larvae must be incubated at around 95°F (35°C).

Buckfast bee   A strain of bees developed by Brother Adam in England and bred for resistance to tracheal mites, disinclination to swarm, hardiness, honey production, and good temperament.

candied honey   Crystallized, granulated, or solidified honey; see **Dyce process**.

capped brood   Larval cells that have been capped over with a brown covering consisting of wax and propolis. Once the cells are capped the larvae spin their cocoons and turn into pupae, the third stage of complete metamorphosis.

cappings   The thin, light-colored wax covering cells full of honey. Once cut from honey frames they are referred to as cappings and make the finest grade of beeswax.

cappings scratcher   A forklike device used to scrape the wax cappings off the sealed honey in order to extract the honey. Also used to remove drone brood to test for the presence of varroa mites.

Carniolan bee   The common name for *Apis mellifera carnica*, a race of honey bee named for Carniola, Austria, but originating in the Balkan region. It is gray-black in color, very gentle, and conserves honey well.

caste   Worker and queen bees are both members of the female gender but differ in both form and function. Drones are male bees with the same form and function, and therefore do not belong to a caste.

Caucasian bee   The common name for *Apis mellifera caucasica*, a race of honey bee originating in the Caucasus Mountains. Grayish in color, they use more propolis than other bee races.

cell   The hexagonal unit-compartment of a comb.

cell bar   A wooden strip on which queen cups are placed for rearing queens.

cellar wintering   The placing of bees in an unheated cellar or special building during the winter months. In some cases these sites are temperature controlled.

chalkbrood   A disease caused by the fungus *Ascosphera* *apis*, which turns larvae into white then gray and black chalk-like mummies.

chilled brood Immature bees that have died from the cold because of insufficient numbers of adult bees to maintain proper incubation temperatures. This may also result from beekeepers inspecting hives during low temperatures or when the forager numbers have been diminished by pesticides.

chorion Membrane or shell covering a bee egg.

chunk comb honey Honey cut from the comb and placed with liquid honey in glass containers.

clarifying The removal of any foreign material or wax from honey usually by allowing the honey to settle, at either room or a higher temperature. Wax and most other material will float to the top and can be skimmed off.

cleansing flight Bees flying out of the hive during winter and early spring when temperatures are sufficient to permit flight. During these flights bees relive themselves of waste products. It is common when snow is on the ground to find yellow spots over the snow.

cluster The form or arrangement in which bees cling together as a swarm assembles. Most swarm clusters are found on the branches of trees and shrubs. In the winter bees form a cluster inside the hive in order to conserve heat.

colony An aggregate of worker bees, drones, and a queen living together in a hive or in some other dwelling (usually a cavity) as one social unit.

comb A back-to-back arrangement of two series of hexagonal cells made of beeswax to hold eggs, brood, pollen, or honey. There are approximately five worker cells to the linear inch and are about four drone cells to the inch.

comb honey Honey in the comb, produced in small, square wooden sections about 4 ¼ × 4 ¼ inches (11 × 11 cm) in size, in plastic rings (Ross Rounds), or plastic boxes.

compound eyes The two large lateral eyes of the adult honey bee composed of many lens elements called ommatidia.

conical escape An escape board made with cones set in a frame to permit a one-way exit for bees, to clear honey supers of bees.

cremed honey See **Dyce process**.

crystallization Granulation of honey; when honey as a supersaturated solution has candied or become solid instead of liquid. Cremed honey is a commercially made soft form of granulated honey.

cut-comb honey Honeycomb cut to fit small plastic boxes or wrapped individually for retail sale.

deep super Hive furniture that holds standard, full-depth frames; the usual depth is 9½ inches or 9 ⅝ inches (24–25 cm).

Demaree The beekeeper, G. Demaree, who invented a popular method of swarm control in 1884; also used as a verb "to demaree," in describing this method. It consists of separating the queen from most of the brood.

dequeen Removal of a queen from a colony.

dextrose See **glucose**.

diastase An enzyme in honey that assists in the conversion of starch to sugar.

disease resistance When strains of bees selected from stock show high survival despite the presence of diseases.

dividing Separating a colony in such a way as to form two or more colonies.

division board A thin vertical board of the same dimensions as a frame; also called dummy board or follower board. It is used to reduce the size of the brood chamber (to a few frames) or to fill up the gap in a hive body using only nine frames. It is also used to permanently divide the hive into two or more parts.

division board feeder A plastic or wood container the shape of a frame, hung in a hive and filled with syrup to feed bees.

drawn combs Honeycombs having the cell walls built up by honey bees from a sheet of foundation base.

drifting The movement of bees into hives other than their own. Young bees who have not yet learned the location of their hive tend to drift more frequently than do older bees. This phenomenon is common when hives are placed in long rows; in such cases bees tend to drift into hives near the end of rows. On days when wind is a factor, drifting also becomes a problem.

drone The male honey bee, developing from unfertilized eggs, which are haploid, or have half the chromosome numbers. The development of individuals from unfertilized eggs is known technically as *parthenogenesis*.

drone brood Brood that is reared in larger cells and produces drone bees. When drone cells are sealed the cappings have the appearance of bullet heads and can therefore be easily distinguished from the cappings over worker brood and those covering honey.

drone comb Comb having cells measuring about 4 cells to the linear inch or about 18.5 cells to the square inch. Such comb is used for the specific purpose of raising drones or is placed in honey supers to facilitate the extraction of honey and to control varroa.

drone-congregating area (DCA) A specific area to which drones fly year after year waiting for opportunity to mate with virgin queens.

drone layer A queen that lays only unfertilized eggs, resulting in only drone offspring, because she is old, is low on sperm, was improperly mated, was not mated at all, or is diseased or injured.

drumming Pounding on the sides of a hive or other bee dwelling to drive the bees upward. This method is used to transfer bees from bee trees into a bee hive or to drive bees from one hive body into another or completely out of a given hive body.

dwindling The rapid dying off of old bees in the spring (often referred to as spring dwindling). It can be caused by nosema and viral diseases or mite infestation.

Dyce process A patented process involving pasteurization and controlled granulation to produce a finely crystallized or granulated honey product that spreads easily at room temperature; also sometimes called "cremed" honey.

dysentery A condition of adult bees resulting from an accumulation of feces on the inner or outer surface of hive furniture. Dysentery is first detected by finding small spots of feces around the entrance and within the hive, and its presence usually occurs during winter. It is caused by unfavorable wintering conditions that restrict cleansing flights and/or low-quality food; can be confused with nosema disease, caused by a microsporidian fungus.

egg The first stage in the bee's life cycle, usually laid by the queen. It is cylindrical, 1/16 inch (1.6 mm) long, and enclosed with a flexible shell or chorion.

embed To force wire into wax foundation by heat, pressure, or both in order to strengthen the foundation. Such additional support to the foundation keeps it intact, especially when honey is being extracted from the comb.

emerging brood Young or teneral bees chewing their way out of the capped brood cells.

entrance reducers A strip of wood notched with different-sized holes to regulate the size of the hive entrance and hence the flow of bees into or out of a hive. Reducers are useful when robbing in an apiary becomes prevalent. Reducers also restrict mice from entering hives; metal reducers are now available for this purpose.

escape board A board having one or more bee escapes in it that is used to remove bees from honey supers.

European foulbrood A bacterial brood disease of bees caused by *Melissococcus* (=*Streptococcus*) *plutonius*.

extender patties Vegetable shortening and sugar patties with antibiotics added (e.g., Terramycin). This patty has been used to suppress American foulbrood disease but is now not recommended as it can lead to resistance of the bacteria to the antibiotic.

extracted honey Honey removed from combs by means of a centrifugal extractor.

extractor A machine used for removing honey from combs by spinning the frames and throwing the honey against the extractor walls; the combs remain intact.

eyelets Metal pieces that fit in the holes of frame end bars; used to prevent reinforcing wires from cutting into the wood and thus slackening the taut wires.

feeders Various types of appliances and containers for feeding bees sugar syrup.

fermentation A chemical breakdown of high-moisture honey by yeasts; in honey, fermentation is caused only by the genus *Zygosaccharomyces*, a yeast able to tolerate the high sugar content of honey. Fermented honey cannot be reversed and is unsalable.

fertile queen A queen inseminated instrumentally or naturally with drone spermatozoa. By either method, the sperm is stored in her spermatheca. Such a queen is capable of laying fertilized eggs.

fertilization Usually refers to eggs laid by queen bees, which become fertilized when sperm stored in the spermatheca is mixed with the egg while it is being laid. Can also refer to the process where a pollen grain grows down a flower's female stigma to fertilize eggs in the ovary, producing seed and fruit.

festooning The activity of young bees engorged with honey and clinging to one another in a suspended form while secreting wax scales. The function of festooning is unknown.

field bees or foragers Worker bees, usually at least 16 days old, that work in the field to collect pollen, nectar (or rob honey from other hives), honeydew, water, and propolis.

flight path Refers to the direction bees fly leaving their hive. If obstructed by a beekeeper standing in front of the hive entrance, bees will collect behind the obstruction and may become defensive.

food chamber A hive body filled with honey; used as extra food or for winter stores.

foulbrood Malignant, contagious bacterial diseases of honey bee brood; see **American foulbrood** and **European foulbrood**.

foundation A thin sheet of beeswax or plastic, embossed or stamped with the base of a normal worker or drone cell on which the bees will construct a complete or drawn comb.

frame Four pieces of wood (or preformed plastic) that form a rectangle, designed to hold wax or plastic foundation/honey comb. A frame consists of one top bar, a bottom bar (of one or two pieces), and two end bars.

frass Excreta of insects, used especially in reference to moths and butterflies; the black droppings found on comb infested with wax moth larvae.

fructose A simple sugar (or monosaccharide); formerly called levulose (fruit sugar). It is a disaccharide found in honey.

fumagillin   An antibiotic made from a fungus used for the treatment of nosema disease.

fume board   A cloth-coated wooden frame with a metal cover, sprinkled with a bee repellent, such as Bee Go, used to remove bees from honey supers.

Fumidil-B   The older trade name (U.S.) for fumagillin; now made in Canada and labeled Fumagilin-B.

Galleria mellonella L.   The scientific name of the greater wax moth. The larvae of this moth chew and destroy honeycomb. They are used for fish bait and for scientific research.

glucose   A simple sugar; formerly called dextrose (grape sugar). It is one of the two main sugars found in honey and forms most of the solid phase in granulated honey.

grafting   A process of removing newly hatched worker larvae from their cells and placing them in artificial queen cups, for the purpose of rearing queens.

granulation   A term applied to crystallized, candied, creamed, or solidified honey.

grease patty   A patty made with vegetable shortening and sugar; used to control tracheal mites.

gynandromorph   Having both male and female organs and/or body parts.

2-heptanone   A chemical substance produced in the mandibular glands of worker honey bees that elicits an alarm reaction.

head   The front (anterior) part of an insect containing the eyes, antennae, and mouthparts.

hive   (*n.*) A home for bees provided by humans, that is, a hive box. (*v.*) To place a swarm into a hive box.

hive body   A wooden (or plastic) box or rim that contains the frames.

hive stand   A structure that serves as a base support for a hive. Such a stand keeps the bottom board off the damp ground.

hive tool   A metal device with a curled scraping surface at one end and a flat blade at the other end; used to separate hive furniture when inspecting bees, to scrape frames, and to remove frames from the hive.

honey   A sweet, viscous liquid composed of sugars. It is made from flower nectar gathered by the bees, ripened or evaporated into honey, and stored in the combs. Well-cured honey contains about 17 percent water.

honey bee   The common name for *Apis mellifera* (honey bearer). The word is written as two words (in American publications) and as one word in Europe. An arthropod in the class Insecta, order Hymenoptera, and superfamily Apoidea that is a social, honey-collecting insect living in perennial colonies. Also known as *A. mellifica* (honey maker).

honey house   A building used for extracting and packaging honey, storing supers, and so on.

honey stomach, crop, or sac   An enlargement of the back or posterior end of the bee's esophagus that lies in the front part of the abdomen. This organ can hold a large amount of liquid due to its invaginated walls; now found to contain *Lactobacillus* and other beneficial bacteria.

honeyflow   Loosely, a time of year when there is a plentiful supply of nectar that bees can collect. Its signs include fresh, white wax and combs filled with liquid. It is a time when bees produce and store surplus honey.

hybrid queen   The result of crossing different bee races or lines to produce a queen of superior qualities.

hypopharyngeal glands   A pair of organs located in the head of a worker bee that produce brood food and royal jelly.

increase   To add to the number of existing colonies in an apiary, by dividing those already on hand or in hiving swarms or packages.

infertile   As in worker bees, not able to lay fertile eggs. In plants, it means unable to reproduce.

inner cover   A lightweight cover with an oblong hole in its center; used under a standard telescoping outer cover on a bee hive.

instrumental insemination   The introduction of drone spermatozoa into the genital organs of a virgin queen by means of special instruments; sometimes called artificial insemination.

introducing or queen cage   A small box made of wire screen and wood or plastic, used in shipping queens or introducing a new queen to a colony.

invertase   An enzyme that speeds the transformation of sucrose (a complex sugar) into the monosaccharides (or simple sugars) fructose and glucose.

Isle of Wight disease   An early name for acarine disease, the infestation of bees by tracheal mites. It is named after the place where these mites were first discovered in the early 1900s.

isopentyl acetate   The alarm odor in honey bees produced in the sting chamber.

Italian bee   The common name for *Apis mellifera ligustica*, the most common race of European bees used commercially. Introduced from Italy in the 1860s, workers have brown and yellow bands on their abdomen; queens have brown or orange abdomens with few or no stripes.

K-wing   The appearance of the bee's wing in the shape of the letter K, in which the hind wing is held in front or over the fore wing; caused by an infestation of tracheal mites or by a virus, or both.

Lactic Acid Bacteria (LAB)   Several species of bacteria able to tolerate acidic conditions and responsible for fermentation, especially of sugars; special strains are commonly used for probiotics (such as in yogurt). New research has discovered LAB living in the honey stomachs of bees and other pollinators, which could help protect the bees from pathogens and help in the preservation and conversion of their food.

larva (*pl.* larvae)   The second stage in the development of an insect, such as the honey bee, that has complete metamorphosis. It is comparable to the caterpillar (or eating) stage of a moth or butterfly.

laying worker   A worker bee that lays unfertilized eggs, which will develop into drones. Laying workers develop usually in colonies that have been queenless for a long time.

levulose   See **fructose**.

local reaction   Bee stings that elicit a brief, sharp pain at the site(s) of the sting(s). The sting site may swell, turn red, and itch but such symptoms will usually subside in a matter of minutes or hours.

mandibles   The jaws of an insect. In the honey bee and most insects, the mandibles move in a horizontal rather than in a vertical plane. The bees use the jaws to form honeycomb, to scrape pollen, and to pick up hive debris.

mating flight   The flight taken by a virgin queen, during which she mates in the air with one or more drones. Normal queens mate ten to twenty times in two or more mating flights.

mead   A wine made with honey as the main source of food for the yeast.

menthol crystals   Crystalline form of the essential oil of the mint plant *Mentha avensis*; used to control tracheal mites.

metamorphosis   The developing process of most insects, in four stages: egg, larva, pupa, and adult; also called complete metamorphosis.

migratory beekeeping   The moving of bee colonies from one locality to another during a single season to pollinate different crops or to take advantage of more than one honeyflow.

mite   An eight-legged creature or acarine (like the tick and spider). At least three species are currently parasitic on bees: *Tropilaelaps*, tracheal, and varroa mites.

movable frame   A wooden or plastic frame containing honeycomb constructed to provide the bee space between frames. When placed in a hive it remains essentially unattached by brace comb and heavy deposits of propolis, thus permitting its removal with ease.

Nasonov gland   The gland associated with the seventh abdominal tergite of the worker honey bee. This gland is commonly called the scent gland because its contents attract bees to gather in a cluster.

nectar   A sweet plant exudation secreted by special nectary glands containing sugars. It is found chiefly in the flowers or reproductive organ of plants. Extra-floral nectaries are found outside of flowers (on leaves or stems).

nectaries   Specialized tissues contained in organs or glands of plants that secrete nectar.

Nosema apis   The scientific name of a microsporidian parasite (now considered a fungus) of honey bees, causing nosema disease. Since the 1990s, a new species, *N. ceranae*, has been identified and is reported to be more virulent that *N. apis* and may have replaced it in many areas. The parasite *Nosema* spp. is also found in other insects, such as locusts and bumble bees.

nosema disease   An abnormal condition of adult bees resulting from the presence of nosema spores in their intestines; often treated with the antibiotic fumagillin.

nucleus   A small colony of bees often used in queen rearing and called a nuc; comes in three to five frame sizes.

nurse bees   Young worker bees with fully functional food glands whose duty is to feed larvae and the queen and to perform particular hive duties. Generally nurse bees are 3 to 10 days old.

observation hive   A hive made largely of glass or Plexiglas with an outside entrance to permit observation of the bees at work from inside a building.

ocellus (*pl.* ocelli)   A simple eye with a single lens. The honey bee has three ocelli on the top of its head that distinguish light from dark.

ommatidium (*pl.* ommatidia)   One of the visual units or lenses comprising the compound eye.

osmophilic yeasts   Yeasts that occur naturally in honey and are responsible for fermentation in honey that has more than 18 percent water. These yeasts belong to the genus *Zygosaccharomyces*.

out apiary   An apiary kept at a distance from the home of the beekeeper; also called an outyard.

outer cover   The top cover that fits over a hive to protect it from the weather. The two most common covers are migratory and telescoping.

ovary   The egg-producing female organ of a plant or animal.

oxytetracycline   The antibiotic sold as Terramycin, registered to control American foulbrood (AFB). Some report that AFB is resistant to this antibiotic and are using tylosin instead.

package   A special wire-screened shipping box containing a quantity of bees (2–5 lb.) with or without a queen.

Paenibacillus The current scientific genus name of the organism (bacterium) that causes American foulbrood disease; formerly known as *Bacillus*.

paradichlorobenzene (PDB) A crystalline chemical used to control wax moths; also called moth crystals. It is not recommended anymore because of its carcinogenic nature.

parthenogenesis The development of young from unfertilized eggs. In honey bees, the unfertilized eggs are laid by virgin queens, laying workers, or mated queens and produce only drone bees.

pheromone A chemical substance that is released externally by one insect or animal and stimulates a response in other insects (or animals) of the same species.

piping A series of sounds, a loud shrill tone followed by shorter ones, made by queens. These sounds are usually made by newly emerged virgin queens to illicit quacking from queens still in their cells, which enables her to locate and destroy them.

play flights Short flights taken in front of the hive and in its vicinity to acquaint the young bees with their immediate surroundings and hive location; also called orientation flights. These may also have other functions and are sometimes mistaken for robbing or preparation for swarming. They are common during late afternoon.

pollen The dustlike male reproductive cell bodies of flowers formed on the anthers and collected by bees. Bees collect pollen and turn it into bee bread by a complex fermentation process; it provides the protein part of the honey bee's diet.

pollen basket or corbicula A flattened depression surrounded by curved spines located on the outside of the tibiae of the bees third set of legs. It is used to carry pollen gathered from flowers back to the hive where the pollen pellets are deposited into cells and packed together as the bees ram their heads against the pellets. This same basket is also used by bees to collect and transport propolis back to the hive.

pollen insert A device inserted into the entrance of a colony into which hand-collected pollen is placed. As the bees leave the hive and pass through the trap, some of the pollen adheres to their bodies. This allows the pollen to be carried to the target blossoms, resulting in cross-pollination.

pollen patty A cake or patty made of pollen pellets and sugar syrup. These patties are fed to stimulate brood rearing.

pollen substitute A food material used to substitute wholly for pollen in the bees' diet; commonly made from soy flour or other products.

pollen supplement A bee food that is mixed with pollen to augment the bees' diet. It can contain brewer's yeast (distiller's soluble from a brewer), soybean flour, natural pollen, and other ingredients formulated in different ways to be digestible to bees.

pollen trap A device for removing pollen from the pollen baskets of bees as they return to their hives.

pollination The transfer of pollen from the anthers (male part) to the stigma (female) of flowers.

pollinator The agent that transmits the pollen.

pollinizer The plant that furnishes pollen. Crab apple trees are often pollinizers in apple orchards.

prime swarm The first swarm to issue from the parent colony; usually contains the old queen.

proboscis A structure formed from the free parts of the bees' maxillae and labium, forming a tube for ingesting nectar, honey, honeydew, and water.

propodeum The first abdominal segment fused to the bee's thorax, which connects the thorax to the abdomen and is typical "wasp waist" of the Apocrita suborder that includes bees, ants, and wasps. The propodeum has a pair of spiracles.

propolis or bee glue A gluey or resinous material that bees collect from trees and plants and use to strengthen the comb or seal cracks. It has antimicrobial activity.

protein Naturally occurring, complex organic macromolecule that contains amino acids. Pollen contains protein, which is an important nutrient to developing bees.

prothoracic/mesothorax spiracle This spiracle lies between the pro- and mesothorax. It is the largest of the spiracles; by way of this spiracle the parasitic mite *Acarapis woodi* enters the tracheal system of the bees.

pupa (*pl.* pupae) The third stage in the development of an insect that is encapsulated in a cocoon. In this stage, the organs of the larva are replaced by those that will be used as an adult.

queen A fully developed mated female bee. The queen is larger and longer than a worker bee and is recognized by workers because of special pheromones.

queen cage A small cage in which a queen and five or six worker bees may be confined; used for shipping queens and usually contains a candy plug.

queen cell A special, elongated cell suspended vertically from honeycomb and resembling a peanut shell in which a queen bee is being raised. It is usually an inch or more in length when fully developed and capped.

queen cup A cup-shaped cell produced by bees and suspended vertically from the honey comb that may eventually develop into a queen cell. These cups can also be obtained commercially or produced by individuals using a mold. Commercial cups are made either of beeswax or plastic. These cups become queen cells when a queen deposits an egg in them or when a queen breeder transfers a young larva into the cup. These cups are also suspended in a vertical position.

queen excluder   A device made of wire, wood and wire, plastic, or punched plastic, having openings of about 0.16 to 0.17 inch (0.41–0.42 cm). This permits workers to pass through but excludes queens and drones. It is used to confine the queen to a specific part of the hive, usually the brood chamber.

queenright   A bee colony having a laying queen.

rabbet   A narrow ledge cut into the top ends of hive bodies on which the frames hang. Some rabbets are cut so that resting frames will be at the right bee space to the top of the box; others are lower, requiring a metal strip to correct the bee space.

rendering wax   The process of melting combs and cappings to separate wax from its impurities and thus refining the beeswax. Wax is usually melted using a hot water tank or solar wax melter.

requeen   To place a new queen into a hive made queenless.

reversing   To exchange places of different hive bodies of the same colony, usually to expand the nest. Done in the spring, the upper hive box full of bees and brood is reversed with the lower, emptier box on the bottom board.

robbing   Applied to bees stealing honey/nectar from other colonies.

round sections   Section comb honey made in plastic rings instead of square wooden or plastic boxes. Also called Ross Rounds.

royal jelly   A highly nutritious glandular secretion of young bees; used to feed the queen and the young queen larvae.

sacbrood   A brood disease of bees caused by a filterable virus.

scent gland   See **Nasonov gland**.

sex attractant   A chemical substance that attracts an animal of the same species, male or female, for the purpose of mating. The sex attractant in *Apis mellifera* is [t]9-oxo-2-enoic acid (9-ODA) from queen mandibular gland.

shallow super   Any one of several super sizes less than the depth of a deep super. Commonly, shallow supers vary from 4¼ to 6¼ inches (11–16 cm) in depth.

skep   A dome-shaped beehive without movable frames, usually made of twisted straw. The use of skeps for keeping bees is illegal in most states in the United States.

slumgum   The refuse from melted comb and cappings after the wax has been rendered or removed.

small hive beetle   The common name for *Aethina tumida*, the beetle first identified in Florida in 1998. Before its discovery in the United States, the beetle was known to exist only in tropical or subtropical areas of Africa. How it found its way to North America is not known.

smoker   A metal container with attached bellows that burns organic fuels to generate smoke; used to control the defensive behavior of bees during routine colony inspections.

solar wax melter   A glass-covered insulated box used to melt wax from combs and cappings, by using the heat of the sun.

spermatheca   A small sac connected with the oviduct of the queen and in which is stored the spermatozoa received during mating with drones or by instrumental insemination.

spermatozoa   Male reproduction cells (gametes) that fertilize eggs.

spiracles   Openings in the body wall connected to the tracheal tubes through which insects breathe.

split   (*v.*) To divide the components of a hive and its population of bees to form a new colony. (*n.*) A colony divided, thus increasing the number of hives.

Starline hybrid   An Italian strain, crossbred for vigor, honey production, and prolific populations of bees. Not commonly found anymore.

sting   An organ of defense of workers and queen bees. It is an egg-laying device (or ovipositor) modified to form a piercing shaft, through which painful organic venom is delivered.

super   A piece of hive furniture in which bees store surplus honey; so called because it is placed over or above the brood chamber.

supering   The act of placing supers on a colony in anticipation of a honey crop.

supersede   To rear a young queen that will replace the mother queen in the same hive. Shortly after the young queen starts to lay eggs, the old queen usually disappears.

surplus honey   The honey removed from a hive, in excess of what bees need for their own use, such as winter food stores.

swarm   The aggregate of worker bees, drones, and a queen that leave the original colony to establish a new one. Swarming is the natural propagation method to form a new bee colony.

swarming season   The time of year, usually late spring, when swarms normally occur.

systemic reaction   A reaction from a bee sting(s) that can be life-threatening and require immediate medical attention. Such a reaction is far more serious than stings that elicit pain at the site of the sting and symptoms include: urticaria (hives), throat tightness, difficulty breathing, and a drop in blood pressure. An EpiPen is often used when someone has a systemic reaction, followed by a trip to the hospital.

Terramycin   The trade name of an antibiotic used to combat European or American foulbrood disease; generic name is oxytetracyline. Research has shown that AFB is now resistant to this antibiotic in many areas.

thin comb-honey foundation   A thin, wireless wax foundation that is used for section comb or chunk honey production.

thorax   The central or second region, between the head and abdomen, of the bee's body that supports the wings and legs.

top bar   The top part of a frame.

trachea (*pl*. tracheae)   The breathing tubes of an insect opening into the spiracles.

tracheal mite   The common name for *Acarapis woodi* (Rennie) (Acari: Tarsonemidae), a mite that inhabits the trachea of adult honey bees. It can be controlled by grease patties and menthol crystals.

travel stain   The darkened or stained surface of comb honey, which is caused by bees walking on its surface over a long time; usually from propolis.

***Tropilaelaps* spp.**   The scientific name for one of two common mites associated with Asian honey bees; the other is *Euvarro*a spp. These mites are not currently present in the New World.

tylosin   Newer antibiotic used for treating colonies with American foulbrood disease, sold as Tylan.

uncapping knife   A knife with a sharp blade used to shave or remove the cappings from combs of sealed honey before extracting. The knives are usually heated by steam, hot water, or electricity.

unfertilized   An ovum (egg) that has not been united with the sperm. Insect eggs not fertilized usually become males.

uniting   Combining two or more weak colonies to form a large one. To prevent fighting between colonies, a sheet of newspaper is placed between the colonies to separate them until they become familiar with each other's scent by means of tearing down the paper.

unripe honey   Honey that is more like nectar, containing over 18 percent water.

unsealed brood or open brood   Immature or larval bees not yet capped over with wax; the term can include cells containing eggs.

***Varroa destructor* Truman and Anderson 2000**   The scientific name for the varroa mite; a destructive parasitic mite (eight-legged) that feeds on brood but is carried by adult bees. Introduced into the United States in 1986 and formerly called *Varroa jacobsoni* Oudemans, this mite is now found worldwide, most recently in Kenya and Uganda.

Varroosis   A disease due to the actions of the varroa mites feeding on honey bees.

ventriculus   The stomach of the bee, located in the abdomen behind the honey stomach but before the hindgut; also called the midgut.

virgin queen   An unmated queen. If she remains unmated due to a variety of circumstances, she will be capable of laying only unfertilized eggs. These eggs will then yield only drone honey bees.

virus (Latin for *toxin*)   An infectious agent (100 times smaller than bacteria) that must grow and reproduce in a host cell; viruses infect all cellular life. Tobacco mosaic virus was the first one discovered, in 1899. Currently 5000 viruses have been described; honey bees have about 18 known viruses.

wasp   A close relative of a honey bee, usually in the genus *Vespula*. Wasps are carnivorous, and some species prey on bees.

wax bloom   A powdery coating forming on the surface of beeswax. It is composed of volatile components of beeswax.

wax glands   The eight glands that secrete beeswax. They are located in pairs on the last four visible ventral abdominal segments (sternites) of young workers.

wax moth   The common name for *Galleria mellonella* L., a moth whose larvae eat comb, pollen, and pupae.

wax scale   A drop of liquid beeswax that hardens into a scale; so named because it has the appearance of a fish scale. These scales, once extruded by the bee's wax glands, harden on contact with the air and serve as the building blocks of honeycomb.

windbreaks   Specifically constructed fences or natural barriers to reduce the force of wind in an apiary during cold weather as well as to reduce drifting in areas of frequent prevailing winds or breezes.

winter cluster   The arrangement or organization of adult bees within the hive during the winter period.

winter hardiness   The ability of some strains of bees to survive long winters by frugal use of honey stores and low bee populations.

wired foundation   Wax foundation containing embedded vertical wires to prevent the finished drawn comb from sagging.

worker bee   A female bee whose organs of reproduction are only partially developed. Workers are responsible for carrying on all the routine tasks of a bee colony.

References

Note: The books and articles listed below are only a small representation of what is available and are just to get you started. Many books are now out of date for some information, and most are out of print. With the Internet technology, many of these citations can be accessed through your computer or at Internet libraries. Also, by using a search engine, such as Google, you can find a plethora of information on each of the topics listed below; however, beware: some of the information may not be very accurate. Read with discrimination.

ABBREVIATIONS

ABJ = American Bee Journal
Ag.= Agriculture
Annu. Rev. Entomol. = Annual Review of Entomology
Ann.= Annals
Apic. Res. = Apicultural Research
ARS = Agriculture Research Service
Bee Wld. = Bee World
CES = Cooperative Extension Service
Dept. Entomol. = Department of Entomology
(ed.) = editor(s)
edn. = edition
Entomol. = Entomology
Exp. Sta. = Experimental Station
Ext. Serv. = Extension Service
Exp. Appl. Acarol. = Experimental and Applied Acarology
IBRA = International Bee Research Association
J. = Journal
JAR = Journal of Apicultural Research
n.d. = no date
Pub. = Publication
Tech. Bull. = Technical Bulletin
Univ. = University

AFRICANIZED BEES

Ambrose, J.T., and D.R. Tarpy. 2007. Africanized honey bees: Some questions and answers. Raleigh: North Carolina CES.

Breed, M.D., E. Guzmán-Novoa, and G.J. Hunt. 2004. Defensive behavior of honey bees: Organization, genetics, and comparisons with other bees. Annu. Rev. Entomol. 49:271–298.

Caron, D.M. 2001. Africanized honey bees in the Americas. Medina, OH: A.I. Root.

Davis, D. 2000. Working safely in areas with Africanized honey bees. Missoula, MT: USDA Forest Service Technology and Development Program.

Guzmán-Novoa, E., G.J. Hunt, J.L. Uribe-Rubio, and D. Prieto-Merlos. 2004. Genotypic effects of honey bee (*Apis mellifera*) defensive behavior at the individual and colony levels: The relationship of guarding, pursuing and stinging. Apidologie 35 (1): 15–24.

Guzman-Novoa, E., and J.L. Uribe-Rubio. 2004. Honey production by European, Africanized and hybrid honey bee (*Apis mellifera*) colonies in Mexico. ABJ 144 (4): 318–320.

Hopkins, J.D. 2005. Africanized honey bees: How to bee-proof your home. Little Rock: Univ. of Arkansas Division of Ag., CES.

Hunt, G.J., E. Guzman-Novoa, J.L. Uribe-Rubio, and D. Prieto-Merlos. 2003. Genotype-environment interactions in honeybee guarding behavior. Animal Behaviour 66 (3): 459–467.

Pinto, M.A., W.S. Sheppard, J.S. Johnston, W.L. Rubink, et al. 2007. Honey bees (Hymenoptera: Apidae) of African origin exist in non-Africanized areas of the southern United States: Evidence from mitochondrial DNA. Ann. Entomol. Society of America 100 (2): 289–295. http://hdl.handle.net/10113/2313.

Schneider, S.S., G. DeGrandi-Hoffman, and D.R. Smith. 2004. The African honey bee: Factors contributing to a successful biological invasion. Annu. Rev. Entomol. 49: 351–376. doi:10.1146/annurev. ento.49.061802.123359.

Websites

Africanized honey bee home page. http://128.194.30.1/ agcom/news/hc/ahb/ahbhome.htm.

Tucson Bee Lab. http://gears.tucson.ars.ag.gov.

USDA map of AHB spread: http://ars.usda.gov/ Research/docs.htm?docid=11059&page=6Dept of Ag., Florida.

BEE BOOK SELLERS

International Bee Research Association (IBRA), 16 North Road, Cardiff CF1 3DY, Wales, UK. http://ibrastore.org.uk/index. php?main_page=index&cPath=1_2.

Northern Bee Books, Scout Bottom Farm, Mytholmroyd, Hebden Bridge, West Yorkshire, HX7 5JS UK. ☎ +44 (0)1422 882751, Fax +44 (0)1422 886157. E-mail: sales@beedata.com.

Wicwas Press, Larry Connor, 1620 Miller Road, Kalamazoo, MI 49001. ☎ (269) 344-8027, Cell: (203) 435-0238. E-mail: LJConnor@aol.com.

Also, bee supply companies sell books.

BEE JOURNALS, ORGANIZATIONS, AND PUBLICATIONS

Abeille de France. 5 Rue de Copenhague, F-75008 Paris, France. http://www.abeille-de-france.com.

ABF Newsletter. American Beekeeping Federation, PO Box 1337, Jesup, GA 31598 USA. http://www.abfnet. net.

American Bee Journal. Dadant & Sons, Hamilton, IL 62341 USA. http://www.dadant.com.

Apiacta. Publication of Apimondia, Corso Vittorio Emanuele II, 101, 1-00186 Rome, Italy. http://www. apimondia.org.

Apidologie. EDP Sciences, 7 av du Hoggar, BP 112, P.A. de Courtaboeuf, 91944 Les Ulis cedex A, France. http://www.edpsciences.org.

The Australasian Beekeeper. P.M.B. 19 Gardiners Road, Maitland 2320, Australia.

Bee Craft. C. Waring (ed.), Stoneycroft, Back ane, Little Addington, Kettering NN14 4AX, UK. http://www. bee-craft.com. (There is also an electronic version available in the United States: Bee Craft America.)

Bee Culture. PO Box 706, Medina, OH 44258 USA. http://www.beeculture.com.

The Beekeepers Quarterly. Northern Bee Books, Scout Bottom Farm, Mytholmroyd, Hebden Bridge, W. Yorkshire HX7 5JS, UK. http://www.beedata.com.

Bees for Development. PO Box 105, Monmouth NP25 9AA, UK. http://www.beesfordevelopment.org.

Entomological Society of America. Publishes periodicals and books. 10001 Derekwood Lane, Suite 100, Lanham, MD 20706 USA. http://www.entsoc.org.

Hivelights. Canadian Honey Council, Suite 236, 234-5149 Country Hills Blvd NW, Calgary AB T3A 5KB, Canada. http://www.honeycouncil.ca.

Insectes Sociaux. Birkhauser Boston, 675 Massachusetts Avenue, Cambridge, MA 02139 USA. http://www. birkhauser.com.

Irish Beekeeper, *An Beachaire*. David Lee, Manager, Scart, Kildorrery, Co. Cork, Ireland.

JAR and Bee World. International Bee Research Association, 16 North Road, Cardiff CF1 3DY, Wales, UK. http://www.ibra.org.uk.

National Honey Market News. 21 N 1st Avenue, Suite 224, Yakima, WA 98902 USA. http://www.beesource. com.

The Speedy Bee. PO Box 1317, Jesup, GA 31598. ☎ (912) 427-4018, Fax (912) 427-8447. Recently converted to being published quarterly rather than monthly. http://thespeedybee.com. E-mail: speedybee@bellsouth.net.

Also many states or local associations have their own newsletters. Check state websites.

BEE LABORATORIES (NATIONAL AND PROVINCIAL)

Arizona: USDA-ARS Carl Hayden Honey Bee Research Center, 2000 E. Allen Rd., Tucson, AZ 85719. ☎ (520) 670-6380. http://gears.tucson.ars.ag.gov.

Canada: Agriculture Canada, Research Branch, Box 29, Beaverlodge, AB, Canada T0H 0C0. ☎ (403) 354-2212.

Canada: Biosystematics Research Centre, Room B149, K.W. Neatby Bldg., Ottawa, ON, Canada K1A OC6. ☎ (613) 996-1665.

Louisiana: UDSA-ARS Honey Bee Breeding and Genetics and Physiology Lab, 1157 Ben Hur Rd., Baton Rouge, LA 70820-5502. ☎ (225) 767-9280. http://www.ars.usda.gov/Main/site_main. htm?modecode=64-13-30-00.

Maryland: USDA-ARS Bee Research Lab, 10300 Baltimore Blvd., Room 100, Bldg. 476, BARC-EAST, Beltsville, MD 20705-2350. ☎ (301) 504-8205. http://www.ars.usda.gov/main/site_main. htm?modecode=12-75-05-00. Sending samples: http://www.ba.ars.usda.gov/psi/brl/directs.htm.

Texas: USDA-ARS Kika de la Garza Subtropical Agricultural Research Center (KSARC), 2413 East Hwy. 83, Bldg. 200, Weslaco, TX 78595. ☎ (956) 969-5005. http://www.ars.usda.gov/main/site_main. htm?modecode=62040000.

Utah: USDA-ARS NPA Pollinating Insect Biology, Management, Systematics Research (Non-*Apis* bees), Natural Resources Biology Bldg., Utah State Univ., Logan, UT 84322-5310. ☎ (435) 797-0530. http://www.ars.usda.gov/main/site_main.htm?modecode=54280500.

BEE LIBRARIES AND MUSEUMS

Cornell Univ., Everett F. Phillips Library (M. Quinby, L.L. Langstroth, and Dyce Collections), Ithaca, NY 14850. http://bees.library.cornell.edu/b/bees/browse.html.

Crane, E. 1979. Directory of the world's beekeeping museums. Bee Wld. 60:9–23.

Dadant and Sons, Hamilton, IL 62341.

Florida Department of Agriculture and Consumer Science, Division of Plant Industry Library, PO Box 1269, Gainesville, FL 32001.

International Bee Research Association, 18 North Road, Cardiff, CF1 3DY Wales, UK.

Michigan State Univ., Special Collections Division, East Lansing, MI 48824.

Smithsonian Institution Libraries, Entomology Branch, Washington, DC 20560.

Univ. of California, Davis, Shields Library, Department of Special Collections (J.S. Harbison, J.E. Eckert Collection) Davis, CA 95616.

Univ. of Connecticut, Storrs (P.J. Hewitt Collection), 32 Hillside Ave., Waterbury, CT 06710.

Univ. of Guelph, McLaughlin Library (B.N. Gates Collection), Guelph, Ontario N1G 2W1, Canada.

Univ. of Massachusetts, Morrill Library (C.C. Crampton Collection), Amherst, MA 01003.

Univ. of Minnesota, Entomology Library (F.F. Jaeger Collection), St. Paul, MN 55108.

Univ. of Wisconsin, Steenbock Memorial Library (C.C. Miller Collection), Madison, WI 53706.

USDA National Agricultural Library, Beneficial Insects Branch, Technical Information Systems, Beltsville, MD 20705.

BEE AND OTHER ORGANIZATIONS, NORTH AMERICAN WEBSITES

4H Clubs. http://www.4-h.org.

American Apitherapy Society. E-mail: aasoffice@apitherapy.org.

American Association of Professional Apiculturists. http://www.masterbeekeeper.org/aapa/.

American Beekeeping Federation. http://www.abfnet.org.

American Honey Producers Association. http://www.americanhoneyproducers.org.

Apiary Inspectors of America. http://www.apiaryinspectors.org.

Canadian Association of Professional Apiculturists. http://www.capabees.ca.

Canadian Honey Council. http://www.honeycounsil.ca.

Eastern Apicultural Society. http://www.easternapiculture.org.

Heartland Apicultural Society. http://www.heartlandbees.com.

International Mead Association. http://www.meadfest.org.

National Council for Agricultural Education. http://www.agedhq.org/councilindex.cfm.

National FFA Organization (originally the Future Farmers of America). http://www.ffa.org.

National Honey Board. http://www.honey.com.

National Honey Packers and Dealers Association. http://www.mytradeassociation/org/nhpda.

Sonoran Arthropod Studies Institute. http://www./'sasionline.org.

State beekeeping organizations. Check with state agriculture department and Cooperative Extension Service (CES).

State Honey and Beekeeping Organizations. http://www.honeyo.com/org-US_State.shtml.

United States Department of Agriculture Cooperative State Research, Education and Extension Service. http://www.csrees.esda.gov.

Western Apicultural Society. http://www.groups.ucanr/WAS.

Xerces Society. http://www.xerces.org.

Young Entomologists' Society. http://members.aol.com/yesbugs/bugclub.html.

BEES, BEEKEEPING, AND BEE MANAGEMENT

Bonney, R.E. 1990. Hive management—A seasonal guide for beekeepers. Charlotte, VT: Garden Way Publishing.

Brown, R. 1998. Honey bees: A guide to management. Marlborough, Wiltshire, UK: Crowood. E-mail: enquiries@crowood.com.

Buchmann, S.L., and B. Repplier. 2005. Letters from the hive: An intimate history of bees, honey, and humankind. New York: Bantam Books.

Caron, D.M. 1999. Honey bee biology and beekeeping. Kalamazoo, MI: Wicwas Press.

Carreck, N., and T. Johnson. 2007. Aspects of sociality in insects. Upminster, Essex, UK: Central Association of Beekeepers. http://www.cabk.org.uk/publications.htm.

Carter, G.A.J. 2004. Beekeeping: A guide to the better understanding of bees, their diseases and the chemistry of beekeeping. Delhi: Biotech Books.

Collison, C.H. 2001. The Fundamentals of Beekeeping. Penn State Univ. College of Ag., Ext. Serv., University Park, PA. 80 pp.

Collison, C.H. 2003. What do you know? Everything you've ever wanted to know about honey bees, beekeeping. Medina, OH: A.I. Root.

Conrad, R. 2007. Natural beekeeping: Organic approaches to modern apiculture. White River Junction, VT: Chelsea Green.

Corona, M., et al. 2007. Vitellogenin, juvenile hormone, insulin signaling, and queen honey bee longevity. Proceedings of the National Academy of Sciences of the United States of America 104:7128–7133.

Crane, E. 1990. Bees and beekeeping: Science, practice, and world resources. Ithaca, NY: Comstock.

Dade, H.A. 2009. Anatomy and dissection of the honeybee. Cardiff, Wales, UK: Bee Research Association.

Danforth, B.N., S. Sipes, J. Fang, and S.G. Brady. 2006. The history of early bee diversification based on five genes plus morphology. Proceedings of the National Academy of Sciences of the United States of America 103:15118–15123.

Delaplane, K. 2007. First lessons in beekeeping. Hamilton, IL: Dadant and Sons.

Delaplane, K.S. 1996. Honey bees and beekeeping: A year in the life of an apiary (3rd edn.) (spiral-bound). Athens: Univ. of Georgia, Georgia Center for Continuing Education, CES.

Erickson, E.H., Jr., S.D. Carlson, and M.B. Garment. 1986. A scanning electron microscope atlas of the honey bee. Ames: Iowa State Univ. Press.

Evans, J. 1989. The complete guide to beekeeping. London, UK: Unwin Hyman Ltd. 192 pp.

Evans, E.C., and C.A. Butler. 2010. Why do bees buzz? New Brunswick, NJ: Rutgers Univ. Press. 229 pp.

Ferrari, S., M. Silva, M. Guarino, and D. Berckmans. 2008. Monitoring of swarming sounds in bee hives for early detection of the swarming period. Computers and Electronics in Agriculture 64 (1): 72–77.

Flottum, K. 2002. Honey as a crop. Small Farm Today 19 (2): 48–49.

Flottum, K. 2008. The backyard beekeeper: An absolute beginner's guide to keeping bees in your yard and garden. Medina, OH: A.I. Root.

Frazier, M., and D. Caron. 2004. Beekeeping basics. Penn. State Univ. College Agric. Sci., Coop. Ext., 98 pp.

Fry, S.N., and R. Wehner. 2002. Honey bees store landmarks in an egocentric frame of reference. J. of Comparative Physiology, A, Sensory, Neural, and Behavioral Physiology 187 (12): 1009–1016.

Goodman, L. 2003. Form and function in the honey bee. Cardiff, UK: IBRA. http://www.ibra.org.uk.

Gould, J.L., and C. Gould. 1995. The honey bee (2nd edn.). New York: Scientific American Library, W.H. Freeman.

Graham, J.M. (ed.). 1992. The hive and the honey bee. Hamilton, IL: Dadant and Sons. (Soon to be updated.)

Guler, A. 2008. The effects of the shook swarm technique on honey bee (Apis mellifera L.) colony productivity and honey quality. JAR 47 (1): 27–34.

Hamdan, K. 2010. The phenomenon of bees bearding. Bee Wld. 87:22–23.

Harris, J.L. 2008. Development of honey bee colonies initiated from package bees on the northern Great Plains of North America. JAR 47 (2): 141–150. Worksheet at http://www.umaniataoba.ca/afs/entomology/links.

Hooper, T. 1997. Guide to bees and honey. Somerset, UK: Marston House, Yeovil.

Horn, T. 2005. Bees in America: How the honey bee shaped a nation. Lexington: Univ. Press of Kentucky.

Jones, J.C., P. Helliwell, M. Beekman, R. Maleszka, B.P. Oldroyd. 2005. The effects of rearing temperature on developmental stability and learning and memory in the honey bee, Apis mellifera. J. Comp Physiol A (2005) 191: 1121–1129. doi: 10.1007/s00359-005-0035-z.

Jones, J.C., M.R. Myerscough, S. Graham, and B.P. Oldroyd. 2004. Honey bee nest thermoregulation: Diversity promotes stability. Science 305 (5682): 402–404.

Kidd, S.M. 2002. The secret life of bees. New York: Penguin Books.

Kritsky, G. 2010. The quest for the perfect hive: A history of innovation in bee culture. Oxford: Oxford University Press.

Langstroth, L.L. 2004. Langstroth's hive and the honeybee: The classic beekeeper's manual. Mineola, NY: Dover.

Luening, R.A., and W.L. Gojmerac. n.d. Beekeeping records. Madison: Univ. of Wisconsin Extension A2655.

Lynn, R.C., and T. Cooney. 2003. Raising healthy honey bees. Seattle: Christian Veterinary Mission.

Martin, E.C. 1980. Beekeeping in the United States. USDA Ag. Handbook 335.

Menzel, R., et al. 2005. Honey bees navigate according to a map-like spatial memory. Proceedings of the National Academy of Sciences of the United States of America 102 (8): 3040–3045.

Milius, S. 2009. Fossil shows first all-American honeybee. ScienceNews. Web edn.: 7/23/09. http://www.sciencenews.org/view/generic/id/45857/title/Fossil_shows_first_all-American_.

Mitchener, C.D. 2000. The bees of the world. Baltimore: Johns Hopkins Univ. Press.

Mitcher, C.D., and D. Grimaldi. 1988. The oldest fossil bee: Apoid history, evolution stasis, and antiquity of social behavior. Proceedings of the National Academy of Sciences of the United States of America 85:6424–6426.

Oldroyd, B.P., S. Wongsiri, and T.D. Seeley. 2006. Asian honey bees: Biology, conservation, and human interactions. Cambridge, MA: Harvard Univ. Press.

Peacock, P. 2008. Keeping Bees: a complete practical guide. Neptune City, NJ: TFH Pub., Inc. 144 pp.

Pierce, A.L., L.A. Lewis, and S.S. Schneider. 2007. The use of the vibration signal and worker piping to influence queen behavior during swarming in honey bees, *Apis mellifera*. Ethology 113:267–275.

Poinar, G.O., Jr., and B.N. Danforth. 2006. A fossil bee from early cretaceous Burmese amber. Science 314:614.

Remolina, S.C., et al. 2007. Senescence in the worker honey bee *Apis mellifera*. J. of Insect Physiology 53:1027–1033.

Ruttner, F. 1988. Biogeography and Taxonomy of Honeybees. Berlin: Springer-Verlag.

Scott, H. 1999. Bee lessons: Think bees, thank natural life, and bee happy. Chapel Hill, NC: Professional Press.

Scott-Dupree, C.D. 1998. The complete beekeeper. Guelph, ON: Univ. of Guelph Press.

Seeley, T.D. 1996. The wisdom of the hive: The social physiology of bee colonies. Cambridge, MA: Harvard Univ. Press.

Seeley, T.D. 2010. Honeybee democracy. Princeton, NJ: Princeton University Press.

Sexton, C.A. 2007. Honey bees. Minneapolis: Bellwether Media.

Shimanuki, H., K. Flottum, and A. Harman (eds.). 2007. The ABC and XYZ of bee culture: An encyclopedia of beekeeping (41st edn.). Medina, OH: A.I. Root.

Skinner, J.A. 2004. Beekeeping in Tennessee. Knoxville: Univ. of Tennessee, Ag. Ext. Serv.

Specht, S. 2009. Secrets of the hive. Resource: Engineering and Technology for Sustainable World 16 (2): 13–15.

Spencer, H. 2007. The honeybee: The most profitable thing you can have. Small Farm Today 24 (1): 39–40.

Strange, J.P., R.P. Cicciaarelli, and N.W. Calderone. 2008. What's in that package? An evaluation of quality of package honey bee (Hymenoptera: Apidae) shipments in the United States. J. of Economics Entomol. 101 (3): 668–673.

Tautz, J. 2008. The buzz about bees. Biology of a superorganism. Berlin: Springer-Verlag.

Tew, J.E. 2001. Beekeeping principles: A manual for the beginner; a guide for the gardener. Clarkson, KY: Walter T. Kelley.

Tew, J.E. 2004. Backyard beekeeping. Auburn: Alabama CES, Alabama A&M Univ. and Auburn Univ.

Thomas, T.D. 2002. Beeing. Guildford, CT: Lyons Press.

Towne, W.F. 1994. Frequency discrimination in the hearing of honey bee (Hymenoptera: Apidae). J. of Insect Behavior 8:281–286.

Traynor, K. 2006. The honey bee's contribution to medicine. ABJ 146 (10): 859–860.

Tsujiuchi, S., E. Sivan-Loukianova, D.F. Eberl, Y. Kitagawa, and T. Kadowaki1. 2007. Dynamic range compression in honey bee auditory system toward waggle dance sounds. PLoS ONE, 2: art. no. e234.

Virtual atlas of the honeybee. On the Web site http://www.neurobiologie.fu-berlin.de/beebrain/. Accessed July 26, 2008.

Waring, A., and C. Waring. 2006. Teach yourself beekeeping. New York: McGraw-Hill.

Whitfield, C.W., et al. 2006. Thrice out of Africa. Ancient and recent expansions of the honey bee, *Apis mellifera*. Science 314:642–645.

Websites

http://www.badbeekeeping.com/weblinks.htm.

Beekeeping software: http://www.beetight.com.

Michigan State Univ. http://cyberbee.net.

http://www.masterbeekeeper.org/aapa/. OhioState: Beekeeping FAQ via WWW. http://www.cis.ohio-state.edu:80/hypertext/faq/usenet/beekeeping-faq/faq.html.

National Honey Board: Bee-l archives/National Honey Board database. IBRA, UK. http://www.cardiff.ac.uk/ibra/index.html.

Yanega, D. Univ. of California, Riverside: http://en.wikipedia.org/wiki/Apis_%28genus%29.

T. Sanford, Bldg. 970, PO Box 110620, Univ. of Florida, Gainesville, FL 32611-0620. E-mail: MTS@gnv.ifas.ufl.edu.

Eric Mussen. California beekeepers newsletter. http://entomology.ucdavis.edu/faculty/mussen/news.cfm.

BEESWAX

Beeswax from the apiary. 1971. Advisory Leaflet 347. Edinburgh: H.M.S.O. Press.

Berthold, R., Jr. 1993. Beeswax crafting. Kalamazoo, MI: Wicwas Press.

Brown, R.H. 1981. Beeswax (2nd edn.). Bee Books New and Old. Burrowbridge, Somerset, UK.

Coggshall, W.L., and R.A. Morse. 1984. Beeswax—Production, harvesting, processing and products. Kalamazoo, MI: Wicwas Press.

Hepburn, H.R. 1986. Honeybees and wax: An experimental natural history. New York: Springer-Verlag.

Websites

http://www.lipidlibrary.co.uk/Lipids/fa_oxy/file.pdf.

http://www.lipidmaps.org/data/get_lm_lipids_dbgif.php?LM_ID=LMFA01010026.

BOOKS ON BEES—HISTORIC

(This list is for collectors of old, out-of-print books)

Alley, H. 1883. The bee-keeper's handy book. Boston: A. Wenham.

Atkins, W., and K. Hawkins. 1924. How to succeed with bees. Watertown, WI: G.B. Lewis.

Bonsels, W. 1922. Adventures of Maya the bee. New York: T. Seltzer.

Casteel, D.B. 1912. The manipulation of the wax scales of the honey bee. Bulletin 161. Washington, DC: USDA.

Coleman, M.L. 1939. Bees in the garden and honey in the larder. New York: Doubleday/Doran.

Comstock, A.B. 1905. How to keep bees. New York: Doubleday/Page.

Cook, A.J. 1888. The beekeeper's guide. East Lansing, MI: Ag. College.

Cowan, T.W. 1908. Wax craft. London, UK: Sampson, Low, Marston.

Dadant, C.P. 1917. First lessons in beekeeping. Hamilton, IL: Dadant and Sons.

Dadant, C.P. 1920. Dadant system of beekeeping. Hamilton, IL: Dadant and Sons.

Dadant, C.P. 1926. Huber's observations on bees. Hamilton, IL: Dadant and Sons.

Dietz, H.F. 1925. Pollination and the honey bee. Indianapolis: W.B. Burford.

Edwards, T. 1923. The lore of the honey-bee. London: Methuen.

Flower, A.B. 1925. Beekeeping up to date. London: Cassell.

Gilman, A. 1929. Practical bee breeding. New York: J.P. Putnam's Sons.

Harrison, C. 1903. The book of the honey bee. London: John Lane/The Bodley Head.

Hawkins, K. 1920. Beekeeping in the south. Hamilton, IL: ABJ.

Herrod-Hempsall, W. 1930. Beekeeping new and old. Vols. I and II. London: British Bee J.

Langstroth, L.L. 1888. Langstroth on the honey-bee. Revised by Dadant. Hamilton, IL: Dadant and Sons.

Latham, A. 1949. Allen Latham's bee book. Hapeville, GA: Hale.

Maeterlinck, M. 1924. Life of the bee. New York: Dodd-Mead.

Manley, R.O.B. 1946. Honey farming. London: Faber and Faber.

Miller, C.C. 1911. Fifty years among the bees. Medina, OH: A.I. Root.

Naile, F. 1942. Life of Langstroth. Ithaca, NY: Cornell Univ. Press.

Pellett, F.C. 1931. The romance of the hive. New York: Abingdon.

Pellett, F.C. 1938. History of American beekeeping. Ames, IA: Collegiate Press.

Phillips, E.F. 1915. Beekeeping. Norwood, MA: Norwood Press.

Phillips, E.F. 1943. Beekeeping. New York: Macmillan.

Root, A.I. 1877. The ABC of bee culture (1st edn.). Medina, OH: A.I. Root.

Sechrist, E.L. 1944. Honey getting. Hamilton, IL: Dadant and Sons.

Snelgrove, L.E. 1946. Swarming, its control and prevention. Weston-Super-Mare, UK: Snelgrove.

Stuart, F.S. 1947. City of bees. New York: McGraw-Hill.

Sturges, A.M. 1924. Practical beekeeping. Philadelphia: David McKay.

Teale, E.W. 1945. The golden throng. New York: Dodd-Mead.

Wedmore, E.B. 1932. A manual of beekeeping for English-speaking beekeepers. London: Longmans, Green.

CANADIAN BEE ASSOCIATIONS AND CONTACTS

Boucher, Claude, poste 302, Provincial Apiarist, Ministère de l'Agriculture, des pêcheries et de l'alimentation du Québec, 675, route Cameron, bureau 101, Ste-Marie, QC G6E 3V7. ☎ (418) 386-8099 télécopieur. E-mail: claude.boucher@mapaq.gouv.qc.ca.

Canadian Honey Council, National Office, 36 High Vale Crescent, Sherwood Park, AB T8A 5J7. ☎ (877) 356-8935 or (780) 570-5930.

Currie, Rob, Univ. Professor, Dept. of Entomology, Univ. of Manitoba, Winnipeg, MB R3T 2N2. ☎ (204) 474-7628. E-mail: rob_currie@umanitoba.ca.

Foster, Leonard J., Univ. Professor, UBC Centre for Proteomics, Dept. of Biochemistry and Molecular Biology, Univ. of British Columbia, Vancouver, BC V6T 1Z4. ☎ (604) 822-2114. E-mail: ljfoster@interchange.ubc.ca.

Giovenazzo, Pierre, Univ. Professor, Département de biologie, Faculté des sciences et de genie, Université Laval, Québec, QC G1K 7P4. ☎ (418) 656-2043. E-mail: pierre.giovenazzo@bio.ulaval.ca.

Guzman, Ernesto, Univ. Professor, Dept. of Environmental Biology, Univ. of Guelph, Guelph, ON N1G 2W1. ☎ (519) 837-0442. E-mail: eguzman@uoguelph.ca.

Houle, Emile, Research Technician, Centre de recherche en sciences animales de Deschambault (CRSAD), 120 a, chemin du Roy Deschambeault, Québec, QC G0A 1S0. ☎ (418) 286-3597. E-mail: emile.houle@crsad.qc.ca.

Jordan, Chris, Provincial Apiarist, Prince Edward Island Dept. of Ag., PO Box 1600, Charlottetown, PEI C1A 7N3. ☎ (902) 368-5729. E-mail: cwjordan@gov.pe.ca.

Kelly, Paul, Research Technician, Dept. of Environmental Biology, Univ. of Guelph, Guelph, ON N1G 2W1. ☎ (519) 837-0442. E-mail: pgkelly@uoguelph.ca.

Kevan, Peter, Univ. Professor, Environmental Biology, Univ. of Guelph, Guelph, ON N1G 2W1. ☎ (510) 837-0442. E-mail: pkevan@uoguelph.ca.

Kozak, Paul, Provincial Apiculturist, Ontario Ministry

of Ag., Food and Rural Affaires, 1 Stone Road, West Guelph, Ontario N1G 4Y2. ☎ (519) 826-3595.

Lafrenière, Rhéal, Provincial Apiarist, Manitoba Ag., Food and Rural Initiatives (MAFRI), 204–545 Univ. Crescent, Winnipeg, MB R3T 5S6. ☎ (204) 945-4327. E-mail: Rheal.Lafreniere@gov.mb.ca.

Marceau, Jocelyn, Research Technician, Ministere de l'Agriculture, des Pecheries et de l'Alimentation du Quebec (MAPAQ), edifice 2, RC-22, l 1665 Blvd. Hamel Ouest, Québec, QC G1N 3Y7. ☎ (418) 644-8263. E-mail: jmarceau@mapaq.gouv.qc.ca.

Maund, Christopher, Provincial Apiarist, Crop Development, New Brunswick Dept. of Ag. and Aquaculture, PO Box 6000, Fredericton, NB E3B 5H1. ☎ (506) 453-7978. E-mail: chris.maund@gnb.ca.

Melathopoulos, Adony, Research Technician, Agriculture and Agri-Food Canada (AAFC), Research Station, PO Box 29, Beaverlodge, AB T0H 0C0. ☎ (780) 354-8171. E-mail: melathopoulosa@agr.gc.ca.

Moran, Joanne, Provincial Apiarist, Nova Scotia Ag., Kentville Agriculture Centre, Kentville, NS B4N 1J5. ☎ (902) 679-6062. E-mail: jmoran@gov.ns.ca.

Nasr, Medhat, Provincial Apiarist, Crop Diversification Centre North, Ag. Research Division, Alberta Ag. and Rural Development, 17507 Fort Road, Edmonton, AB T5Y 6H3. ☎ (780) 422-6096. E-mail: medhat.nasr@gov.ab.ca.

Ostermann, David, Provincial Apiarist, MAFRI, 204–545 Univ. Crescent, Winnipeg, MB R3T 5S6. ☎ (204) 945-4327. E-mail: dostermann@gov.mb.ca.

Otis, Gard, Univ. Professor, Dept. of Environmental Biology, Univ. of Guelph, Guelph, ON N1G 2W1. ☎ (519) 837-0442. E-mail: gotis@uoguelph.ca.

Pernal, Stephen, Research Scientist, AAFC Research Station, PO Box 29, Beaverlodge, AB T0H 0C0. ☎ (780) 354-8171. E-mail: pernals@agr.gc.ca.

Rogers, Richard, Consultant, Wildwood Labs Inc., 53 Blossom Drive, Kentville, NS B4N 3Z1. Consulting Entomologist / Apiculturist, Wildwood Labs Inc., 53 Blossom Drive, Kentville, Nova Scotia. Canada B4N 3Z1 ☎ (902) 679-2818 Fax 902-679-0637 E-mail: rrogers@wildwoodlabs.com.

Scott-Dupree, Cynthia, Univ. Professor, Dept. of Environmental Biology, Univ. of Guelph, Guelph, ON N1G 2W1. ☎ (519) 837-0442. E-mail: csdupree@uoguelph.ca.

Skinner, Alison, Tech Transfer Specialist, Ontario Beekeepers' Association Research Office, Orchard Park Office Centre, 5420 Highway 6 North, Guelph, ON N1H 6J2. ☎ (519) 836-3609. E-mail: alison_bee@yahoo.com.

Tam, Janet, Tech Transfer Specialist, Ontario Beekeepers' Association Research Office, Orchard Park Office Centre, 5420 Highway 6 North, Guelph, ON N1H 6J2. E-mail: shrewless@yahoo.com.

van Westendorp, Paul, Provincial Apiarist, British Columbia Ministry of Ag. and Lands, 1767 Angus Campbell Road, Abbotsford, BC V3G 2M3. ☎ (604) 556-3030. E-mail: paul.vanwestendorp@gov.bc.ca; vanwestendorp@telus.net.

White, Jane, Natural Resources Specialist, Dept. of Natural Resources Newfoundland and Labrador, Agrifoods Development Branch, PO Box 2006, Corner Brook, NL A2H 6J8. ☎ (709) 637-2591. E-mail: jane-white@gov.nl.ca.

Wilson, Geoffrey, Provincial Apiculturist, Saskatchewan Dept. of Agriculture & Food, McIntosh Mall, Box 3003, 800 Central Avenue, Prince Albert, SK S6V 6G1.

Winston, Mark, Univ. Professor, Morris J. Wosk Centre for Dialogue, Simon Fraser Univ., Harbour Centre, 3309–515 W. Hastings Street, Vancouver, BC V5B 5K3. ☎ (778) 782-7892. E-mail: winston@sfu.ca.

CHILDREN'S BOOKS ON BEES

Clayboujrne, A. How do bees make honey. Starting Point Science Series. Educational Development Corporation is the United States trade publisher of a line of children's books produced in the United Kingdom by Usborne Publishing Limited. PO Box 470663, Tulsa, OK 74147-0663.

Cole, J., and B. Degen. 1996. The magic school bus. Inside a beehive. New York: Scholastic.

Hodgson, N.B. 1973. Children's books on bees and beekeeping. Bridgnorth, Salop, UK: Astley Abbotts.

Houghton, G. 2004. Bees, inside and out. New York: PowerKids Press.

Kalman, B., A. Larin, and N. Walker. 1998. Hooray for beekeeping! New York: Crabtree.

Krebs, L., and M. Iwai. 2008. The bee man. Cambridge, MA: Barefoot Books.

Rustad, M.E.H. 2003. Honey bees. Mankato, MN: Pebble Books.

Schaefer, L.M. 1999. Honey bees and flowers. Mankato, MN: Pebble Books.

Schaefer, L.M., and G. Saunders-Smith. 1999. Honey bees and hives. Mankato, MN: Pebble Books.

Stockton, F.R. 2005. Bee-man of Orn. Maurice Sendak (Illustrator). New York: HarperCollins. http://search.barnesandnoble.com/Honey-Bees/Joyce-Milton/e/9780448428468 (Barnes and Noble locator).

COLONY COLLAPSE DISORDER

Chauzat, M.-P., M. Higes, R. Martín-Hernández, A. Meana, N. Cougoule, and J.-P. Faucon. 2007. Presence of *Nosema ceranae* in French honey bee colonies. JAR 46 (2): 127–128.

Chen, Y., J.D. Evans, I.B. Smith, and J.S. Pettis. 2008.

Nosema ceranae is a long-present and wide-spread microsporidian infection of the European honey bee (*Apis mellifera*) in the United States. J. of Invertebrate Pathology 97 (2): 186–188.

Chen, Y. 2002. Phylogenetic analysis of acute bee paralysis virus strains. Applied and Environmental Microbiology 68 (12): 6446–6450.

Dobbelaere, W., D.C. de Graaf, W. Reybroeck, E. Desmedt, J.E. Peeters, and F.J. Jacobs. 2001. Disinfection of wooden structures contaminated with *Paenibacillus larvae* subsp. larvae spores. J. of Applied Microbiology 91 (2): 212–216.

Ellis, J. 2007. Colony collapse disorder (CCD) in honey bees. Gainesville: Univ. of Florida Dept. of Entomology and Nematology, Florida CES.

Fries, I., R. Martín, A. Meana, P. García-Palencia, and M. Higes. 2006. Natural infections of *Nosema ceranae* in European honey bees. JAR 45 (4): 230–233.

Gliński, Z., and J. Jarosz. 2001. Infection and immunity in the honey bee *Apis mellifera*. Apiacta 36 (1): 12–24.

Higes, H., R Martín-Hernández, E. Garrido-Bailón, A.V. González-Porto, P. García-Palencia, et al. . . . 2009. Honeybee colony collapse due to *Nosema ceranae* in professional apiaries. Environmental Microbiology Reports (1): 110–113. doi: 10.1111/j.1758-2229.2009.00014.x.

Higes, M., R. Martín, and A. Meana. 2006. *Nosema ceranae*, a new microsporidian parasite in honeybees in Europe. J. of Invertebrate Pathology 92 (2): 81–83.

Higes, M., R. Martín-Hernández, C. Botías, et al. 2008. How natural infection by *Nosema ceranae* causes honeybee colony collapse. Environmental Microbiology 10 (10): 2659–2669.

Higes, M., R. Martín-Hernández, E. Garrido-Bailón, P. García-Palencia, and A. Meana. 2008. Detection of infective *Nosema ceranae* (Microsporidia) spores in corbicular pollen of forager honeybees. J. of Invertebrate Pathology 97 (1): 76–78.

Honey Bee Health, Special Issue. 2010. Apidologie 41 (3).

Huang, W.F., M. Bocquet, K.C. Lee, I.H. Sung, et al. 2008. The comparison of rDNA spacer regions of *Nosema ceranae* isolates from different hosts and locations. J. of Invertebrate Pathology 97 (1): 9–13.

Huang, W.F., J.H. Jiang, Y.W. Chen, and C.H. Wang. 2007. A *Nosema ceranae* isolate from the honeybee *Apis mellifera*. Apidologie 38 (1): 30–37.

Jacobsen, R. 2008. Fruitless fall: The collapse of the honey bee and the coming agricultural crisis. New York: Bloomsbury.

Johnson, R., J.D. Evans, G.E. Robinson, M.R. Berenbaum. 2009. Changes in transcript abundance relating to colony collapse disorder in honey bees (*Apis mellifera*) Proceedings of the National Acad. Sciences (106): 14790-14795. doi: 10.1073/pnas.0906970106.

Kashmir bee virus. 2004. Apiculture factsheet 230. Victoria, BC: Ministry of Ag. and Lands, Government of British Columbia. http://www.agf.gov.bc.ca/apiculture/factsheets/230_kashmir.htm.

Klee, J., A.M. Besana, E. Genersch, S. Gisder, A. Nanetti, et al. 2007. Widespread dispersal of the microsporidian *Nosema ceranae*, an emergent pathogen of the western honey bee, *Apis mellifera*. J. of Invertebrate Pathology 96 (1): 1–10.

Klee, J., W. Tek Tay, and R.J. Paxton. 2006. Specific and sensitive detection of *Nosema bombi* (Microsporidia: Nosematidae) in bumble bees (*Bombus* spp.; Hymenoptera: Apidae) by PCR of partial rRNA gene sequences. J. of Invertebrate Pathology 91 (2): 98–104.

Maori, E., N. Paldi, S. Shafir, H. Kalev, et al. 2009. IAPV, a bee-affecting virus associated with Colony Collapse Disorder can be silenced by dsRNA ingestion. Insect Molecular Biology. (18): 55–60. doi: 10.1111/j.1365-2583.2009.00847.x.

Pajuelo, A.G., C. Torres, and Fco. J.O. Bermejo. 2008. Colony losses: A double blind trial on the influence of supplementary protein nutrition and preventative treatment with fumagillin against *Nosema ceranae*. JAR 47 (1): 84–86.

Paxton, R.J., J. Klee, S. Korpela, and I. Fries. 2007. *Nosema ceranae* has infected *Apis mellifera* in Europe since at least 1998 and may be more virulent than *Nosema apis*. Apidologie 38 (6): 558–565.

Ribière, M., J.-P. Faucon, and M. Pépin. 2000. Detection of chronic honey bee (*Apis mellifera* L.) paralysis virus infection: Application to a field survey. Apidologie 31:567–577.

Rodriguez-Saona, C. 2007. Honey bees, colony collapse disorder, and cranberries. Plant and Pest advisory 13 (2). http://njaes.rutgers.edu/pubs/plantandpestadvisory/2007/cb0426.pdf.

Schacker, M., and B.V. McKibben. 2008. A spring without bees. Stonington, CT: Globe Pequot Press.

Stankus, T. 2008. A review and bibliography of the literature of honey bee colony collapse disorder: A poorly understood epidemic that clearly threatens the successful pollination of billions of dollars of crops. American J. of Agricultural and Food Information 9 (2): 115–143. 34.

Tentcheva, D., and L. Gauthier. 2004. Prevalence and seasonal variations of six bee viruses in *Apis mellifera* L. and *Varroa destructor* mite populations in France. Applied and Environmental Microbiology 70 (12): 7185–7191.

vanEngelsdorp, D., D. Cox-Foster, M. Frazier, N. Ostiguy, and J. Hayes. 2006. Colony collapse disorder preliminary report. Mid-Atlantic Apiculture Research and Extension Consortium (MAAREC) CCD Working Group, Penn State Univ. State College, PA. http://maarec.cas.psu.edu/bkCD/Bee_Diseases/disease_index.html.

vanEngelsdorp, D., J.D. Evans, C. Saegerman, C. Mullin, E. Haubruge, B.K. Nguyen, M. Frazier, J. Frazier, D. Cox-Foster, Y. Chen, R. Underwood, D.R. Tarpy, J.S. Pettis. 2009. Colony collapse disorder: a descriptive study. Online: PloS one, 4 (8).

vanEngelsdorp, D., J. Hayes Jr., R.M. Underwood, and J. Pettis. 2008. A survey of honey bee colony losses in the U.S., Fall 2007 to Spring 2008. PLoS one, 3 (12): art. no. e4071.

Williams, G.R., M.A. Sampson, D. Shutler, and R.E.L. Rogers. 2008. Does fumagillin control the recently detected invasive parasite Nosema ceranae in western honey bees (Apis mellifera)? J. of Invertebrate Pathology 99 (3): 342–344.

Williams, G.R., A.B.A. Shafer, R.E.L. Rogers, D. Shutler, and D.T. Stewart. 2008. First detection of Nosema ceranae, a microsporidian parasite of European honey bees (Apis mellifera), in Canada and central USA. J. of Invertebrate Pathology 97 (2): 189–192.

Zoet, K. 2008. Dutch bee breeding and Nosema ceranae. ABJ 148 (3): 185–186.

Websites

Penn State Univeristy: CCD Working Group. http://maarec.cas.psu.edu/bkCD/Bee_Diseases/disease_index.html.

American Assoc. of Professional Apiculturists. http://www.masterbeekeeper.org/aapa/pdf/position_papers/Final_AAPA_POSITION_STATEMENT_COLONY_HEALTH-1.pdf.

COURSES

East Sydney College of Technical and Further Education, Darlinghurst, New South Wales, Australia. http://pip.com.au/~abestuds.

Univ. of Minnesota. http://www.extension.umn.edu/honeybees/components/shortcourse.htm.

Hot Courses, UK. http://www.hotcourses.com/uk-courses/Bee-Keeping-courses/hc2_browse.pg_loc_tree/16180339/0/p_type_id/1/p_bcat_id/4160/page.htm.

North Carolina State Beekeepers. http://www.ncbeekeepers.org/courses.htm.

The AgricultureB2B.com Resource. http://www.agricultureb2b.com/biz/e/Bee-Industry/Education/.

Cooperative Extension System, eXtension is an educational partnership of 74 universities in the United States. http://www.extension.org/pages/Bee_Health_is_Focus_of_New_National_Web_Resource.

Note there are many universities, state organizations, and private companies that offer beekeeping courses.

DISEASES AND PESTS

Akratanakul, P. 1987. Honeybee diseases and enemies in Asia: A practical guide. Rome: Food and Agriculture Organization of the United Nations.

Bailey, L., and B.V. Ball. 1991. Honey bee pathology (2nd edn.). London: Academic Press.

Chen, Y.P., J.D. Evans, I.B. Smith, and J.S. Pettis. 2008. Nosema ceranae is a long-present and wide-spread microsporidian infection of the European honey bee (Apis mellifera) in the United States. J. of Invertebrate Pathology 97 (2): 186–188.

Chen, Y P., J.S. Pettis, A. Collins, and M.F. Feldlaufer. 2006. Prevalence and transmission of honeybee viruses. Applied and Environmental Microbiology 72 (1): 606–611.

Chen, Y.P., and R. Siede. 2007. Honey bee viruses. Advances in Virus Research 70: 33–80.

Currie, R.W. 2001. Control of parasitic mites in honey bees. Final Report: September 25, 2001 Saskatchewan Agri-Food Innovation Fund, Canada, Project 97000002. 10 pp.

Currie, R.W. 2001. Chalkbrood and disease control in honey bees. Univ. of Manitoba. Saskatchewan Agri-Food Innovation Fund, Canada.

Devillers, J., and M.H. Pham-Delègue. 2002. Honey bees: Estimating the environmental impact of chemicals. London: Taylor and Francis.

Dobbelaere, W., D.C. de Graaf, W. Reybroeck, E. Desmedt, J.E. Peeters, and F.J. Jacobs. 2001. Disinfection of wooden structures contaminated with Paenibacillus larvae subsp. larvae spores. J. of Applied Microbiology 91 (2): 212–216.

Evans, J.D., and M. Spivak. 2010. Socialized medicine: Individual and communal disease barriers in honey bees. J. of Invertebrate Pathology 103 (Suppl. 1): S62–S72.

Flores, A. 2007. Honey bee genetics vital in disease resistance. Ag. Research 5 (1): 14. http://www.ars.usda.gov/is/AR/archive/jan07/bee0107.htm.

Foul brood disease of honey bees: Recognition and control. 2001. London, UK: Central Science Laboratory National Bee Unit, Department for Environment, Food and Rural Affairs (DEFRA).

Fries, I., F. Feng, A. Da Silva, S.B. Slemenda, and N.J. Pieniazek. 1996. Nosema ceranae n. sp. (Microspora, Nosematidae), morphological and molecular characterization of a microsporidian parasite of the Asian honey bee Apis cerana (Hymenoptera, Apidae). European J. of Protistology 32 (3): 356–365.

Fries, I., R. Martín, A. Meana, P. García-Palencia, and M. Higes. 2006. Natural infections of Nosema ceranae in European honey bees. JAR 45 (4): 230–233.

Higes, M., R. Martín, and A. Meana. 2006. Nosema cera-

nae, a new microsporidian parasite in honeybees in Europe. J. of Invertebrate Pathology 92 (2): 93–95.

Higes, M., R. Martín-Hernández, C. Botías, E.G. Bailón, A.V. González-Porto, L. Barrios, et al. 2008. How natural infection by *Nosema ceranae* causes honeybee colony collapse. Environmental Microbiology 10 (10): 2659–2669.

Hive equipment sterilization. 2009. The Pennsylvania State Beekeeper Newsletter 1 (Jan.): 13. Pennsylvania State Beekeepers Association. http://www.pastatebee-keepers.org or newsletter editor: pabee@epix.net.

Kashmir bee virus. 2004, July. Apiculture factsheet 230. Victoria, BC: Ministry of Agriculture and Lands, Government of British Columbia, Canada. http://www.agf.gov.bc.ca/apiculture/factsheets/230_kashmir.htm.

Matheson, A. 1996. The conservation of bees. IBRA. London: Academic Press.

Miksha, R. 2004. Bad beekeeping. Victoria, BC: Trafford.

Monck, M., and D. Pearce. 2007. Mite pests of honey bees in the Asia-Pacific region. Canberra: Australian Centre for International Agricultural Research.

Morse, R.A., and R. Nowogrodzki (eds.). 1990. Honey bee pests, predators and diseases (3rd edn.). Ithaca, NY: Comstock.

Oldroyd, B.P., and S. Wongsiri. 2006. Asian honey bees: Biology, conservation, and human interactions. Cambridge, MA: Harvard Univ. Press.

Powell, G. 2006, Jan. Cleaning up American foul-brood. Buzz Newsletter. Iowa Honey Producers Association.

Protecting honey bees from pesticides. 2005. Oklahoma City: Oklahoma Department of Agriculture, Food and Forestry.

Ritter, W., and P. Akratanakul. 2006. Honey bee diseases and pests: A practical guide. Rome: Food and Agriculture Organization of the United Nations.

Shimanuki, H., and D. Knox. 1991. Diagnosis of honey bee diseases. USDA Ag. Handbook AH-690.

Suszkiw, J. 2005. New antibiotic approved for treating bacterial honey bee disease. USDA-ARS, Washington, DC. http://ars.usda.gov/IS/pr/2005/051219.htm.

Szabo, T.I., and D.C. Szabo. 2006. Honey bees in bear territory. ABJ 146 (6): 512–514.

Tentcheva, D., and L. Gauthier. 2004. Prevalence and seasonal variations of six bee viruses in *Apis mellifera* L. and *Varroa destructor* mite populations in France. Applied and Environmental Microbiology 70 (12): 7185–7191.

Tropilaelaps: Parasitic mites of honey bees. 2005. London: Department for Environment, Food and Rural Affairs.

Websites

Arizona Cooperative Extension: http://www.extension.org/pages/Bee_Health_is_Focus_of_New_National_Web_Resource.

Department for Environment, Food and Rural Affairs (Defra), London: http://www.wasba.org/SHB.pdf.

Ohio State Univ.: http://www.biosci.ohio-state.edu/~acarolog/sum2k1.htm.

Texas A & M: http://insects.tamu.edu/research/collection/hallan/acari/Mesostigmata1.htm.

Tree of Life Web Project: http://tolweb.org/tree?group=Acari.

Univ. of British Columbia: http://www.zoology.ubc.ca/~srivast/mites/; http://www.zoology.ubc.ca/~srivast/mites/scans/partspred.jpg.

Univ. of Georgia: http://www.ent.uga.edu/bees/Disorders/Nosema.htm.

Univ. of Michigan: http://insects.ummz.lsa.umich.edu/beemites/.

Univ. of Minnesota: http://www.extension.umn.edu/Honey bees/components/pdfs/poster 167 nosema sp. ores 24x33.pdf by G.C. Reuter, K. Lee, and M. Spivak; testing for nosema spores using a hemacytometer. St. Paul: Univ. of Minnesota.

USDA-ARS: http://www.sel.barc.usda.gov/acari/frames/collaborators.html.

DRONES

Ble Kanga, L.H., W.A. Jones, and R.R. James. 2003. Field trials using the fungal pathogen, *Metarhizium anisopliae* (Deuteromycetes: Hyphomycetes) to control the ectoparasitic mite, *Varroa destructor* (Acari: Varroidae) in honey bee, *Apis mellifera* (Hymenoptera: Apidae) colonies. J. Economic Entomology 96 (4): 1091–1099.

Burley, L.M., R.D. Fell, and R.G. Saacke. 2008. Survival of honey bee (Hymenoptera: Apidae) spermatozoa incubated at room temperature from drones exposed to miticides. J. Economic Entomology 101 (4): 1081–1087.

Calderón, R.A., L.G. Zamora, and J.W. Van Veen. 2007. The reproductive rate of *Varroa destructor* in drone brood of Africanized honey bees. JAR 46 (3): 140–143.

Calderón, R.A., L.G. Zamora, J.W. Van Veen, and M.V. Quesada. 2007. A comparison of the reproductive ability of *Varroa destructor* (Mesostigmata: Varroidae) in worker and drone brood of Africanized honey bees (*Apis mellifera*). Experimental and Applied Acarology 43 (1): 25–32.

Calderone, N.W. 2005. Evaluation of drone brood removal for management of *Varroa destructor* (Acari: Varroidae) in colonies of *Apis mellifera* (Hymenoptera: Apidae) in the northeastern United States. J. Economic Entomology 98 (3): 645–650.

Charriére, J.-D., A. Imdorf, B. Bachofen, and A. Tschan. 2003. The removal of capped drone brood: An effective means of reducing the infestation of varroa in honey bee colonies. Bee Wld. 84 (3): 117–124.

Collins, A.M. 2004. Sources of variation in the viability of honey bee, *Apis mellifera* L., semen collected for artificial insemination. Invertebrate Reproduction and Development 45 (3): 231–237.

Collins, A.M., T.J. Caperna, V. Williams, W.M. Garrett, and J.D. Evans. 2006. Proteomic analyses of male contributions to honey bee sperm storage and mating. Insect Molecular Biology 15 (5): 541–549.

Colonello, N.A., and K. Hartfelder. 2003. Protein content and pattern during mucus gland maturation and its ecdysteroid control in honey bee drones. Apidologie 34 (3): 257–267.

Cruz-Landim, C., and R.P. Dallacqua. 2005. Morphology and protein patterns of honey bee drone accessory glands. Genetics and Molecular Research (GMR) 4 (3): 473–481.

De Oliveira Tozetto, S., M.M.G. Bitonde, R.P. Dallaqua, and Z.L.P. Simões. 2007. Protein profiles of testes, seminal vesicles and accessory glands of honey bee pupae and their relation to the ecdysteroid titer. Apidologie 38 (1): 1–11.

Es'kov, E.K. 2004. Variability of honey bee (*Apis Mellifera*) drones. Zoologicheskii Zhurnal 83 (3): 367–370.

Gençer, H.V., and Ç. Firatli. 2005. Reproductive and morphological comparisons of drones reared in queenright and laying worker colonies. JAR 44 (4): 163–167.

Hayworth, M.K., N.G. Johnson, M.E. Wilhelm, R.P. Gove, J.D. Metheny, and O. Rueppell. 2009. Added weights lead to reduced flight behavior and mating success in polyandrous honey bee queens (*Apis mellifera*). Ethology 115 (7): 698–706.

Herrmann, M., T. Trenzcek, H. Fahrenhorst, and W. Engels. 2005. Characters that differ between diploid and haploid honey bee (*Apis mellifera*) drones. Genetics and Molecular Research 4 (4): 624–641.

Hrassnigg, N., R. Brodschneider, P.H. Fleischmann, and K. Crailsheim. 2005. Unlike nectar foragers, honeybee drones (*Apis mellifera*) are not able to utilize starch as fuel for flight. Apidologie 36 (4): 547–557.

Jensen, A.B., K.A. Palmer, N. Chaline, N.E. Raine, A. Tofilski, S.J. Martin, B.V. Pedersen, J.J. Boomsma, and F.L.W. Ratnieks. 2005. Quantifying honey bee mating range and isolation in semi-isolated valleys by DNA microsatellite paternity analysis. Conservation Genetics 6 (4): 527–537.

Koeniger, N., and G. Koeniger. 2007. Mating flight duration of *Apis mellifera* queens: As short as possible, as long as necessary. Apidologie 38 (6): 606–611.

Latshaw, J.S. 2010. Survivor drone project dispersing honey bee genetic diversity. ABJ 150 (2): 157–159.

Maggi, M., N. Damiani, S. Ruffinengo, D. de Jong, J. Principal, and M. Eguaras. 2010. Brood cell size of *Apis mellifera* modifies the reproductive behavior of *Varroa destructor*. Experimental and Applied Acarology 50 (3): 269–279.

Mattila, H.R., and T.D. Seeley. 2007. Genetic diversity in honey bee colonies enhances productivity and fitness. Science 317 (5836): 362–364.

Richard, F.-J., D.R. Tarpy, and C.M. Grozinger. 2007. Effects of insemination quantity on honey bee queen physiology. PLoS ONE 2 (10), art. no. e980.

Rueppell, O., M.K. Fondrk, and R.E. Page Jr. 2005. Biodemographic analysis of male honey bee mortality. Aging Cell 4 (1): 13–19.

Santomauro, G., N.J. Oldham, W. Boland, and W. Engels. 2004. Cannibalism of diploid drone larvae in the honey bee (*Apis mellifera*) is released by odd pattern of cuticular substances. JAR 43 (2): 69–74.

Shafir, S., L. Kabanoff, M. Duncan, and B.P. Oldroyd. 2009. Honey bee (*Apis mellifera*) sperm competition in vitro—Two are no less viable than one. Apidologie 40 (5): 556–561.

Underwood, R.M., M.J. Lewis, and J.F. Hare. 2004. Reduced worker relatedness does not affect cooperation in honey bee colonies. Canadian J. of Zoology 82 (9): 1542–1545.

Wantuch, H.A., and D.R. Tarpy. 2009. Removal of drone brood from *Apis mellifera* (Hymenoptera: Apidae) colonies to control *Varroa destructor* (Acari: Varroidae) and retain adult drones. J. Economic Entomology 102 (6): 2033–2040.

Woyke, J. 2008. Why the eversion of the endophallus of honey bee drone stops at the partly everted stage and significance of this. Apidologie 39 (6): 627–636.

Z°ółtowska, K., Z. Lipiński, and M. Dmitryjuk. 2007. Effects of *Varroa destructor* on sugar levels and their respective carbohydrate hydrolase activities in honey bee drone prepupae. JAR 46 (2): 110–113.

EQUIPMENT

Blume, K.R. 2010. Building, managing and using a top bar hive. Bee Culture 138:55–59.

Bérubé, C., Jr. 1989. The Kenya top-bar hive as a better hive in developing countries. ABJ 129:525–527. See also Bérubé's websites.

Celia, C. 2006. A livestock trailer for bees. ABJ 146 (6): 507–508. http://www.beesource.com/plans/index.htm.

Gentry, C. 1982. Small scale beekeeping. Washington, DC: Peace Corps, Information Collection and Exchange Division. Search at http://www.eric.ed.gov.

Goodwin, M., and H. Haine. 1998. Using paraffin wax

and steam chests to sterilise hive parts that have been in contact with colonies with American Foulbrood Disease. New Zealand Beekeeper 5:4, 21.

Great Lakes IPM: company that sells IPM products such as Varroa mite (sticky) boards, and bee lures. http://www.greatlakesipm.com.

Griffiths, G.L. 1992, March. Protocol for wax dipping bee equipment. In: Preservation of wooden hive equipment. Miscellaneous Pub. 4/92. Compiled by L. Allan, Senior Apiculturalist, Department of Ag., Western Australia.

Hot wax dipping of beehive components for preservation and sterilization. 2001. Pub. 01/051, Project DAV-167A. Kingston, Australia: Rural Industries Research and Development Corporation and the Department of Natural Resources and Environment. http://www.rirdc.gov.au or http://www.queenrightcolonies.com/uploads/HotWax DippingofBeehives.pdf.

Kalnins, M.A., and B.F. Detroy. 1984. Effects of wood preservative treatment on honey bees and hive products. J. of Ag. and Food Chemistry 32:1176–1180. http://dx.doi.org/10.1021/jf00125a060.

Mangum, Wyatt A. 1987. Building a Regular or Observation Kenya Top Bar Hive. Gleanings in Bee Culture 115: 646–648.

Matheson, A. 1980. Easily constructed paraffin wax dipper. New Zealand Beekeeper 41 (4): 11–12.

Northern Tool and Equipment. n.d. Tool's bench scale 19333. http://NorthernTool.com.

Plans and dimensions for a 10-frame bee hive. n.d. Pub. CA 33-24. Beltsville, MD: USDA-ARS, Entomol. Research. Division.

Robinson, P.J., and J.R.J. French. 1984. Beekeeping and wood preservation in Australia. Proceedings of the 21st forest products research conference, vol. 1.

Robinson, P.J., and J.R.J. French. 1986. Beekeeping and wood preservation in Australia. Australian Bee J. 67 (1): 8–10.

Sanford, M.T. 1998. Preserving woodenware in beekeeping operations. ENY-125. Gainesville, Florida: CES, Institute of Food and Agricultural Sciences, Univ. of Florida. http://edis.ifas.ufl.edu.

Sperling, D., and D.M. Caron. 1980. The moveable-comb frameless hive: "appropriate technology" alternative to the Langstroth hive? ABJ 120:284–289.

Tew, J. 1984. Paraffin wax dipping of hive equipment. Gleanings in Bee Culture 112 (8): 422, 426.

Weatherhead, T.F., and M.J. Kennedy. 1980. Adhesives and wood preservatives for the beekeeper in the 1980s. Australasian Beekeeper 82 (4): 88–94.

Williams, D. 1980. Preserving beehive timber. New Zealand Beekeeper 41 (4): 13.

Websites

Detecto scale Model 4510. http://www.detectoscale.com.

Free hive beetle DIY trap. http://www.greenbeehives .com/.

Louisiana State Univ.: http://www.Lsuagcenter.com/en/our offices/departments/Biological Ag Engineering/Features/Extension/Building Plans/small animal/bees/Honey+Extractor.htm.

Paraffin wax dipping. http://www.bobsbeekeeping.com .au/uploads/tips/Hot%20Wax%20Dipping%20of%20 Beehives.pdf.

Top bar hive websites:

Food and Agriculture Organization of the United Nations (FAO): http://www.fao.org/docrep/t0104e/T0104E07.htm; and a picture at http://www.fao.org/docrep/t0104e/T0104E02.GIF.

Peace Corps:

http://www.beekeeping.com/articles/us/small_beekeeping/index.htm.

http://www.ibiblio.org/pub/academic/agriculture/entomology/beekeeping/general/management/top_bar_faqs/tbhf.html.

http://www.top-bar-hive.com/my-beehive/searching-for-the-right-top-bar-hive-design/.

FEEDING BEES

Avni, D., A. Dag, and S. Shafir. 2009. The effect of surface area of pollen patties fed to honey bee (*Apis mellifera*) colonies on their consumption, brood production and honey yields. JAR 48 (1): 23–28.

De Jong, D., E.J. Da Silva, P.G. Kevan, and J.L. Atkinson. 2009. Pollen substitutes increase honey bee haemolymph protein levels as much as or more than does pollen. JAR 48 (1): 34–37.

DeGrandi-Hoffman, G., G. Wardell, F. Ahumada-Segura, T. Rinderer, R. Danka, and J. Pettis. 2008. Comparisons of pollen substitute diets for honey bees: Consumption rates by colonies and effects on brood and adult populations. JAR 47 (4): 265–270.

Dufault, R., B. LeBlanc, R. Schnoll, C. Cornett, et al. 2009. Mercury from chlor-alkali plants: Measured concentrations in food product sugar. Environmental Health 8:2.

Mattila, H.R., and B.H. Smith. 2008. Learning and memory in workers reared by nutritionally stressed honey bee (*Apis mellifera* L.) colonies. Physiology and Behavior 95 (5): 609–616.

Rosendale, D.I., I.S. Maddox, M.C. Miles, M. Rodier, M. Skinner, and J. Sutherland. 2008. High-throughput microbial bioassays to screen potential New Zealand functional food ingredients intended to manage the growth of probiotic and pathogenic gut bacteria. International J. of Food Science and Technology 43 (12): 2257–2267.

Somerville, D. 2005. Fat bees, skinny bees: A manual on honey bee nutrition for beekeepers. Barton, Goulburn NSW 2580, Australia: Rural Industries Research and Development Corporation. http://docs.ksu.edu.sa/PDF/Articles18/Article180069.pdf.

HONEY AND HONEY PRODUCTS

Balayiannis, G., and P. Balayiannis. 2008. Bee honey as an environmental bioindicator of pesticides' occurrence in six agricultural areas of Greece. Archives of Environmental Contamination and Toxicology 55 (3): 462–470.

Bishop, H. 2006. Robbing the bees: A biography of honey, the sweet liquid gold that seduced the world. New York: Free Press.

Blasa, M., M. Candiracci, A. Accorsi, M.P. Piacentini, and E. Piatti. 2007. Honey flavonoids as protection agents against oxidative damage to human red blood cells. Food Chemistry 104 (4): 1635–1640. http://dx.doi.org/10.1016/j.foodchem.2007.03.014.

Bogdanov, S. 2006. Contaminants of bee products. Apidologie 37 (1): 1–18.

Bogdanov, S., T. Jurendic, T. Sieber, and P. Gallmann. 2008. Honey for nutrition and health: A review. J. of the American College of Nutrition 27 (6): 677–689.

Cabeza de Vaca, F.G., and A.E. Macias. 2008. Wound care for salvaging diabetic foot. J. of Hospital Infection 70 (4): 386–387.

Chesson, L., and B. Tipple. 2010. The isotope waggle dance. Bee Culture 138 (8): 30–32.

Conti, M.E., and F. Botre. 2001. Honeybees and their products as potential bioindicators of heavy metals contamination. Environmental Monitoring and Assessment 69 (3): 267–282. http://www.kluweronline.com/issn/1420-2026/contents.

Cooper, J. 2009. Wound management following orbital exenteration surgery. British J. of Nursing 18 (6): S4, S6, S8, passim. http://www.ncbi.nlm.nih.gov/pubmed/19374038.

Crane, E. 1980. A book of honey. Oxford, UK: Oxford Univ. Press.

Crane, E. 1999. The world history of beekeeping and honey hunting. New York: Routledge.

Cutting, K.F. 2008. Honey and contemporary wound care: An overview. Wound Management. OWM 53 (11): 49–54. http://www.o-wm.com/article/8058.

Eddy, J.J., M.D. Gideonsen, and G.P. Mack. 2008. Practical considerations of using topical honey for neuropathic diabetic foot ulcers: A review. Wisconsin Medical J. 107 (4): 187–190.

Edgar, J.A., E. Roeder, and R.J. Molyneux. 2002. Honey from plants containing pyrrolizidine alkaloids: A potential threat to health. J. of Ag. and Food Chemistry 50 (10): 2719–2730.

Fessenden, R., and M. McInnes. 2009. The honey revolution, restoring the health of future generations. Colorado Springs: World ClassEmprise. http://www.thehoneyrevolution.com.

Field, O. 1983. Honey by the ton. London: Barn Owl Books.

Forte, G., S. D'Ilio, and S. Caroli. 2001. Honey as a candidate reference material for trace elements. J. of AOAC International 84 (6): 1972–1975.

Frankel, S., G.E. Robinson, and M.R. Berenbaum. 1998. Antioxidant capacity and correlated characteristics of 14 unifloral honeys. JAR 37:27–31.

Gheldof, N., and N.J. Engeseth. 2002. Antioxidant capacity of honeys from various floral sources based on the determination of oxygen radical absorbance capacity of inhibition of in vitro lipoprotein oxidation in human serum samples. J. of Ag. and Food Chemistry 50:3050–3055.

Gheldorf, N., XH Wang, NJ Engeseth. 2003. Buckwheat honey increases serum antioxidant capacity in humans. J. Ag. and Food Chemistry 51:1500–1505.

Goltz, L. 2001. Honey color: How important? ABJ 141 (1): 33–35.

Granja, R.H.M.M., A.M.M. Niño, R.A.M. Zucchetti, R.E.M. Niño, R. Patel, and A.G. Salerno. 2009. Determination of streptomycin residues in honey by liquid chromatography-tandem mass spectrometry. Analytica Chimica Acta 637 (1-2): 64–67.

Guzman-Novoa, E., and J.L. Uribe-Rubio. 2004. Honey production by European, Africanized and hybrid honey bee (Apis mellifera) colonies in Mexico. ABJ 144 (4): 318–320.

Honey Market News. Ongoing. Fruit and Vegetable Division, Ag. Market Service, Washington, DC.

Iurlina, M.O., A.I. Saiz, R. Fritz, and G.D. Manrique. 2009. Major flavonoids of Argentinean honeys. Optimisation of the extraction method and analysis of their content in relationship to the geographical source of honeys. Food Chemistry 115 (3): 1141–1149.

Jull, A.B., A. Rodgers, and N. Walker. 2008. Honey as a topical treatment for wounds. Cochrane Database of Systematic Reviews (4), art. no. CD005083.

Kevan, P.G., M.A. Hannan, N. Ostiguy, and E. Guzman-Novoa. 2006. A summary of the varroa-virus disease complex in honey bees. ABJ 146 (8): 694–696.

Killion, C.E. 1981. Honey in the comb. Paris, IL: Killion and Sons.

McInnes, M., and S. McInnes. 2006. The hibernation diet (with Maggie Stanfield). London, UK: Souvenir Press.

Molan, P.C. 2006. The evidence supporting the use of honey as a wound dressing. Lower Extremity Wounds 5:40–54.

Morse, R.A., and M.L. Morse. n.d. Honey shows: Guidelines for exhibitors, superintendents, and judges. Reprinted. Kalamazoo, MI: Wicwas Press.

Nicholls, J., and A.M. Miraglio. 2003. Honey and healthy diets. Cereal Foods World 48 (3): 116–119.

Pascoe, T. 1996. Beekeeping and back pain. Bee Biz (2): 10–11.

Penner, L.R. 1980. The honey book. New York: Hastings House.

Pieper, B. 2009. Honey-based dressings and wound care: An option for care in the United States. J. of Wound, Ostomy, and Continence Nursing 36 (1): 60–66.

Przybylowski, P., and A. Wilczynska. 2001. Honey as an environmental marker. Food Chemistry 74 (3): 289–291.

Robson, V., S. Dodd, and S. Thomas. 2009. Standardized antibacterial honey (Medihoney™) with standard therapy in wound care: Randomized clinical trial. J. of Advanced Nursing 65 (3): 565–575.

Subrahmanyam, M. 2007. Topical application of honey for burn wound treatment—An overview. Ann. of Burns and Fire Disasters 20:137–139.

Tananaki, C., A. Thrasyvoulou, E. Karazafiris, and A. Zotou. 2006. Contamination of honey by chemicals applied to protect honeybee combs from wax-moth (*Galleria mellonela* L.). Food Additives and Contaminants 23 (2):159–163.

Tosi, E., M. Ciappini, E. Re, and H. Lucero. 2002. Honey thermal treatment effects on hydroxymethylfurfural content. Food Chemistry 77 (1): 71–74.

Turski, M.P., M. Turska, W. Zgrajka, D. Kuc, and W.A. Turski. 2009. Presence of kynurenic acid in food and honeybee products. Amino Acids 36 (1): 75–80.

Visavadia, B.G., J. Honeysett, and M. Danford. 2008. Manuka honey dressing: An effective treatment for chronic wound infections. British J. of Oral and Maxillofacial Surgery 46 (8): 696–697.

Wijesinghe, M., M. Weatherall, K. Perrin, and R. Beasley. 2009. Honey in the treatment of burns: A systematic review and meta-analysis of its efficacy. New Zealand Medical J. 122 (1295): 47–60.

Websites

Airborne Honey: Technical Information for Manufactures: http://www.airborne.co.nz/manufacturing.shtml.

Honey: Bee Product Service: http://www.bee-hexagon.net (S. Bogdanov); http://www.bee-hexagon.net/files/file/fileE/Honey/4PhysicalPropertiesHoney.pdf.

National Honey Board: http://www.honey.com.

HONEY COOKBOOKS

Bass, L.L. 1983. Honey and spice. Ashland, OR: Coriander Press.

Berto, H. 1972. Cooking with honey. New York: Gramercy.

Charlton, J., and J. Newdick. 1995. A taste of honey. Edison, NJ: Chartwell Books.

Davenport, M. 1992. Cooking with honey! Tigard, OR: Paddlewheel Press.

Geiskopf, S. 1979. Putting it up with honey: A natural foods canning and preserving cookbook. Ashland, OR: Quicksilver Productions.

Opton, G., and N. Highes. 2000. Honey: A connoisseur's guide with recipes. Berkeley, CA: Ten Speed Press.

HONEY PLANTS

Ayers, G.S. 2002. Honey quality from a unifloral source. ABJ 142 (9): 657–660.

Burgett, D.M., B.A. Stringer, and L.D. Johnston. 1989. Nectar and pollen plants of Oregon and the Pacific Northwest. Blogett, OR: Honeystone Press.

Chittka, L., and J.D. Thomson (eds.). 2001. Cognitive ecology of pollination: Animal behavior and floral evolution. Cambridge, UK: Cambridge Univ. Press.

Crompton, C.W., and W.A. Wojtas. 1993. Pollen grains of Canadian honey plants. Pub. 892/E. Ottawa, ON: Agriculture Canada.

Dalby, R. 2004. A honey of a tree: Black locust. ABJ 144 (5): 382–384.

Delaplane, K. 1998. Bee conservation in the Southeast. CES Bulletin 1164. Univ. of Georgia, College of Agricultural and Environmental Sciences.

Hooper, T., and M. Taylor. 1988. The beekeepers' garden. London: Alphabooks.

Matheson, A. (ed.). 1994. Forage for bees in an agricultural landscape. Cardiff, UK: IBRA.

O'Neal, R.J. and G.D. Waller. 1984. On the Pollen Harvest by the Honey Bee (*Apis mellifera* L.) near Tucson, AZ (1976-1981). Desert Plants 6 (2): 81-110.

Ramsay, J. 1987. Plants for beekeeping in Canada and the Northern U.S. Cambridge, UK: IBRA, Cardiff UK and Burlington Press. 198 pp.

Sawyer, R. 1988. Honey identification. Cardiff, UK: Cardiff Academic Press.

Tew, J. 1998. Nectar and pollen producing plants of Alabama: A guide for beekeepers. Wooster: Ohio State Univ., Ohio Agricultural Research and Development Center (OARDC).

Tew, J. 2000. Some Ohio nectar and pollen producing plants. Extension fact sheet. Wooster: Ohio State Univ.

Websites

Alabama Cooperative Extension System: http://www.aces.edu/department/ipm/npt1.htm; http://www.aces.edu/department/ipm/npt2.htm.

Penn State Univ.: http://MAAREC.cas.psu.edu.

Texas A & M Univ.: http://uvalde.tamu.edu/herbarium//index.html; http://www.vftn.org/projects/bryant/navbar_pages/texas_regions.htm.

Univ. of Arizona: http://ag.arizona.edu/pima/gardening/aridplants/.

Wikipedia: http://en.wikipedia.org/wiki/List_of_crop_plants_pollinated_by_bees.

INTEGRATED PEST MANAGEMENT (IPM)

Integrated Pest Management (IPM) for beekeepers. 2000. Mid-Atlantic Apiculture Research and Extension Consortium (MAAREC), Penn State Univ. Pub. 4.8. http://maarec.psu.edu/pdfs/IPM_FOR_.PDF.

Spivak, M. 2008. Bee health: Putting control in last place. ABJ 148 (11): 979–980.

VanderDussen, D. 2008. Varroa/tracheal mite IPM program for Canada using the alternative management option available in Canada as of 2005. Hive Lights (20): 23–25.

Websites

Penn State Univ.: http://MAAREC.cas.psu.edu.

Univ. of Minnesota: http://extension.umn.edu/honeybees.

USDA: http://www.reeis.usda.gov/web/crisprojectpages/186610.html.

Virginia Apiculture Program: http://web.ento.vt.edu/ento/project.jsp?projectID=12.

INTERNATIONAL BEEKEEPING

Adjare, S. 1990. Beekeeping in Africa. Service Bulletin 68/6. Rome: Food and Agriculture Organization of the United Nations.

Bradbear, N. 2004. Beekeeping and sustainable livelihoods. Rome: Agricultural Support Systems Division, Food and Agriculture Organization of the United Nations.

Crane, E. 1999. The world history of beekeeping and honey hunting. Routledge, New York.

IBRA, 16 North Road, Cardiff CF1 3DY, Wales, UK. http://www.ibra.org.uk.

Ransome, H.M. 2004. The sacred bee in ancient times and folklore. Mineola, NY: Dover.

Sammataro, D. 1980. Lesson plans for beekeeping. Washington, DC: US Peace Corps. Also online at http://www.beekeeping.com/articles/us/philippines/index.htm.

Websites

Bees for Development. http://www.beesfordevelopment.org/info/info/topbar/practical-beekeeping-topb.shtml.

INTERNATIONAL HONEY AND BEEKEEPING ASSOCIATIONS

http://www.honeyo.com/org-International.shtml.
IBRA: http://www.ibra.org.uk/.

MEAD

Morse, R.A. 1980. Making mead. Ithaca, NY: Wicwas Press.

Schramm, K. 2003. The compleat meadmaker: Home production of honey wine from your first batch to award-winning fruit and herb variations. Boulder, CO: Brewers.

Spence, P. 2002. Mad about mead! Nectar of the gods. Woodbury, MN: Llewellyn Worldwide.

MITES

Adamczyk, S., R. Lázaro, C. Pérez-Arquillué, S. Bayarri, and A. Herrera. Impact of the use of fluvalinate on different types of beeswax from Spanish hives. 2010. Archives of Environmental Contamination and Toxicology 58 (3): 733–739.

Aggarwal, K., and R.P. Kapil. 1988. Observations on the effect of queen cell construction on *Euvarroa sinhai* infestation in drone brood of *Apis florea*. In: G.R. Needham, R.E. Page Jr., M. Delfinado-Baker, and C.E. Bowman (eds.), Africanized Honey Bees and Bee Mites, pp. 404–408. Chichester: Ellis Horwood.

Ali, M.A., M.D. Ellis, J.R. Coats, and J. Grodnitzky. 2002. Laboratory evaluation of 17 monoterpenoids and field evaluation of two monoterpenoids and two registered acaricides for the control of *Varroa destructor* Anderson & Trueman (Acari: Varroidae). American Bee Journal 142 (1): 50–53.

Aliano, N.P., and M.D. Ellis. 2005. A strategy for using powdered sugar to reduce varroa populations in honey bee colonies. J. of Apicultural Research 44 (2): 54–57.

Anderson, D.L., and M.J. Morgan. 2007. Genetic and morphological variation of bee-parasite *Tropilaelaps* mites (Acair: Laelapidae): new and re-defined species. Experimental and Applied Acarology 43:1–24.

Anderson, D.L., and J.W.H. Trueman. 2000. *Varroa jacobsoni* (Acari: Varroidae) is more than one species. Experimental and Applied Acarology 24:165–189.

Aumeier, P. 2001. Bioassay for grooming effectiveness towards *Varroa destructor* mites in Africanized and Carniolan honey bees. Apidologie 32:81–90.

Berry, J.A., W.B. Owens, and K.S. Delaplane. 2010. Small-cell comb foundation does not impede Varroa mite population growth in honey bee colonies. Apidologie 41 (1): 40–44.

Booppha, B., S. Eittsayeam, K. Pengpat, and P. Chantawannakul. 2010. Development of bioactive ceramics to control mite and microbial diseases in bee farms. Advanced Materials Research 93–94:553–557.

Branco, M.R., N.A.C. Kidd, and R.S. Pickard. 2006. A comparative evaluation of sampling methods for *Varroa destructor* (Acari: Varroidae) population estimation. Apidologie 37:452–461.

Büchler, R., S. Berg, and Y. Le Conte. 2010. Breeding for resistance to *Varroa destructor* in Europe. Apidologie 41 (3): 393–408.

Çakmak, I. 2010. The over wintering survival of highly *Varroa destructor* infested honey bee colonies determined to be hygienic using the liquid nitrogen freeze killed brood assay. J. of Apicultural Research 49 (2): 197–201.

Çakmak, I., L. Aydin, and H. Wells. 2006. Walnut leaf smoke versus mint leaves in conjunction with pollen traps for control of *Varroa destructor*. Bulletin of the Veterinary Institute in Pulawy 50 (4): 477–479.

Calderón, R.A., J.W. van Veen, M.J. Sommeijer, and L.A. Sanchez. 2010. Reproductive biology of *Varroa destructor* in Africanized honey bees (*Apis mellifera*). Experimental and Applied Acarology 50 (4): 1–17.

Charriere, J., A. Imdorf, B. Bachofen, and A. Tschan. 2003. The removal of capped drone brood: an effective means of reducing the infestation of Varroa in honey bee colonies. Bee World 84 (3): 117–124.

Currie, R.W. 2001. Management of varroa mites in honey bees. University of Manitoba. Saskatchewan Agri-Food Innovation Fund, Canada.

Currie, R.W., and P. Gatien. 2006. Timing acaricide treatments to prevent *Varroa destructor* (Acari: Varroidae) from causing economic damage to honey bee colonies. Canadian Entomologist 138 (2): 238–252.

Danka, R.G., J.D. Villa, et al. 1995. Field test of resistance to *Acarapis woodi* (Acari: Tarsonemidae) and of colony production by four stocks of honey bees (Hymenoptera: Apidae). J. of Econom. Entomol. 88:584–591.

Delaplane, K.S., J.D. Ellis, and W.M. Hood. 2010. A test for interactions between *Varroa destructor* (Acari: Varroidae) and *Aethina tumida* (Coleoptera: Nitidulidae) in colonies of honey bees (Hymenoptera: Apidae). Ann. Entomol. Soc. America 103 (5): 711–715.

Delfinado-Baker, M., and K. Aggarwal. 1987. A new Varroa (Acari: Varroidae) from the nest of *Apis cerana* (Apidae). International J. of Acarology 13:233–237.

Donzé, G., et al. 1996. Effect of mating frequency and brood cell infestation rates on the reproductive success of the honeybee parasite *Varroa jacobsoni*. Ecological Entomol. 21:17–26.

Eickwort, G.C. 1988. The origins of mites associated with honey bees. In: G.R. Needham, R.E. Page Jr., M. Delfinado-Baker, and C. Bowman (eds.), Africanized honey bees and bee mites, pp. 327–338. New York: J. Wiley & Sons.

Eickwort, G.C. 1994. Evolution and life-history patterns of mites associated with bees. In: M.A. Houck (ed.), Mites: Ecological and evolutionary analyses of life-history patterns, pp. 218–251. New York: Chapman and Hall.

Ellis, A.M., G.W. Hayes, and J.D. Ellis. 2009. The efficacy of dusting honey bee colonies with powdered sugar to reduce varroa mite populations. J. of Apicultural Research 48 (1): 72–76.

Ellis, J.D., Jr., K.S. Delaplane, and W.M. Hood. 2001. Efficacy of a bottom screen device, Apistan™, and Apilife VAR™, in controlling *Varroa destructor*. American Bee Journal 141 (11): 813–816.

Ellis, J.D., Jr., J.D. Evans, and J. Pettis. 2010. Colony losses, managed colony population decline, and Colony Collapse Disorder in the United States. J. of Apicultural Research 49 (1): 134–136.

Elzen, P.J., J.R. Baxter, D. Westervelt, D. Causey, C. Randall, L. Cutts, and W.T. Wilson. 2001. Acaricide rotation plan for control of varroa. American Bee Journal 141 (6): 412.

Elzen, P.J., R.L. Cox, and W.A. Jones. 2004. Evaluation of food grade mineral oil treatment for varroa mite control. American Bee Journal 144 (12): 921–923.

Elzen, P.J., F.A. Eischen, J.R. Baxter, G.W. Elzen, and W.T. Wilson. 1999. Detection of resistance in U.S. *Varroa jacobsoni* Oud. (Mesotigmata: Varroidae) to the acaricide fluvalinate. Apidologie 30:13–18.

Elzen, P.J., R.D. Stipanovic, and R. Rivera. 2001. Activity of two preparations of natural smoke products on the behavior of *Varroa jacobsoni* Oud. American Bee Journal 141 (4): 289–291.

Elzen, P.J., and D. Westervelt. 2002. Detection of coumaphos resistance in *Varroa destructor* in Florida. American Bee Journal 142 (4): 291–292.

Emsen, B., and A. Dodologlu. 2009. The effects of using different organic compounds against honey bee mite (*Varroa destructor* Anderson and Trueman) on colony developments of honey bee (*Apis mellifera* L.) and residue levels in honey. J. of Animal and Veterinary Advances 8 (5): 1004–1009.

Fakhimzadeh, K. 2000. Potential of super-fine ground, plain white sugar dusting as an ecological tool for the control of varroasis in the honey bee (*Apis mellifera*). American Bee Journal 140 (6): 487–491.

Fassbinder, C., J. Grodnitzky, and J. Coats. 2002. Monoterpenoids as possible control agents for *Varroa destructor*. J. of Apicultural Research 41 (3–4): 83–88.

Floris, I., A. Satta, P. Cabras, V.L. Garau, and A. Angioni. 1997. A sequential sampling technique for female adult mites of *Varroa jacobsoni* Oudemans in the sealed worker brood of *Apis mellifera ligustica*. Apidologie 28:63–70.

Fuchs, S. 1990. Preference for drone brood cells by *Varroa jacobsoni* Oud in colonies of *Apis mellifera carnica*. Apidologie 21 (3): 193–199.

Gashout, H.A., and E. Guzmán-Novoa. 2009. Acute toxicity of essential oils and other natural compounds to the parasitic mite, *Varroa destructor*, and to larval and adult worker honey bees (*Apis mellifera* L.). J. of Apicultural Research 48 (4): 263–269.

Glenn, G.M., A.P. Klamczynski, D.F. Woods, B. Chiou, W.J. Orts, and S.H. Imam. 2010. Encapsulation of plant oils in porous starch microspheres. J. of Agricultural and Food Chemistry 58 (7): 4180–4184.

Harbo, J.R., and J.W. Harris. 2004. Effect of screen floors on populations of honey bees and parasitic mites (*Varroa destructor*). J. of Apicultural Research 43 (3): 114–117.

Harris, J.W., R.G. Danka, and J.D. Villa. 2010. Honey bees (Hymenoptera: Apidae) with the trait of varroa sensitive hygiene remove brood with all reproductive stages of varroa mites (Mesostigmata: Varroidae). Ann. Entomol. Soc. America 103 (2): 146–152.

Harz, M., F. Müller, and E. Rademacher. 2010. Organic acids: Acute toxicity on *Apis mellifera* and recovery in the haemolymph. J. of Apicultural Research 49 (1): 95–96.

Huang, Z. 2001. Mite zapper: A new and effective method for Varroa mite control. American Bee Journal 141 (10): 730–732.

Imdorf, A., and J-D. Charriere. Alternate varroa control. Bern: Swiss Bee Research Center, Dairy Research Station Liebefed. http://www.alp.admin.ch/themen/00502/00550/index.html?lang=en.

Jacobson, S. 2010. Locally adapted, varroa resistant honey bees: Ideas from several key studies. American Bee Journal 150 (8): 777–781.

James, R.R., G. Hayes, and J.E. Leland. 2006. Field trials on the microbial control of varroa with the fungus *Metarhizium anisopliae*. American Bee Journal 146 (11): 968–972.

Johnson, R.M., Z.Y. Huang, and M.R. Berenbaum. 2010. Role of detoxification in *Varroa destructor* (Acari: Parasitidae) tolerance of the miticide tau-fluvalinate. International J. Acarology 36 (1): 1–6.

Kanga, L.H.B., R.R. James, and D.G. Boucias. 2002. *Hirsutella thompsonii* and *Metarhizium anisopliae* as potential microbial control agents of *Varroa destructor*, a honey bee parasite. J. of Invertebrate Pathology 81 (3): 175–184.

Le Conte, Y., M. Ellis, and W. Ritter. 2010. Varroa mites and honey bee health: Can Varroa explain part of the colony losses? Apidologie 41 (3): 353–363.

Lee, K.V., R.D. Moon, E.C. Burkness, W.D. Hutchison, and M. Spivak. 2010. Practical sampling plans for *Varroa destructor* (Acari: Varroidae) in *Apis mellifera* (Hymenoptera: Apidae) colonies and apiaries. J. of Economic Entomol. 103 (4): 1039–1050.

Maggi, M.D., S.R. Ruffinengo, L.B. Gende, E.G. Sarlo, M.J. Eguaras, P.N. Bailac, and M.I. Ponzi. 2010. Laboratory evaluations of *Syzygium aromaticum* (L.) Merr. et Perry essential oil against *Varroa destructor*. J. of Essential Oil Research 22 (2): 119–122.

Mangum, W.A. 2007. Honey bee biology. A coexistence between varroa mites and bees, without any miticide treatments. American Bee Journal 147 (6): 489–491.

Martin, S.J., B.V. Ball, and N.L. Carreck. 2010. Prevalence and persistence of deformed wing virus (DWV) in untreated or acaricide-treated *Varroa destructor* infested honey bee (*Apis mellifera*) colonies. J. of Apicultural Research 49 (1): 72–79.

Mattheson, A. (ed.). 1994. New perspectives on varroa. Cardiff, UK: IBRA.

Meikle, W.G., G. Mercadier, N. Holst, C. Nansen, and V. Girod. 2008. Impact of a treatment of *Beauveria bassiana* (Deuteromycota: Hyphomycetes) on honeybee (*Apis mellifera*) colony health and on *Varroa destructor* mites (Acari: Varroidae) Apidologie 39 (2): 247–259.

Melathopoulos, A.P., M.L. Winston, R. Whittington, H. Higo, and M. le Doux. 2000. Field evaluation of neem and canola oil for the selective control of the honey bee (Hymenoptera: Apidae) mite parasites *Varroa jacobsoni* (Acari: Varroidae) and *Acarapis woodi* (Acari: Tarsonemidae). J. of Econom. Entomol. 93 (3): 559–567.

Milani, N. 1999. The resistance of *Varroa jacobsoni* Oud. to acaricides. Apidologie 30:229–234.

Mitchell, D., and D. Vanderdussen. 2010. Mite-away quick strip™ mid honey flow efficacy trial. American Bee Journal 150 (5): 487–489.

Mullin, C.A., M. Frazier, J.L. Frazier, S. Ashcraft, R. Simonds, D. vanEngelsdorp, and J.S. Pettis. 2010. High levels of miticides and agrochemicals in North American apiaries: Implications for honey bee health. PLoS ONE 5 (3): e9754. doi:10.1371/journal.pone.0009754.

Navajas, M., D.L. Anderson, L.I. De Guzman, Z.Y. Huang, J. Clement, T. Zhou, and Y. Le Conte. 2010. New Asian types of *Varroa destructor*: A potential new threat for world apiculture. Apidologie 41 (2): 181–193.

O'Meara, J. 2005. Walnut leaf smoke: A thrifty control of varroa mites. American Bee Journal 145 (1): 60–62.

Otis, G.W., and J. Kralj. 2001. Mites of economic importance not present in North America. In: T.C. Webster and K.S. Delaplane (eds.), Mites of the honey bee, pp. 251–272. Hamilton, IL: Dadant & Sons.

Peng, C.Y.S., X. Zhou, and H.K. Kaya. 2002. Virulence and site of infection of the fungus, *Hirsutella thompsonii*, to the honey bee ectoparasitic mite, *Varroa destructor*. J. of Invertebrate Pathology 81 (3): 185–195.

Pettis, J.S. 2001. Biology and life history of tracheal mites. In T.C. Webster and K.S. Delaplane (eds.), Mites of the honey bee, pp. 29–41. Hamilton, IL: Dadant & Sons.

Rath, W. 1992. The key to Varroa: the drones of *Apis cerana* and their call cap. American Bee Journal 132 (5): 329–331.

Rinderer, T.E., L.I. De Guzman, V.A. Lancaster, G.T. Delatte, and J.A. Stelzer. 1999. Varroa in the mating yard: I. The effects of *Varroa jacobsoni* and Apistan[(R)] on drone honey bees. American Bee Journal 139 (2): 134–139.

Rinderer, T.E., J.W. Harris, G.J. Hunt, and L.I. De Guzman. 2010. Breeding for resistance to *Varroa destructor* in North America. Apidologie 41 (3): 409–424.

Rosenkranz, P., P. Aumeier, and B. Ziegelmann. 2010. Biology and control of *Varroa destructor*. J. of Invertebrate Pathology 103 (Suppl. 1): S96–S119.

Ruffinengo, S.R., M. Maggi, S. Fuselli, I. Floris, G. Clemente, N.H. Firpo, P.N. Bailac, and M.I. Ponzi. 2006. Laboratory evaluation of *Heterothalamus alienus* essential oil against different pests of *Apis mellifera*. J. of Essential Oil Research 18 (6): 704–707.

Sammataro, D. 1996. Tracheal mites can be suppressed by oil patties. American Bee Journal 136:279–282.

Sammataro, D. 2006. An easy dissection technique for finding tracheal mites (Acari: Tarsonemidae) in honey bees (with Video link). International J. of Acarology 32:339–343. Video: http://www.ars.usda.gov/pandp/docs.htm?docid=14370.

Sammataro, D., J. Finley, B. Leblanc, G. Wardell, F. Ahumada-Segura, and M.J Carroll. 2009. Feeding essential oils and 2-heptanone in sugar syrup and liquid protein diets to honey bees (*Apis mellifera* L.) as potential varroa mite (*Varroa destructor*) controls. J. of Apicultural Research 48 (4): 256–262.

Sammataro, D., U. Gerson, and G. Needham. 2000. Parasitic mites of honey bees: Life history, implications, and impact. Annu. Rev. Entomol. 45:519–548.

Sammataro, D., and J.A. Yoder (eds.). In press. Recent investigations focused on the problems of honey bee pollinators.

Satta, A., I. Floris, M. Eguaras, P. Cabras, V.L. Garau, and M. Melis. 2005. Formic acid-based treatments for control of *Varroa destructor* in a Mediterranean area. J. of Econom. Entomol. 98 (2): 267–273.

Schäfer, M.O., W. Ritter, J.S. Pettis, and P. Neumann. 2010. Winter losses of honeybee colonies (Hymenoptera: Pidae): The role of infestations with *Aethina tumida* (Coleoptera: Nitidulidae) and *Varroa destructor* (Parasitiformes: Varroidae). J. of Econom. Entomol. 103 (1): 10–16.

Spivak, M., and G.S. Reuter. 1998. Performance of hygienic honey bee colonies in a commercial apiary. Apidologie 29:291–302.

Tabor, K.L., and J.T. Ambrose. 2001. The use of heat treatment for control of the honey bee mite, *Varroa destructor*. American Bee Journal 141 (10): 733–736.

Tu, S., X. Qiu, L. Cao, R. Han, Y. Zhang, and X. Liu. 2010. Expression and characterization of the chitinases from *Serratia marcescens* GEI strain for the control of *Varroa destructor*, a honey bee parasite. J. of Invertebrate Pathology 104 (2): 75–82.

Underwood, R.M., and R.W. Currie. 2003. The effects of temperature and dose of formic acid on treatment efficacy against *Varroa destructor* (Acari: Varroidae), a parasite of *Apis mellifera* (Hymenoptera: Apidae). Exp. App. Acarol. 29:303–313.

Walter, D.E. 2006. Invasive mite identification. Fort Collins, CO, and Raleigh, NC: Colorado State University and USDA/APHIS/PPQ Center for Plant Health Science and Technology. http://www.lucidcentral.org/keys/v3/mites/Invasive_Mite_Identification/key/Whole_site/Home_whole_key.html.

Walter, D.E., G. Krantz, and E. Lindquist. Acari. The mites. Tree of Life Web Project. http://tolweb.org/Acari/2554.

Wantuch, H.A., and D.R. Tarpy. 2009. Removal of drone brood from *Apis mellifera* (Hymenoptera: Apidae) colonies to control *Varroa destructor* (Acari: Varroidae) and retain adult drones. J. of Econom. Entomol. 102 (6): 2033–2040.

Warrit, N., D.R. Smith, and C. Lekprayoon. 2006. Genetic subpopulations of Varroa mites and their *Apis cerana* hosts in Thailand. Apidologie 37 (1): 19–30.

Wilkinson, D., and G.C. Smith. 2002. Modeling the efficiency of sampling and trapping *Varroa destructor* in the drone brood of honey bees (*Apis mellifera*). American Bee Journal 142 (3): 209–212.

Wilson, W.T., J.S. Pettis, C.E. Henderson, and R.A. Morse. 1997. Tracheal mites. In: Honey bee pests, predators and diseases. 3rd ed. Medina, OH: A.I. Root.

Websites

Barlow, V.M. 2006. Sampling methods for varroa mites on the domesticated honeybee, http://www.ext.vt.edu/pubs/entomology/444-103/444-103.html.

British Columbia apiculture fact sheet: http://www.agf.gov.bc.ca/apiculture/factsheets/222_vardetect.htm.

Drone uncapping: http://bees.tennessee.edu/ipm/uncapping.htm.

Great Lakes IPM sells varroa sticky boards: http://glipm@greatlakesipm.com.

Kansas Statue University: http://docs.ksu.edu.sa/PDF/Articles18/Article180069.pdf.

Victoria, Australia: http://new.dpi.vic.gov.au/notes/agg/bees--and--wasps/ag000-sugar-shake-test-detection-of-varroa-mite.

Virginia Cooperative Extension: http://pubs.ext.vt.edu/444/444-103/444-103.html.

http://www.alp.admin.ch/themen/00502/00515/00516/index.html?lang=en&download=M3wBPgDB/.

http://www.bioone.org/doi/abs/10.1603/00220493 (2005)098%5B0267:FATFCO%5D2.0.CO%3B2.

NON-APIS BEE POLLINATORS

Batra, S.W.T. 1989. Japanese hornfaced bees, gentle and efficient new pollinators. Pomona 22:3–5.

Buchmann, S.L., and G.P. Nabhan. 1996. Forgotten pollinators. Washington, DC: Island Press.

Canto-Aguilar, M.A., and V. Parra-Tabla. 2000. Importance of conserving alternative pollinators: Assessing the pollination efficiency of the squash bee, *Peponapis limitaris* in *Cucurbita moschata* (Cucurbitaceae). J. of Insect Conservation 4 (3): 203–210.

Costa, L., and G.C. Venturieri. 2009. Diet impacts on *Melipona flavolineata* workers (Apidae, Meliponini). JAR 48 (1): 38–45.

Dogterom, M. 2002. Pollination with mason bees: A gardener and naturalists' guide to managing mason bees for fruit production. Coquitlam, BC, Canada: Beediverse Publishing.

dos Santos, C.G., F.L. Megiolaro, J.E. Serrão, and B. Blochtein. 2009. Morphology of the head salivary and intramandibular glands of the stingless bee *Plebeia emerina* (Hymenoptera: Meliponini) workers associated with propolis. Ann. Entomol. Society of America 102 (1): 137–143.

Kremen, C., N.M. Williams, and R.W. Thorp. 2002. Crop pollination from native bees at risk from agricultural intensification. Proceedings of the National Academy of Sciences of the United States of America 99 (26): 16812–16816.

Richards, K.W. 1996. Effect of environment and equipment on productivity of alfalfa leafcutter bees (Hymenoptera: Megachilidae) in southern Alberta, Canada. Canadian Entomologist 128 (1): 47–56.

Sampson, B.J., S.J. Stringer, J.H. Cane, and J.M. Spiers. 2004. Screenhouse evaluations of a mason bee *Osmia ribifloris* (Hymenoptera: Megachilidae) as a pollinator for blueberries in the southeastern United States. Fruits Review 3 (3–4): 381–392.

Shuler, R.E., T.A.H. Roulston, and G.E. Farris. 2005. Farming practices influence wild pollinator populations on squash and pumpkin. J. of Economic Entomol. 98 (3): 790–795.

Southwick, E.E. 1992. Estimating the economic value of honey bees (Hymenoptera: Apidae) as agricultural pollinators in the United States. J. of Economic Entomol. 85:621–633.

Stephen, W.P., and S. Rao. 2005. Unscented color traps for non-*Apis* bees (Hymenoptera: Apiformes). J. of Kansas Entomol. Society 78 (4): 373–380.

Stubbs, C.S., F.A. Drummond, and E.A. Osgood. 1994. *Osmia ribifloris biedermannii* and *Megachile rotundata* (Hymenoptera: Megachilidae) introduced into the lowbush blueberry agroecosystem in Maine. J. of Kansas Entomol. Society 67:173–185.

Tepedino, V.J. 1981. The pollination efficiency of the squash bee (*Peponais pruinosa*) and the honey bee (*Apis mellifera*) on summer squash (*Cucurbita pepo*). J. of Kansas Entomol. Society 54:359–377.

Torchio, P.F. 1991. Use of *Osmia lignaria propinqua* (Hymenoptera: Megachilidae) as a mobile pollinator of orchard crops. Environmental Entomol. 20:590–596.

Tuell, J.K., J.S. Ascher, and R. Isaacs. 2009. Wild bees (Hymenoptera: Apoidea: Anthophila) of the Michigan highbush blueberry agroecosystem. Ann. Entomol. Society of America 102 (2): 275–287.

Weinberg, D., and C.M.S. Plowright. 2006. Pollen collection by bumblebees (*Bombus impatiens*): The effects of resource manipulation, foraging experience, and colony size. JAR 45 (2): 22–27.

Willis, D.S. 1995. Foraging dynamics of *Peponapis pruinosa* (Hymenoptera: Anthophoridae) on pumpkin (*Cucurbita pepo*) in southern Ontario. Canadian Entomol. 127:167–175.

Websites

Agriculture and Agri-Food Canada: Steven Javorek, Kenna MacKenzie, Ken Richards. http://www.agr.gc.ca.

American Farm Bureau Federation: http://www.fb.org.

American Museum of Natural History: Jerry Rozen, John Ascher. http://www.amnh.org.

National Pollination/pollinator groups: http://www.pollinator.org/; http://pollinator.com/.

North American Pollinator Protection Campaign: http://www.nappc.org/. Go to NAAPPC site and click on Partners.

World Wildlife Fund: Jeff England, Taylor Ricketts. http://www.panda.org,

The Xerces Society: http://www.xerces.org/.

PESTICIDES

Alix, A., and C. Vergnet. 2007. Risk assessment to honey bees: A scheme developed in France for non-sprayed systemic compounds. Pest Management Science 63 (11): 1069–1080.

Chauzat, M.P., and J.P. Faucon. 2007. Pesticide residues in beeswax samples collected from honey bee colonies (*Apis mellifera* L.) in France. Pest Management Science 63 (11): 1100–1106.

Chauzat, M.P., J.P. Faucon, A.C. Martel, J. Lachaize, N. Cougoule, and M. Aubert. 2006. A survey of pesticide residues in pollen loads collected by honey bees in France. J. of Economic Entomol. 99 (2): 253–262.

Choudhary, A., and D.C. Sharma. 2008. Dynamics of pesticide residues in nectar and pollen of mustard (*Brassica juncea* (L.) Czern.) grown in Himachal Pradesh (India). Environmental Monitoring and Assessment 144 (1–3): 143–150.

Colin, M.E., J.M. Bonmatin, I. Moineau, C. Gaimon, S. Brun, and J.P. Vermandere. 2004. A method to quantify and analyze the foraging activity of honey bees: Relevance to the sublethal effects induced by systemic insecticides. Archives of Environmental Contamination and Toxicology 47 (3): 387–395.

Conti, M.E., and F. Botre. 2001. Honeybees and their products as potential bioindicators of heavy metals contamination. Environmental Monitoring and Assessment 69 (3): 267–282. http://www .kluweronline.com/issn/1420-2026/contents.

Cutler, G.C., and C.D. Scott-Dupree. 2007. Exposure to clothianidin seed-treated canola has no long-term impact on honey bees. J. of Economic Entomol. 100 (3): 765–772.

Desneux, N., A. Decourtye, and J.M. Delpuech. 2007. The sublethal effects of pesticides on beneficial arthropods. Annu. Rev. Entomol. 52:81–106.

Devillers, J., A. Decourtye, H. Budzinski, M.H. Pham-Delègue, S. Cluzeau, and G. Maurin. 2003. Comparative toxicity and hazards of pesticides to *Apis* and non-*Apis* bees. A chemometrical study. Environmental Research 14 (5–6): 389–403.

Frazier, M., C. Mullin, J. Frazier, and S. Ashcraft. 2008. What have pesticides got to do with it? ABJ 148 (6): 521–523.

Johnson, R.M., M.D. Ellis, C.A. Mullin, and M. Frazier. 2010. Pesticides and honey bee toxicity—USA. Apidologie 41 (3): 312–331.

Johnson, R.M., Z.Y. Huang, and M.R. Berenbaum. 2010. Role of detoxification in *Varroa destructor* (Acari: Parasitidae) tolerance of the miticide tau-fluvalinate. International J. of Acarology 36 (1): 1–6.

Karazafiris, E., C. Tananaki, U. Menkissoglu-Spiroudi, and A. Thrasyvoulou. 2008. Residue distribution of the acaricide coumaphos in honey following application of a new slow-release formulation. Pest Management Science 64 (2): 165–171.

Martel, A.-C., S. Zeggane, C. Aurières, P. Drajnudel, J.-P. Faucon, and M. Aubert. 2007. Acaricide residues in honey and wax after treatment of honey bee colonies with Apivar® or Asuntol® 50. Apidologie 38 (6): 534–544.

Mineau, P., K.M. Harding, M. Whiteside, et al. 2008. Using reports of bee mortality in the field to calibrate laboratory-derived pesticide risk indices. Environmental Entomol. 37 (2): 546–554.

Mullin C.A., M. Frazier, J.L. Frazier, S. Ashcraft, R. Simonds, D. vanEnglesdorp, and J. Pettis. 2010. High levels of miticides and agrochemicals in North American apiaries: Implications for honey bee health. PLoS ONE 5(3): e9754. doi:10.1371/journal.pone.0009754.

Romaniuk, K., W. Witkiewicz, and A. Spodniewska. 2004. Residues of HCH and DDT in linden flowers, bees, drones as well as maggots and *Varroa destructor* females [Pozostałości HCH i DDT w kwiatach lipy, pszczołach, trutniach oraz czerwiu i samicach *Varroa destructor*]. Medycyna Weterynaryjna 60 (12): 1352–1353.

Rortais, A., G. Arnold, M.-P. Halm, and F. Touffet-Briens. 2005. Modes of honeybees exposure to systemic insecticides: Estimated amounts of contaminated pollen and nectar consumed by different categories of bees. Apidologie 36: 71–83.

Tew, J.E. 1996. Protecting honey bees from pesticides. Extension fact sheet. Wooster: Ohio State Univ. Extension, Ohio Ag. Research and Development Center (OARDC).

Thompson, H.M., and C. Maus. 2007. The relevance of sublethal effects in honey bee testing for pesticide risk assessment. Pest Management Science 63 (11): 1058–1061.

Websites

USGS pesticide usage maps: http://water.usgs.gov/ nawqa/pnsp/usage/maps/compound_listing. php?year=02. Contact for pesticide samples.

POLLEN

Hodges, D. 1974, 1994. The pollen loads of the honeybee. London: IBRA.

Kesseler, R., and M.M. Harley. 2004. Pollen: The hidden sexuality of flowers (2nd edn.). London: Papadakis Publisher. 264 pp.

Kirk, W.D.J. 2006. A colour guide to pollen loads of the honey bee (2nd revised edn., spiral-bound). Cardiff, UK: IBRA.

LeBlanc, B.W., O.K. Davis, S. Boue, A. DeLucca, and T. Deeby. 2009. Antioxidant activity of Sonoran Desert bee pollen. Food Chemistry 115 (4): 1299–1305.

Lipiński, Z., M. Farjan, K. Zóltowska, and B. Polaczek. 2008. Effects of dietary transgenic *Bacillus thuringiensis* maize pollen on hive worker honeybees. Polish J. of Environmental Studies 17 (6): 957–961.

Pleasants, J., R.L. Hellmich, G.P. Dively, M.K.D. Sears, E. Stanley-Horn, H.R. Mattila, J.E. Foster, P. Clark, and G.D. Jones. 2001. Corn pollen deposition on milkweeds in and near cornfields. Proceedings of the National Academy of Sciences of the United States of America 98 (21): 11919–11924. doi: 10.1073/pnas.211287498.

Price, L.D., F. Chukwuma, and J.J. Adamczyk Jr. 2004. Honey bee (Hymenoptera: Apidae) pollen load rate based on pollen grain size. J. of Entomol. Science 39 (4): 677–678.

Wenning, C.J. 2003. Pollen and the honey bee. ABJ 143(5): 394–397.

POLLINATION

Abrol, D.P. 2007. Honeybees and rapeseed: A pollinator-plant interaction. Advances in Botanical Research 45: 337–367.

Barth, F.G. 1991. Insects and flowers. Princeton, NJ: Princeton Univ. Press.

Blanke, M.M. 2008. Perspectives of fruit research and apple orchard management in Germany in a changing climate. Acta Horticulturae 772:441–446.

Capaldi, E., and F. Dyer. 1999. The role of orientation flights on homing performance in honeybees. J. of Experimental Biology 202:1655–1666.

Chittka, L., and N.E. Raine. 2006. Recognition of flowers by pollinators. Current Opinion in Plant Biology 9:428–435.

Chittka, L., J.D. Thompson, and N.M. Waser. 1999. Flower constancy, insect psychology, and plant evolution. Naturwissenschaften 86:361–377.

Dafni, A., P.G. Kevan, and B.C. Husband (eds.). 2005. Practical pollination biology. ON, Canada: Enviroquest.

Dag, A., R.A. Stern, and S. Shafir. 2005. Honey bee (*Apis mellifera*) strains differ in apple (*Malus domestica*) pollen foraging preference. JAR 44 (1): 15–20.

David, A., and S. Bucknall. 2004. Plants and honey bees: An introduction to their relationships. Mytholmroyd, West Yorkshire, UK: Northern Bee Books.

Dedej, S., and K.S. Delaplane. 2003. Honey bee (Hymenoptera: Apidae) pollination of rabbiteye blueberry *Vaccinium ashei* var. 'Climax' is pollinator density-dependent. J. of Economic Entomol. 96 (4): 1215–1220.

DeGrandi-Hoffman, G., and J.C. Watkins. 2000. The foraging activity of honey bees *Apis mellifera* and non-*Apis* bees on hybrid sunflowers (*Helianthus annuus*) and its influence on cross-pollination and seed set. JAR 39 (1–2): 37–45.

Delaplane, K.S., and D.F. Mayer. 2000. Crop pollination by bees. New York: CABI. 344 pp.

Dryer, A.G., C. Neumeyer, and L. Chittka. 2005. Honeybee (*Apis mellifera*) vision can discriminate between and recognize images of human faces. J. of Experimental Biology 208:4709–4714.

Dryer, A.G., M.G.P. Rosa, and D.H. Reser. 2008. Honeybees can recognize images of complex natural scenes for use as potential landmarks. J. of Experimental Biology 211:1180–1186.

Ellis, A., and K.S. Delaplane. 2008. Effects of nest invaders on honey bee (*Apis mellifera*) pollination efficacy. Agriculture, Ecosystems and Environment 127 (3–4): 201–206.

Free, J.B. 1993. Insect pollination of crops (2nd edn.). London: Academic Press.

Goodman, R., G. Hepworth, P. Kaczynski, B. McKee, S. Clarke, and C. Bluett. 2001. Honeybee pollination of buckwheat (*Fagopyrum esculentum* Moench) cv. Manor. Australian J. of Experimental Ag. 41 (8): 1217–1221.

Gould, J. 2004. Animal navigation. Current Biology 14:R221–R224.

Gruter, C., M.S. Balbuena, and W.M. Farnia. 2008. Informational conflicts created by the waggle dance. Proceedings of the Royal Society, B, Biological Sciences 275:1321–1327.

Gunduz, A., H. Bostan, S.Turedi, İ. Nuhoğlu, and T. Patan. 2007. Wild flowers and mad honey. Wilderness and Environmental Medicine 18:69–71. doi: 10.1580/06-WEME-LE-042R.1.

Higo, H.A. 1995. Mechanisms by which honey bee (Hymenoptera: Apidae) queen pheromone sprays enhance pollination. Ann. Entomol. Society of America 88:366–373.

Higo, H.A., N.D. Rice, M.L. Winston, and B. Lewis. 2004. Honey bee (Hymenoptera: Apidae) distribution and potential for supplementary pollination in commercial tomato greenhouses during winter. J. of Economic Entomol. 97 (2): 163–170. http://archives.cnn.com/2000/ NATURE/05/05/pollinators.peril/.

James, R.R., and T.L. Pitts-Singer. 2008. Bee pollination in agricultural ecosystems. Oxford, UK: Oxford Univ. Press. 232 pp.

Kerns, C.A., and D.W. Inouye. 1993. Techniques for pollination biologists. Niwot: Univ. Press of Colorado. 583 pp.

Klein, B.A. 2006. Caste-dependent sleep of worker honey bees. J. of Experimental Biology 211:3028–3040.

Ledford, H. 2007. Plant biology: The flower of seduction. Nature 445 816–817.

McGregor, S.E. 1976. Insect pollination of crops. Washington, DC: USDA-ARS. Online at http://gears.tucson.ars.ag.gov/book/.

National Research Council. 2007. Status of pollinators in North America. Washington, DC: National Academies Press.

Nicolson, S.W., M. Nepi, and E. Pacini (eds.). 2007. Nectaries and nectar. Doetinchem, The Netherlands: Springer.

Pierce, A.L., L.A. Lewis, and S.S. Schneider. 2007. The use of the vibration signal and worker piping to influence queen behavior during swarming in honey bees, *Apis mellifera*. Ethology 113:267–276.

Proctor, M., P. Yeo, and A. Lack. 1996. The natural history of pollination. Portland, OR: Timber Press. 479 pp.

Sauer, S., E. Hermann, and W. Kaiser. 2004. Sleep deprivation in honey bees. J. of Sleep Research 13:145–152.

Seeley, T.D., and P.K. Visscher. 2003. Choosing a home: How the scouts in a honey bee swarm perceive the completion of their group decision making. Behavioral Ecology and Sociobiology 54:511–520.

Shemesh, Y., M. Cohen, and G. Block. 2007. The natural plasticity in circadian rhythms is mediated by reorganization in the molecular clockwork in honeybees. FASEB J. 21:2304–2311.

Shivanna, K.R., and V.K. Sawhney (eds.). 1997. Pollen biotechnology for crop production and improvement. Cambridge, UK: Cambridge Univ. Press.

Skinner, J.A. n.d. Making a pollination contract. Ag. Ext. Serv. Knoxville: Univ. of Tennessee.

Tallamy, D.W. 2008. Bringing nature home. How you can sustain wildlife with native plants. Portland, OR: Timber Press. http://bringingnaturehome.net/.

Thom, C., D.C. Gilley, and J. Tautz. 2004. Working piping in honey bees (*Apis mellifera*): The behavior of piping nectar foragers. Behavioral Ecology and Sociobiology 53:199–205.

Traynor, J. 1993. Almond pollination handbook. Bakersfield, CA: Kovak Books.

Waser, N.M., and J. Ollerton (eds.). 2007. Plant-pollinator interactions: from specialization to generalization. Chicago: Univ. of Chicago Press.

Wehner, R. 1986. Visual navigation in insects: Coupling of egocentric and geocentric information. J. of Experimental Biology 199:129–140.

Whynott, D. 1991. Following the bloom: Across America with the migratory beekeepers. Boston: Beacon Press.

Williams, S.K., and A.G. Dryer. 2007. A photographic simulation of insect vision. J. of Ophthalmis Photography 29:10–14.

Williams, S.K., D. Reiser, and A.G. Dyer. 2008. A biologically inspired mechano-optical imaging system based on insect vision. J. of Biocommunication 34: e3–e7.

Websites

Hanlon, M. A bee's-eye view: How insects see flowers very differently to us. http://www.dailymail.co.uk/pages/livearticles/technology/technology.html?in_article_id=47389.

North American Pollinator Protection Campaign: http://www.nappc.org/.

NYC Beewatchers, a great pollinator project: http://www.nycbeewatchers.org.

Orchard mason bee: http://www.accessone.com/~knoxclr/omb.htm.

Pollinator Conservation Digital Library: http://libraryportals.com/PCDL/plants.

Pollinator Conservation Program Xerces Society: http://www.xerces.org/Pollinaotr_Insect_Conservation/.

Pollinator Partnership: http://www.pollinator.org.

Send for sample pollination contract: Penn State Univ., Dept. Entomology, 501 ASI Bldg., State College, PA 16802.

PRODUCTS OF THE HIVE (OTHER THAN HONEY)

Bankova, V. 2005. Recent trends and important developments in propolis research. Complementary and Alternative Medicine 2 (1): 29–32. http://ecam.oxfordJ.s.org/cgi/content/full/2/1/29.

Bowling, A.C. 2006. Complementary and alternative medicine and multiple sclerosis. New York: Dmos Medical.

Brätter, C., M. Tregel, C. Liebenthal, and H.D. Volk. 1999. Prophylactic effectiveness of propolis for immunostimulation: A clinical pilot study. Forsch Komplementarmed 6 (5): 256–260. http://www.ncbi.nlm.nih.gov/sites/entrez?uid=10575279.

Croci, A.N., B. Cioroiu, et al. 2009. HPLC evaluation of phenolic and polyphenolic acids from propolis. Farmacia 57 (1): 52–57.

da Silva, F.B., J.M. Almeida, and S.M. Sousa. 2004. Natural medicaments in endodontics—A comparative study of the anti-inflammatory action. Brazilian Oral Research 18 (2): 174–179. http://www.scielo.br/scielo.php?pid=S1806-83242004000200015&script=sci_arttext&tlng=en.

Duarte, S., P.L. Rosalen, M.F. Hayacibara, J.A. Cury, W.H. Bowen, et al. 2006. The influence of a novel propolis on mutans streptococci biofilms and caries development in rats. Archives of Oral Biology 51 (1): 15–22. http://www.ncbi.nlm.nih.gov/sites/entrez?uid=16054589.

Fearnley, J. 2001. Bee propolis: Natural healing from the hive. London: Souvenir Press.

Gambichler, T., S. Boms, and M. Freitag. 2004. Contact dermatitis and other skin conditions in instrumental musicians. BMC Dermatology 4:3.

Gregory, S.R., N. Piccolo, M.T. Piccolo, M.S. Piccolo, and J.P. Heggers. 2002. Comparison of propolis skin cream to silver sulfadiazine: A naturopathic alternative to antibiotics in treatment of minor burns. J. of Alternative and Complementary Medicine 8 (1): 77–83. http://www.ncbi.nlm.nih.gov/sites/entrez?uid=11890438.

Hocking, B., and F. Matsumura. 1960. Bee brood as food. Bee Wld. 41 (5): 113–120.

Koo, H., J.A. Cury, P.L. Rosalen, G.M. Ambrosano, M. Ikegaki, and Y.K. Park. 2002. Effect of a mouth-rinse containing selected propolis on 3-day dental plaque accumulation and polysaccharide formation. Caries Research 36 (6): 445–448. http://www.ncbi.nlm.nih.gov/sites/entrez?uid=12459618.

Krell, R. 1996. Value-added products from beekeeping. Service Bulletin 124. Rome: Food and Agriculture Organization of the United Nations.

Lensky, Y., and A. Mizrahi (eds.). 1996. Bee products—Properties, applications and apitherapy: Proceedings of an international conference held in Tel Aviv, Israel, May 26–30, 1996.

Marcucci, M.C. 1995. Propolis: Chemistry, composition, biological properties and therapeutic activity. Apidologie 26:83–99.

Mizrahi, A., and Y. Lensky. 1997. Bee products: Properties, applications, and apitherapy. New York: Springer.

Munn, P. (ed.). 1998. Beeswax and propolis for pleasure and profit. Cardiff, UK: IBRA. http://www.ibra.org.uk.

Munstedt, K., and M. Zygmunt. 2001. Propolis—Current and future medical uses. ABJ 141 (7): 507–510.

Mustafa, F.B., F.S.P. Ng, T.H. Nguyen, and L.H.K. Lim. 2008. Honeybee venom secretory phospholipase A, induces leukotriene production but not histamine release from human basophils. Clinical and Experimental Immunology 151 (1): 94–100. http://dx.doi.org/10.1111/j.1365-2249.2007.03542.x.

Ocakci, A., M. Kanter, M. Cabuk, and S. Buyukbas. 2006. Role of caffeic acid phenethyl ester, an active component of propolis, against NAOH-induced esophageal burns in rats. International J. of Pediatric Otorhinolaryngology 70 (10): 1731–1739. http://www.ncbi.nlm.nih.gov/sites/entrez?uid=16828884.

Orsi, R.O., J.M. Sforcin, V.L.M. Rall, S.R.C. Funari, et al. 2005. Susceptibility profile of salmonella against the antibacterial activity of propolis produced in two regions of Brazil. J. of Venomous Animals and Toxins Including Tropical Diseases 11 (2): 109–116. http://www.doaj.org/doaj?func=abstract&id=115293.

Park Y.K., S.M. Alencar, and C.L. Aguiar. 2005. Botanical origin and chemical composition of Brazilian propolis. J. of Ag. and Food Chemistry 50:2502–2506.

Park, Y.K., M.H. Koo, J.A. Abreu, M. Ikegaki, J.A. Cury, and P.L. Rosalen. 1998. Antimicrobial activity of propolis on oral microorganisms. Curr. Microbiol. 36 (1): 24–28. doi:10.1007/s002849900274; http://www.ncbi.nlm.nih.gov/sites/entrez?uid=9405742.

Quiroga, E.N., D.A. Sampietro, et al. 2006. Propolis from the northwest of Argentina as a source of antifungal principles. J. of Applied Microbiology 101 (1): 103–110. http://dx.doi.org/10.1111/j.1365-2672.2006.02904.

Sanford, MT. 1994. Producing pollen. Gainesville: Univ. of Florida, Institute of Food and Agricultural Sciences. Document ENY118. Revised February 1, 1995.

Sehn, E., L. Hernandes, S.L. Franco, C.C.M. Gonçalves, and M.L. Baesso. 2009. Dynamics of re-epithelialisation and penetration rate of a bee propolis formulation during cutaneous wounds healing. Analytica Chimica Acta 635 (1): 115–120.

Sforcin, J.M. 2007. Propolis and the immune system: A review. J. of Ethnopharmacology 113 (1): 1–14. http://dx.doi.org/10.1016/j.jep.2007.05.012.

Taylor, R.L. 1975. Butterflies in my stomach. Santa Barbara, CA: Woodbridge Press.

Taylor, R.L., and B.J. Carter. 1976. Entertaining with insects; or, The Original Guide to Insect Cookery. Santa Barbara, CA: Woodbridge Press.

Townsend, L. n.d. Rearing wax worms, ENTFACT-011 pamphlet. Univ. of Kentucky College of Ag. http://www.ca.uky.edu/entomology/entfacts/ef011.asp; http://www.ca.uky.edu/entomology/entfacts/entfact-pdf/ef011.pdf.

Trusheva, B., M. Popova, V. Bankova, S. Simova, M.C. Marcucci, et al. 2006. Bioactive constituents of Brazilian red propolis. Evidence-based Complementary and Alternative Medicine 3 (2): 249–254.

Zhou, J., X. Xue, Y. Li, J. Zhang, F. Chen, L. Wu, L. Chen, and J. Zhao. 2009. Multiresidue determination of tetracycline antibiotics in propolis by using HPLC-UV detection with ultrasonic-assisted extraction and two-step solid phase extraction. Food Chemistry 115 (3): 1074–1080.

Websites

Ohio State Univ., rearing waxmoth: http://www.ncbuy.com/flowers/articles/01_10233.html.

The Food Insects Newsletter was published from 1988 to 2000, articles about edible insects from all over the world, including instructions to raise insects, their nutritional properties, recipes, medicinal uses, and so forth. All thirteen volumes are now available as a single book. Can be ordered at http://www.hollowtop.com/finl_html/finl.html.

QUEENS

Al-Lawati, H., G. Kamp, and K. Bienefeld. 2009. Characteristics of the spermathecal contents of old and young honeybee queens. J. of Insect Physiology 55 (2): 117–122. http://dx.doi.org/10.1016/j.jinsphys.2008.10.010.

Cobey, S. 2003. The extraordinary honey bee mating strategy and a simple field dissection of the spermatheca: A three-part series, Part 1. ABJ 143 (1): 67–69.

Collins, A.M. 2004. Functional longevity of honey bee, *Apis mellifera*, queens inseminated with low viability semen. JAR 43 (4): 167–171.

Collins, A.M. 2005. Insemination of honey bee, *Apis mellifera*, queens with non-frozen stored semen: Sperm concentration measured with a spectrophotometer. JAR 44 (4): 141–145.

Doolittle, G.M. 2008. Scientific queen rearing. Kalamazoo, MI: Wicwas Press.

Flores, A. 2007. Honey bee genetics vital in disease resistance. Ag. Research 55 (1): 14. http://www.ars.usda.gov/is/AR/archive/jan07/bee0107.htm.

Frake, A.M., L.I. De Guzman, and T.E. Rinderer. 2009. Comparative resistance of Russian and Italian honey bees (Hymenoptera: Apidae) to small hive beetles (Coleoptera: Nitidulidae). J. of Economic Entomol. 102 (1): 13–19.

Harris, J.L. 2008. Effect of requeening on fall populations of honey bees on the northern Great Plains of North America. JAR 47 (4): 271–280.

Laidlaw, H.H., and R. Page. 1998. Queen rearing and bee breeding. Kalamazoo, MI: Wicwas Press.

Mackensen, O., and K.W. Tucker. 1970. Instrumental insemination of queen bees. USDA Ag. Handbook 390.

Marterre, B. 2009. Foolproof requeening. ABJ 149 (3): 227–231.

Mattila, H.R., and T.D. Seeley. 2007. Genetic diversity in honey bee colonies enhances productivity and fitness. Science 317:362–364.

Morse, R.A. 1979. Rearing queen honey bees. Ithaca, NY: Wicwas Press.

Morse, R.A. 1994. Rearing queen honey bees (2nd edn.). Kalamazoo, MI: Wicwas Press.

Page, R.E., and H.H. Laidlaw. 1985. Closed population honeybee breeding program. Bee Wld. 66 (2): 63–72.

Pettis, J.S., A.M. Collins, R. Wilbanks, and M.F. Feldlaufer. 2004. Effects of coumaphos on queen rearing in the honey bee, *Apis mellifera*. Apidologie 35 (6): 605–610.

Rinderer, T.E. (ed.). 1986. Bee genetics and breeding. Mytholmroyd, Hebden Bridge, UK: Northern Bee Books. Reprinted 2009.

Rothenbuhler, W.C. 1964. Behaviour genetics of nest cleaning in honey bees. I. Responses of four inbred lines to disease-killed brood. Animal Behaviour 12:578–583.

Rothenbuhler, W.C. 1964. Behaviour genetics of nest cleaning in honey bees. IV. Responses of F1 and backcross generations to disease-killed brood. American Zoologist 4:111–128.

Spivak, M., and G.S. Reuter. 1994. Successful queen rearing. St. Paul: Minnesota Ext. Serv., Univ. of Minnesota.

Villa, J.D., and T.E. Rinderer. 2008. Inheritance of resistance to *Acarapis woodi* (Acari: Tarsonemidae) in crosses between selected resistant Russian and selected susceptible U.S. honey bees (Hymenoptera: Apidae). J. of Economic Entomol. 101 (6): 1756–1759.

Websites

Arizona Cooperative Extension: http://www.extension.org/events/1097.

The Development of Honey Bee Artificial Insemination, on the website Apiculture Program, Dept. of Entomol., North Carolina State Univ. http://www.cals.ncsu.edu/entomology/apiculture/PDF%20files/2.14.pdf.

Glenn Apiaries: Principles of Honey Bee Genetics, on their website: http://www.glenn-apiaries.com/genetics.html.

Honey bee insemination instrument, on the website: Chung Jin Biotech Co. Ltd. http://younan99.en.ecplaza.net/catalog.asp?DirectoryID=81469&CatalogID=197772.

Univ. of California Davis: http://entomology.ucdavis.edu/courses/beeclasses/queenrearing.html.

Univ. of Minnesota: http://www.extension.umn.edu/Honeybees/components/publiccourses.htm.

SMALL HIVE BEETLES

Arbogast, R.T., B. Torto, and P.E.A. Teal. 2009. Monitoring the small hive beetle *Aethina tumida* (Coleoptera: Nitidulidae) with baited flight traps: Effect of distance from bee hives and shade on the numbers of beetles captured. Florida Entomologist 92 (1): 165–166.

Benda, N.D., D. Boucias, B. Torto, and P.E.A. Teal. 2008. Detection and characterization of *Kodamaea ohmeri* associated with small hive beetle *Aethina tumida* infesting honey bee hives. JAR 47 (3): 194–201.

Caron, D.M., A. Park, J. Hubner, R. Mitchell, and I.B. Smith. 2001. Small hive beetle in the Mid-Atlantic States. ABJ 141 (11): 776–777.

Ellis, J.D., and K.S. Delaplane. 2008. Small hive beetle (*Aethina tumida*) oviposition behaviour in sealed brood cells with notes on the removal of the cell contents by European honey bees (*Apis mellifera*). JAR 47 (3): 210–215.

Hood, M.W. 2004. The small hive beetle, *Aethina tumida*: A review. Bee Wld. 85 (3): 51–59.

Lundie, A.E. 1940. Small Hive Beetle *Aethina tumida*. South African Dept. Agri. and Forestry Entomolo. Series, Science Bulletin 220, 30 pp.

Neumann, P., and D. Hoffmann. 2008. Small hive beetle diagnosis and control in naturally infested honeybee colonies using bottom board traps and CheckMite+ strips. J. of Pest Science 81 (1): 43–48. http://dx.doi.org/10.1007/s10340-007-0183-8.

Neumann, P., and J.D. Ellis. 2008. The small hive beetle (*Aethina tumida* Murray, Coleoptera: Nitidulidae) distribution, biology and control of an invasive species. JAR 47 (3): 181–183.

Nolan, M.P., IV, and W.M. Hood. 2008. Comparison of two attractants to small hive beetles, *Aethina tumida*, in honey bee colonies. JAR 47 (3): 229–233.

Sanford, M.T. 2005. Small hive beetle, *Aethina tumida* (Murray). Featured creatures. EENY-94. Gainesville, Univ. of Florida. http://entomology.ifas.ufl.edu/creatures/misc/bees/small_hive_beetle.htm.

Tew, J.E. 2001. The small hive beetle: A new pest of honey bees. Auburn: Alabama CES, Alabama A & M and Auburn Universities.

Westervelt, D.A. 2005. Small hive beetles in the USA—What we've learned in nine years. ABJ 145 (10): 805–807.

Websites

Penn State, MAAREC: http://maarec.psu.edu/pdfs/Small_Hive_Beetle_-_PMP.pdf.Div. Plant Industry, Florida, small hive beetle: http://www.doacs.state.fl.us/pi/enpp/ento/aethinanew.html.

SOCIAL INSECTS

Batra, S.W.T. 1966. Nests and social behavior of halictine bees of India (Hymenoptera: Halictidae). Indian J. of Entomol. 28:375–393.

Beshers, S.N., and J.H. Fewell. 2001. Models of division of labor in social insects. Annu. Rev. Entomol. 46:413–440.

Brian, M.V. 1983. Social insects: Ecology and behavioural biology. New York: Chapman and Hall.

Burda, H., R.L. Honeycutt, S. Begall, O. Locker-Grutjen, and A. Scharff. 2000. Are naked and common mole-rats eusocial and if so, why? Behavioral Ecology and Sociobiology 47 (5): 293–303.

Costa, J.T., and T.D. Fitzgerald. 2005. Social terminology revisited: Where are we ten years later? Annales Zoologici Fennici 42:559–564.

Free, J.B. 1987. Pheromones of social bees. London: Chapman and Hall.

Gadagkar, R. 1993. And now . . . eusocial thrips! Current Science 64 (4): 215–216.

Hughes, W.O.H., B P. Oldroyd, M. Beekman, and F.L.W. Ratnieks. 2008. Ancestral monogamy shows kin selection is key to the evolution of eusociality. Science 320 (5880): 1213–1216. http://www.sciencemag.org/cgi/content/abstract/320/5880/1213.

Michener, C.D., and D.J. Brothers. 1974. Were workers of eusocial Hymenoptera initially altruistic or oppressed? Proceedings of the National Academy of Sciences of the United States of America 68:1242–1245.

Moritz, R.F.A., and E.E. Southwick. 1992. Bees as super-organisms: An evolutionary reality. Berlin: Springer-Verlag.

Papaj, D.R., and A.C. Lewis (eds.). 1993. Insect learning. New York: Chapman and Hall.

Reeve, H.K., and B. Hölldobler. 2007. The emergence of a superorganism through intergroup competition. Proceedings of the National Academy of Sciences of the United States of America 104:9736–9740.

Seeley, T.D. 1995. The wisdom of the hive. Cambridge, MA: Harvard Univ. Press.

Seeley, T.D. 1985. Honeybee ecology: A study of adaptation in social life. Princeton, NJ: Princeton Univ. Press.

Von Frisch, K. 1971. Bees, their vision, chemical senses and language. Ithaca, NY: Cornell Univ. Press.

Von Frisch, K. 1973. Animal architecture. New York: Harcourt, Brace, Jovanovich.

Wheeler, W.M. 1918. A study of some ant larvae with a consideration of the origin and meaning of social habits among insects. Proceedings of the American Philosophical Society 57:293–343.

Wilson, E.O. 1971. The insect societies. Cambridge, MA: Harvard Univ. Press.

Wilson, E.O. 1975. Sociobiology. Cambridge, MA: Belknap/Harvard Univ. Press.

Wilson, E.O., and B. Hölldobler. 2005. Eusociality: Origin and consequences. Proceedings of the National Academy of Sciences of the United States of America 102 (38): 13367–13371. http://www.pnas.org/content/102/38/13367.full.pdf+html.

Yanega, D. Univ. of California, Riverside: http://en.wikipedia.org/wiki/Apis_%28genus%29).

SUPPLIERS, FOREIGN

B.J. Sherriff, Carclew Road, Mylor Downs, Falmouth, Cornwall TR11 5UN, UK. http://www.bjsherriff.co.uk.

Maisemore Apiaries, Old Road, Maisemore, Gloucester GL2 8HT, UK. http://www.bees-online.co.uk.

National Bee Supplies, Merrivale Road, Exeter Road Industrial Estate, Okehampton, Devon EX20 1UD, UK. http://www.beekeeping.co.uk.

Saf Natura s.r.l., 36015 Schio (VI) Italia - Via Lago di Misurina, 26 Z.I. E-mail: safnatura@witcom.com.

Swienty A/S, Hortoftvej 16, Ragebol, DK-6400 Sonderborg, Denmark. http://www.swienty.com.

Thomas, 86 rue Abbé Thomas, BP 2, F-45450, Fay aux Loges, France. http://www.thomas-apiculture.com.

Thorne, E.H. (Beehives) Ltd. Beehive Business Park, Rand, Nr. Market Rasen LN8 5NJ, UK. http://www.thorne.co.uk.

SUPPLIERS, MAJOR U.S.

Betterbee, Inc., 8 Meader Road, Greenwich, NY 12834. http://www.betterbee.com.

Brushy Mountain Bee Farm, Inc., 610 Bethany Church Road, Moravian Falls, NC 28654. http://www.brushy-mountainbeefarm.com.

Dadant and Sons, 51 S. 2nd Street, Hamilton, IL 62341. http://www.dadant.com.

The Walter T. Kelley Co., 807 West Main Street, Clarkson, KY 42726. http://www.kelleybees.com.

Mann Lake Ltd., 501 1st Street, S. Hackensack, MN 56452. http://www.mannlakeltd.com.

Note: Check bee magazines for other suppliers.

VENOM

Allan, S. 2008. Allergy: Bee-keepers hold clues for T-cell tolerance. Nature Reviews Immunology 8 (12): 910.

Balit, C., G. Isbister, and N. Buckley. 2003. Randomized controlled trial of topical aspirin in the treatment of bee and wasp stings. J. of Toxicology and Clinical Toxicology 41 (6): 801–808.

Bowling, A.C. 2006. Complementary and alternative medicine and multiple sclerosis. New York: Demos Medical.

de Graaf, D.C., M. Aerts, E. Danneels, and B. Devreese. 2009. Bee, wasp and ant venomics pave the way for a component-resolved diagnosis of sting allergy. J. of Proteomics 72 (2): 145–154.

Garcia, D.P. 2009. Bee aware . . . of what to do. J. of the Kentucky Medical Association 107 (1): 23–24.

Georgieva, D., K. Greunke, N. Genov, and C. Betzel. 2009. 3-D Model of the bee venom acid phosphatase: Insights into allergenicity. Biochemical and Biophysical Research Communications 378 (4): 711–715.

Hellner, M., D. Winter, R. Von Georgi, and K. Munstedt. 2008. Apitherapy: Usage and experience in German beekeepers. Evidence-based Complementary and Alternative Medicine 5 (4): 475–479.

Kokot, Z.J., and J. Matysiak. 2009. Simultaneous determination of major constituents of honeybee venom by LC-DAD. Chromatographia 69 (11–12): 1401–1405.

Lehnert, T. 1980. Hymenopterous insect stings. In: Beekeeping in the United States, pp. 141–143. USDA Ag. Handbook 335.

Martínez-Gómez, J.M., P. Johansen, I. Erdmann, G. Senti, R. Crameri, and T.M. Kündig. 2009. Intralymphatic injections as a new administration route for allergen-specific immunotherapy. International Archives of Allergy and Immunology 150 (1): 59–65.

Meier, J., and J. White. 1995. Clinical toxicology of animal venoms and poisons. Boca Raton, FL: CRC Press.

Mraz, C. 1995. Health and the honeybee. Burlington, VT: Queen City.

Müller, U.R., N. Johansen, A.B. Petersen, J. Fromberg-Nielsen, and G. Haeberli. 2009. Hymenoptera venom allergy: Analysis of double positivity to honey bee and *Vespula* venom by estimation of IgE antibodies to species-specific major allergens *Api m1* and *Ves v5*. Allergy: European J. of Allergy and Clinical Immunology 64 (4): 543–548.

Peiren, N., D.C. de Graaf, F. Vanrobaeys, E.L. Danneels, B. Devreese, J. Van Beeumen, and F.J. Jacobs. 2008. Proteomic analysis of the honey bee worker venom gland focusing on the mechanisms of protection against tissue damage. Toxicon 52 (1): 72–83.

Ramya, J., and D. Rajagopal. 2008. Morphology of the sting and its associated glands in four different honey bee species. JAR 47 (1): 46–52.

Resiman, R. 1994. Insect stings. New England J. of Medicine 26:523–527.

Ruëff, F., M.B. Bilò, M. Jutel, H. Mosbech, U. Müller, and B. Przybilla. 2009. Sublingual immunotherapy with venom is not recommended for patients with Hymenoptera venom allergy. J. Allergy and Clinical Immunology 123 (1): 272–273.

Sanford, M. 2003. Bee stings and allergic reaction. ENY122. One of a series of the Entomology and Nematology Department, Florida CES, Institute of Food and Agricultural Sciences. Gainesville: Univ. of Florida. http://edis.ifas.ufl.edu.

Scherer, K., J.M. Weber, T.M. Jermann, A. Krautheim, E. Tas, E.V. Ueberschlag, et al. 2008. Cellular in vitro assays in the diagnosis of hymenoptera venom allergy. International Archives of Allergy and Immunology 146 (2): 122–132.

Schmidt, J.O. 1992. Allergy to venomous insects. In: J.M. Graham (ed.), The hive and the honey bee, pp. 1209–1269. Hamilton, IL: Dadant and Sons.

Smorawska-Sabanty, E., and M.L. Kowalski. 2008. Step-wise approach to the diagnosis of hymenoptera venom allergy. Alergia Astma Immunologia 13 (4): 227–241.

Sturm, G.J., C. Schuster, B. Kranzelbinder, M. Wiednig, A. Groselj-Strele, and W. Aberer. 2009. Asymptomatic sensitization to hymenoptera venom is related to total immunoglobulin E levels. International Archives of Allergy and Immunology 148 (3): 261–264.

Traynor, K. 2006. The honey bee's contribution to medicine. ABJ 146 (10): 859–860.

Visscher, P., R. Vetter, and S. Camazine. 1996. Removing bee stings. Lancet 348 (9023): 301–302.

VIRUS

Chen, Y.P., J.S. Pettis, A. Collins, and M.F. Feldlaufer. 2006. Prevalence and transmission of honeybee viruses. Applied and Environ. Microbiol. 72:606–611. doi:10.1128/AEM.72.1.606-611.2006.

Chen, Y.P., and R. Siede. Honey bee viruses. 2007. Advances in Virus Research 70: 33–80.

Cox-Foster, D.L., S. Conlan, E.C. Holmes, G. Palacios, J.D. Evans, N.A. Moran, P.-L. Quan, T. Briese, et al. 2007. A metagenomic survey of microbes in honey bee colony collapse disorder. Science 318 (5848): 283–287.

de Miranda, J.R., and I. Fries. 2008. Venereal and vertical transmission of deformed wing virus in honeybees (Apis mellifera L.) J. Invertebrate Pathology 98 (2): 184–189.

de Miranda, J.R., G. Cordoni, and G. Budge. 2010. The acute bee paralysis virus-Kashmir bee virus-Israeli acute paralysis virus complex. J. Invertebrate Pathology 103 (Suppl. 1): S30–S47.

de Miranda, J.R., and E. Genersch. 2010. Deformed wing virus. J. of Invertebrate Pathology 103 (Suppl. 1): S48–S61.

Eyer, M., Y.P. Chen, M.O. Schäfer, J.S. Pettis, and P. Neumann. 2009. Honey bee sacbrood virus infects adult small hive beetles, Aethina tumida (Coleoptera: Nitidulidae). JAR 48 (4): 296–297.

Kevan, P.G., M.A. Hannan, N. Ostiguy, and E. Guzman-Novoa. 2006. A summary of the varroa-virus disease complex in honey bees. ABJ (8): 694–696.

Nielsen, S.L., M. Nicolaisen, and P. Kryger. 2008. Incidence of acute bee paralysis virus, black queen cell virus, chronic bee paralysis virus, deformed wing virus, Kashmir bee virus and sacbrood virus in honey bees (Apis mellifera) in Denmark. Apidologie 39 (3): 310–314, 45.

Palacios, G., J. Hui, P.L. Quan, A. Kalkstein, K.S. Honkavuori, A.V. Bussetti, S. Conlan, J. Evans, Y.P. Chen, D. VanEngelsdorp, H. Efrat, J. Pettis, D. Cox-Foster, E.C. Holmes, T. Briese, and W.I. Lipkin. 2008. Genetic analysis of Israel acute paralysis virus: Distinct clusters are circulating in the United States. J. of Virology 82 (13): 6209–6217.

Ribière, M., V. Olivier, and P. Blanchard. 2010. Chronic bee paralysis: A disease and a virus like no other? J. Invertebrate Pathology 103 (Suppl. 1): S120–S131.

WEBSITES FOR BEE PRODUCTS AND APITHERAPY

American Apitherapy Society: http://www.apitherapy.org.

Apitherapy for multiple sclerosis: http://www.pacificrim.net/~jwolf.

Apitronic Services, bee venom collector devices: http://www.beevenom.com/collectordevices.htm#COLL.

Bee-L list, discussion logs: http://www.internode.net:80/~allend/index.html.

Glaser, D. Are wasp and bee stings alkali or acid and does neutralising their pH give sting relief? http://www.insectstings.co.uk.

Univ. of Florida: http://gnv.ifas.ufl.edu/~ent1/software/det_bees.htm.

WEBSITES, MISCELLANEOUS

The following is a list of websites on the Internet. Many of these sites change very fast, so explore the Internet for anything to do with honey bees; you will be surprised by the results.

Entomology, in General

Great Lakes Entomology Resources: http://insects.ummz.lsa.umich.edu/entostuff.html.

WINTERING

Akyol, E., H. Yeninar, N. Sahinler, and A. Guler. 2006. The effects of additive feeding and feed additives before wintering on honey bee colony performances, wintering abilities and survival rates at the East Mediterranean region. Pakistan J. of Biological Sciences 9 (4): 589–592.

Amdam, G.V., K. Hartfelder, K. Norberg, A. Hagen, and S.W. Omholt. 2004. Altered physiology in worker honey bees (Hymenoptera: Apidae) infested with the mite Varroa destructor (Acari: Varroidae): A factor in colony loss during overwintering? J. of Economic Entomol. 97 (3): 741–747.

Lalonde, T., and A. Dziadyk. 2004. Wintering bees using "hard" wraps. ABJ 144 (3): 221–223.

Mattila, H.R., and G.W. Otis. 2006. Influence of pollen diet in spring on development of honey bee (Hymenoptera: Apidae) colonies. J. of Economic Entomol. 99 (3): 604–613.

Szabo, T.I., and D.C. Szabo. 2004. Observations on the importance of hive top entrances during winter. ABJ 144 (12): 936–938.

Underwood, R.M., and R.W. Currie. 2004. Indoor winter fumigation of Apis mellifera (Hymenoptera: Apidae) colonies infested with Varroa destructor (Acari: Varroidae) with formic acid is a potential control alternative in northern climates J. of Economic Entomol. 97 (2): 177–186.

Underwood, R.M., and R.W. Currie. 2007. Effects of release pattern and room ventilation on survival of varroa mites and queens during indoor winter fumigation of honey bee colonies with formic acid. Canadian Entomologist 139 (6): 881–893.

Underwood, R.M., and R.W. Currie. 2008. Indoor winter fumigation with formic acid does not have a long-term impact on honey bee (Hymenoptera: Apidae) queen performance. JAR 47 (2): 108–112.

Wineman, E., Y. Lensky, and Y. Mahrer. 2003. Solar heating of honey bee colonies (*Apis mellifera* L.) during the subtropical winter and its impact on hive temperature, worker population and honey production. ABJ 143 (7): 565–570.

Index